Ending Famine in India

Global Connections: Routes and Roots

Global Connections: Routes and Roots explores histories that challenge existing demarcations between and within local, regional, and interregional arenas. The series encompasses single-site and vernacular histories as much as studies of long-distance connection.
This series seeks to bridge early modern and modern history. By taking a wide timeframe of c. 1200 to the present, we embrace the many and shifting nodal points, key regions, modes of transportation and other forms of connectivity that together form the "routes" and "roots" of global history. This includes the making and unmaking of power in different manifestations, as well as the intellectual genealogies and trajectories of the ideas that did so.
We welcome all work that explores the global as method. We stress the need to recover local primary sources as a way of investigating both the individual and the collective agency of all those involved in the making of the global.

Other titles in this series:

Bérénice Guyot-Réchard and Elisabeth Leake, *South Asia Unbound: New International Histories of the Subcontinent*, 2023

Carolien Stolte and Su Lin Lewis (eds), *The Lives of Cold War Afro-Asianism*, 2022

Harald Fischer-Tiné and Nico Slate (eds), *The United States and South Asia from the Age of Empire to Decolonization: A History of Entanglements*, 2022

Neilesh Bose (eds), *India after World History: Literature, Comparison, and Approaches to Globalization*, 2022

Jos Gommans and Ariel Lopez (eds), *Philippine Confluence: Iberian, Chinese and Islamic Currents, c. 1500–1800*, 2020

Michele Louro, Carolien Stolte, Heather Streets-Salter and Sana Tannoury-Karam (eds), *The League Against Imperialism: Lives and Afterlives*, 2020

ENDING FAMINE IN INDIA

A Transnational History of Food Aid and Development, c. 1890–1950

Joanna Simonow

Leiden University Press

The Open Access edition was made possible by a grant from the Swiss National Science Foundation (SNF) for the promotion of scientific research.

Global Connections: Routes and Roots, volume 7

Cover design: Andre Klijsen
Cover illustration: Plates and mugs stored in place, from Siloam Girls' Boarding School, Tirukoilur, South India, 1974. Photo: Petersen, Erna. Danmission Photo Archive. https://creativecommons.org/licenses/by/3.0/
Lay-out: Crius Group

ISBN 978 90 8728 404 6
e-ISBN 978 94 0060 449 0
https///doi.org/10.24415/9789087284046
NUR 692

Printed and bound by CPI Group (UK) Ltd, Croydon, CR0 4YY

Table of Contents

Figures and Tables

Acknowledgements

One of the most significant and yet basic insights that came out of writing this monograph is that there are no such things as single-authored books. This book builds on decades of research by other scholars who paved the way for my engagement with the topic, and it is also the result of numerous conversations during public and private gatherings of colleagues, friends and family members who generously shared their views. I benefitted from the helpful comments, constructive criticism and encouragement of a long list of persons in and outside academia who helped refining my arguments. Without these long and intense conversations, this book would not have taken shape. The labour of archivists and librarians in India, Britain and the US was equally essential to this book and I am thankful for their guidance. I am indebted to the Swiss National Science Foundation, the main sponsor of my research, as well as the British Association for South Asian Studies and the Marsilius Kolleg in Heidelberg for their generous support. As time went by, the list of universities and research groups that hosted my work grew; all of them deserve a note of gratitude. I began my book in the Department of History of the Modern World of ETH Zurich and finished revising the manuscript in the Department of History of the South Asia Institute Heidelberg. I am particularly grateful for the mentoring, collegial support and infrastructure provided in both places. I thank the editors of the book series and the staff of LUP who guided me through the publication process. Despite all this support, I take responsibility for unintentional misconstructions and flaws that readers might detect in the following pages.

Among the people I wish to profusely thank for helping me to get this book off the ground, for commenting on earlier drafts of chapters or creating occasions to test my ideas, are (in no particular order) Harald Fischer-Tiné, Ravi Ahuja, Johannes Paulmann, Maria Framke, Elena Valdameri, Katharina Stornig, Bernhard Schär, Soni, Michael Brunner, Philipp Krauer, Andreas Greiner, Sujeet George, Nikolay Kamenov, Anindita Nag, Margarete Pratschke, Robert Kramm, Martina Andermatt, Monique Ligtenberg, Sönke Bauck, Judith Große, Stephanie Willi, Vasudha Bharadwaj, Srimanjari, Nitin Sinha, Prakash Kumar, Kama Maclean, Benjamin Siegel, Aditya Ramesh, Nico Slate and Sonam Dechen Gurung. Last but not least, I can only fail to adequately express my appreciation to Tim, whose genius and generosity are immeasurable and contributed to this book in countless ways.

Abbreviations

ABCFM	American Board of Commissioners for Foreign Missions
ABP	Amrita Bazar Patrika
AFSC	American Friends Service Committee
AIIH&PH	All-India Institute of Hygiene and Public Health
AIWC	All-India Women's Conference
AMM	American Marathi Mission
BMCC	Bengal Medical Coordination Committee
BRC	Bengal Relief Committee
BWRS	British War Relief Society
CFTRI	Central Food Technological Research Institute
CPA	Communist Party of Australia
CPGB	Communist Party of Great Britain
CPI	Communist Party of India
FAU	Friends Ambulance Unit
IFEC	India Famine Emergency Committee
IFRC	India Famine Relief Committee
IPTA	Indian People's Theatre Association
IRFA	Indian Research Fund Association
IVS	India Village Service
MARS	Mahila Atma Raksha Samiti
MFM	Meals for Millions
MPF	Multi-Purpose Food
NAC	Nutrition Advisory Committee
NCCI	National Christian Council of India
NMML	Nehru Memorial Museum and Library
NPC	National Planning Committee
PHM	People's History Museum
PRC	People's Relief Committee
PSS	Poona Sarvajanik Sabha
SVM	Student Volunteer Movement
TWI	Training Within Industry

Introduction

Ending Famine in India illuminates the panoply of historical actors who promoted scientific, religious and political solutions to famine in colonial and early postcolonial India: including medical practitioners, nutritional scientists, social reformers, agricultural experts, missionaries, politicians and colonial administrators. The study of this wide web of actors and agendas locates Indian famines in the intersecting histories of humanitarianism, development, science and (anti)colonialism. It also pushes the geographical boundaries of the history of Indian famines beyond the (future) Indian nation and the British Empire. This book shows that many of the activities geared towards ending hunger in the subcontinent arose in the tripartite relationship of India, Britain and the United States.[1] Our understanding of the history of famine in India has been shaped by national and imperial frames.[2] Tracing the early and persistent ideological and material investments of North Americans in ending famine in South Asia and the resulting entanglements between Indian and US societies breaks new ground. The book is also novel in terms of its temporal scope, linking periods of time, and hence famines, which are commonly studied separately. The mitigation of famine in India between the "late Victorian Holocausts" and the Bengal Famine of the 1940s has been dealt with only peripherally.[3] In this intervening period, famines still occurred regularly on a district level, prompting Indian social service organisations, nutritional scientists, missionaries and colonial officers to undertake and debate anti-famine measures. These minor famines (in terms of their geographical scope and the number of people affected) have received less scholarly attention than earlier and later famines.[4] *Ending Famine in India* treats them as missing links to tease out continuities in the responses of elites to famine from the late nineteenth to the twentieth century and across the colonial and postcolonial divide. Without claiming to be exhaustive, the book provides a selective account of key moments and actors to highlight historical developments and continuities. Three themes structure the book and are developed in the respective book sections. These are the interplay of famine, nutritional science and food aid; the expansion of American missionary activity in South Asia through famine relief and rural reform; and Indian political mobilisation against the backdrop of famines. Before providing further explanations on the aims and content of the book, some preliminary remarks are offered on the meaning of the term "famine" and the way it is used in this book.

I do not delve deeply into definitional questions, nor offer new insights into the causes and nature of famines in colonial India, which have been studied and extensively debated by economic, political, social and environmental historians.[5]

This is not to say that definitions are futile. The definition of famine carried (and still carries) weight, because it guided political stakeholders and humanitarian aid.[6] This book, however, traces the policies, relief measures, and scientific solutions that were generated in response to famines, mostly with no agreement among historical actors on a singular definition of the phenomenon. Therefore, rather than taking a univocal definition of famine as a starting point, the book foregrounds ambivalences and conflicts over the claimed existence of famine conditions in India and illuminates the activities taken to mitigate them. In the period covered in this book, the meaning of famine was fiercely debated. The colonial administration in India, aiming to keep relief expenses at a minimum and avoid even larger investments into welfare, differentiated between endemic hunger and famine. By the late nineteenth century, the colonial administration understood famines as exceptional periods that demanded the state to intervene, but viewed India's endemic hunger a burden too great to carry. The understanding of famine in India was shaped by the institutionalisation of colonial famine relief, which began with the drafting of the Indian famine codes in the 1880s and refined the indicators used to monitor the food situation in India.[7] Rainfall, crop failure, food prices, mortality, crime rates and migration were observed to determine the right moment to set colonial (anti-)famine policies in motion: not too early to avoid offsetting the market but not too late either to prevent the loss of life. The success of the colonial early warning system of famine depended on the accuracy of information and the timeliness of communication across administrative levels. The Indian famine codes further distinguished between scarcity and famine, but how exactly they differed remained unclear. On paper, scarcity existed when paupers began to wander, private charity and credit contracted, grain-trade showed "feverish activity", crime rates rose and people as well as cattle migrated in search for food and fodder.[8] To determine the tipping point at which scarcity turned into famine, the famine codes prescribed further tests of need once scarcity was evident. So-called test-works offered employment at an outrageously low wage. The rush of agricultural labourers to the test-works was considered further proof of an impending famine. In this case, the responsible local official was instructed to report to the provincial government, which in turn decided to declare a famine or postpone the declaration until further notice. Since the Indian famine codes failed "to fix in formal language exactly the point where conditions of scarcity cease, and where conditions of famine begin" the decision to declare famine rested with the provincial governments.[9] The absence of a clear-cut definition of famine allowed provincial governments to weigh indicators very differently. Mortality had long been the only criterion to distinguish between scarcity and famine in Punjab.[10] In the early twentieth century, after the experience of major epidemics in India, excess mortality during scarcities was no longer considered a reliable indicator of famine. Now evidence had to be produced that people were dying of hunger and not diseases. But establishing food deprivation as a cause of death was a difficult undertaking, prone to manipulation and error.[11]

Conflicting views and counter-narratives complicated how famine was understood, debated and responded to in British India. Indian writers, economists and politicians revealed flaws in anti-famine policies and the measurement of famine. They also expressed their dissent with the colonial approach to famine that limited state responsibilities to short-term hunger relief, at the expense of welfare policies that eliminated endemic hunger and poverty. "'Famine' had not been officially 'declared' in any part of India when we were there, but if famine means hunger and want, the masses of the people of India are never free from it."[12] Published in the report of a delegation spearheaded by the India League's leader V.K. Krishna Menon (1896–1974) in 1933, the statement echoed long-standing criticism of the colonial government for ignoring, if not producing, poverty, mal- and undernutrition outside of famines. Beyond the confines of colonial administrative language, the word famine denoted a range of different phenomena. Its inflationary use was often intentional and served the purpose of challenging colonial anti-famine policies and drawing the attention of donors. Although famine decreased in scope in the early twentieth century, the promise of ending famine in India still held power to release resources and political support. The numerous activities explored in this book came about not least because the importance of freeing India from famine had become widely accepted, with famine in India not only being a popular cause of humanitarianism but framed as a danger to political stability and economic development in and beyond South Asia.

Nutritional Science, Famine and Food Aid in South Asia

The advance of chemical science and dietetics in mid-nineteenth century India was followed by the emergence of nutritional science as a discipline in the early twentieth century. Famines in and outside India drove nutritional innovation, because they offered physicians the opportunity to study alimentary requirements and the bodily effects of food deprivation on a large scale. The results of the scientific study of food and food consumption were a new body of knowledge and a language of nutrition.[13] Their impact on food aid and anti-famine policies in India is the focus of this first part of the book. Despite the vast body of literature on food, nutrition, and science in colonial South Asia, the (lack of) influence of nutritional scientific opinion on the historical genesis of food aid in South Asia is largely unexplored.[14] An exception is the work of Nadja Durbach whose study of British institutional feeding, including the management of famine relief in nineteenth-century India, revealed the limited impact of nutritional knowledge on state practice.[15] Chapter 1 shows that the discrepancy between nutritional standards and food allocation in state-controlled institutions studied by Durbach for the nineteenth century remained visible in the colonial management of famine in the later period. Although nutritional studies again and again pointed to the insufficiency of relief rations in terms of their size and content, the economic rationale that guided colonial famine relief discouraged its reform. In order to understand to what extent colonial famine relief violated nutritional standards of the time, chapter 1 reviews the famine codes, the rules and guidelines that governed the administrative responses to famine in colonial India from 1883 onwards.[16] The introduction to the principles of colonial famine relief also provides essential context for the second and third parts of the book that shift attention to missionary interventions and Indian nationalist mobilisation. My review of the famine codes extends beyond 1901, when the last all-India famine commission published its report, and which commonly marks the end point of historical studies of colonial famine relief. In the decades that followed, colonial anti-famine policies developed even more unevenly across India as changes to them were introduced on a provincial level. Although not generalisable, the findings of the individual studies discussed in the chapter exemplify some of the ongoing debates and negotiations over famine relief in the early twentieth century. They also allow me to trace the influence of Indian social reformers on the revision of colonial famine relief. In Madras in the 1920s and 1930s, for instance, the Servants of India Society drew on nutritional studies to deplore the inadequacies of colonial famine relief and to insist on its reform.

The discussion on the intersection of nutrition science and famine relief is expanded to the 1940s and 1950s in chapter 2 of the book. Against the backdrop of famine in Bengal, World War II and India's quest for food security after independence, the collaboration between Indian and American nutritionists, philanthropists and political activists grew and led to new initiatives to meet India's food needs. Economic and political rationales contributed to the rising popularity of food supplements and blended food that were increasingly applied to mitigate starvation in India. Remedies for hunger developed and tested in South Asia in these decades radiated further, demonstrating South Asia's role as a hub for humanitarian engagement and knowledge production in nutrition. The chapter looks beyond the role of India in the development of British imperial famine relief to help anchor South Asia in the global history of food aid.[17]

From Famine Relief to Community Development: The American Missionary Movement in South Asia

How famines allowed American missionary expansion in South Asia is the focus of the second part of the book. It takes American famine relief for India in the late nineteenth century as its starting point and ends with the contribution of missionaries to Indian Community Development after independence. (North) American humanitarianism of the 1890s had important historical precursors, but it visibly picked up steam in this decade. Sharing one's wealth to assist foreign populations became popular in the United States at this time. Fundraising committees and relief providers mushroomed during wars and famines to put American money to work.[18] The urge of Americans to assist foreign populations gained considerable momentum in the World War I era, which constituted another turning point in the history of American foreign relief.[19] The history of US-sponsored famine relief in colonial South Asia follows a similar timeline. It began with the mobilisation of unprecedented amounts of US grain and money to assist the work of missionaries in India in 1896. The geography of international disasters and US foreign policy directed American humanitarianism towards specific regions of the world. After World War I, American humanitarianism focussed on Europe and the Near East, but in the shadow of this aid drive, Americans also opened their pockets to relieve famine in colonial South Asia.[20]

Historians have demonstrated that American humanitarianism was a complex phenomenon. It was nourished by imperial ambition, tied to economic interests and steeped in religious rhetoric. It drew strength from the missionary movement that provided personnel and ideological support, particularly in the nineteenth century.[21] The participants and motivations of American international disaster relief grew even more diverse as time progressed. Missionaries continued to provide relief abroad in the interwar period, however, secular-minded do-gooders outstripped the missionary involvement in humanitarianism at the end of the Progressive Era.[22] With the secular Anglo-American humanitarian movement focussing on Europe, missionaries remained the primary vehicles of US humanitarianism in South Asia during and after World War I. Chapter 3 of the book examines American famine relief in India against this historical context. Without neglecting the domestic changes in the United States that fanned the growth of US humanitarian involvement in South Asia, it seeks to deviate from the common approach of tracing the roots of American humanitarianism primarily and exclusively in the United States. I draw from the findings of historians, most notably Ian Tyrrell and David Hollinger, who studied how missionary work in (what came to be known as) foreign mission fields profoundly shaped American society.[23] With this purpose in mind, the chapter studies the famine relief work of the American Marathi Mission (AMM) which spearheaded American humanitarianism in the Bombay province. The focus on the work of a single relief provider in one region of South Asia may seem small on first sight, but it allows for drawing out larger processes. I shift back and forth between the American East Coast and western India to examine the link between the growth of humanitarian spending in the United States and the work of American missionaries in India. I detail the efforts of missionaries to gear the humanitarian spending of Americans to colonial South Asia; I illustrate how the reliance on Indian mission members and the partial integration of American aid in British colonial structures undermined the branding of US missionary work as uniquely American.

Chapter 4 broadens the book's perspective on the history of the American protestant response to famine in India both with regard to the temporal scope of the study and the historical actors populating it. It examines how the wish of the American protestant mission movement to demonstrate its capacity to help the prevention of famines in India contributed to its growing involvement in rural development in the interwar period. Recent literature on rural reconstruction and agricultural education in India has broken new ground. It has highlighted, on the one hand, the long historical genealogy of post-war and post-independence community development and, on the other hand, the early American involvement.[24] In his study of the Young Men's Christian Association in India, Harald Fischer-Tiné has illustrated the contributions of this American missionary institution to rural reform in the interwar period and its influence on later secular development work in South Asia and beyond.[25] Prakash Kumar has used the Allahabad Agricultural Institute as a case study to gain new insights into the American character of rural reform, which he argues coexisted and interacted with British colonial and Indian approaches to agricultural education and reform.[26] Chapter 4 builds on this literature and seeks to contribute to its debates. I foreground the historical continuity of interwar and post-independence missionary rural work, examine the intersections of famine and rural reform and explore the gendering of agricultural education. The chapter also offers a modest contribution to the history of the global food system by showing how missionaries contributed to reframing food security as a matter requiring a global framework and international coordination. Historians have demonstrated that the disruption of agricultural production and the return of famine to Europe during World War I lifted the deterrence of mass starvation onto the agenda of European and North American politicians and economists. Instead of considering food provision as a national and regional matter, global food imbalances were foregrounded and international coordination and cooperation were considered important remedies in the 1930s and 1940s.[27] This shifting response to food shortages between the world wars has not been associated with the American mission movement, and is seldom studied in relation to South Asia.[28] In this context, chapter 4 demonstrates that framing food security as a problem of global dimensions undergirded the efforts of missionaries to internationalise rural mission work.

Anticolonial Famine Relief: Mobilising Against Hunger and Colonialism

How Indian social and political elites employed famines to promote visions of the Indian nation is the central theme of part III of the book that is divided into three chapters. Whereas the first part of the book has shed some light on the intersection of Indian nationalism, food and nutrition, chapter 5 demonstrates how Indian nationalists took on the task of documenting famines and organising for relief in the first two decades of the twentieth century. The book section thus explicates how ending famine became central to Indian nationalist politics. But it also moves beyond this to study how famine relief became an anticolonial and anti-imperial activity, used by a broader section of activists to express political solidarity with Indian political demands. The anticolonial roots of humanitarianism are commonly understudied.[29] Despite earlier conflations of nationalism and famine, the first two decades of the twentieth century deserve particular reflection. In these decades, famine relief acquired a central position in the Indian social service movement, while Indians outside South Asia began to take part in famine relief.[30] Driven by patterns of South Asian migration and diaspora formation, Indian activists in North America and Canada began to write about famine and raise money in support of Indian-led relief efforts.[31] By shifting the geographical focus to North America but tying it to South Asia, Chapter 5 highlights how famines were employed to bridge distance and forge connections between Indians at home and abroad. At the same time, preventing future famines in India even became a concern for opponents of Asian immigration in the United States, who considered famines a cause of Indian migration and political radicalism.[32]

Given the richness of historical sources and the importance of the famine in the history of India, chapter 6 focusses exclusively on the Bengal Famine of 1942–44.[33] It revisits the history of the famine to gain a deeper insight into its "transformative effect on Indian politics and national aspirations" that Benjamin Siegel has recently flagged in his seminal book *Hungry Nation*.[34] Famine relief in Bengal became an ideological battleground for Indian political forces that vied for significance by relieving hunger. While Hindu/Muslim communalism has become the main framework for examining the 1940s in Bengal, the range of ideologies and political forces that came into play during the famine shows that communalism does not suffice to understand the complexity of the relief effort in Bengal or its political impact.[35] Indian social and political movements had diversified in the interwar period.[36] On the eve of independence famine relief promoted very different visions of India's future, now also put forward by Indian women's organisations, Hindu nationalists and Indian communists.

Clearly, independence did not end starvation nor Indian criticism of the government's response to famine. Food became central to Indian politics in the first two decades of independence.[37] Offering a missing perspective on India's quest for sustenance, the last chapter of the book examines the continuity of Indian activism in the United States in the 1940s and early 1950s. Political alliances between Indians and Americans had grown in the interwar period. They expanded decisively in the 1940s when anti-British and anti-imperial sentiment was rising in the United States, giving impetus to Indian political mobilisation. Against the backdrop of the Bengal famine, the "Indian food crisis" of 1946 and the famine in Bihar and Madras in 1951,

Americans and Indians lobbied the US government to start food aid and mobilised non-governmental aid themselves. Thus, Indian and American philanthropists, politicians and scientists already collaborated in the field of nutrition and food aid before the official Indo-US food aid agreements were signed into existence in the postcolonial period.

The book ends in the 1950s, right before the onset of the Green Revolution once again changed the historical setting. However, the history of famine in India continues well into the present. Many of the debates and contestations traced in this book are topical. Claiming its ability to provide sustenance to its population continues to be of existential importance to the Indian government and has invited new debates on the adequate measurement of hunger.[38] This book does not provide an answer to the puzzle as to why hunger prevails in India, but it offers a historical perspective on debates and conflicts that marked the fight against famine, food insecurity and starvation in South Asia.

PART I

Nutritional Science, Famine and Food Aid in South Asia

CHAPTER 1

The Limits of Famine Relief: Colonialism, Nutritional Science, and the Indian Social Service Movement, 1890s–1930s

Abstract

The regulation of the amount and type of food to be consumed in state institutions during famines is a part of the history of famine relief in colonial South Asia that has so far received only marginal attention. The chapter traces the evolution of famine rations and wages from the late nineteenth century to the interwar period against the backdrop of advances in nutritional science and the increasing involvement of Indian social service organisations in the mitigation of famine.

Keywords: Nutritional Science, Famine Relief, Social Service, Wallace R. Aykroyd

1.1 Introduction: Colonial Famine Relief in British India

The British Encounter of Famine in India

Famines convoyed the East India Company's advance into South Asia. In the hundred-plus years that passed between the Bengal famine of 1770 and the South Indian Famine of 1876–78, colonial responses to famine in India were ad-hoc and varied greatly.[1] Until the draft of the Provisional Famine Code of 1883, British administrators had no rulebook to consult when dealing with famines. Instead, they relied on a body of assumptions and theories. After the famine in Bengal of 1770, when the regulation of the grain trade had exacerbated rather than staved off the famine, the free market paradigm won influence among British administrators. Driven by the conviction that is was unadvisable to interfere with the Indian food economy in times of famine, administrators placed their trust in the self-regulatory forces of the market, theorised by political economists Adam Smith (1723–90) and David Ricardo (1772–1823).[2] Keeping the interference in famine to a minimum was also advocated by the adherents of Thomas Robert Malthus (1766–1834), whose writings on population growth cast famines as a necessary check against overpopulation.[3] *Laissez-faire* governed colonial administrative reactions to famine for the remainder of the colonial period, but the administrative response to famine widened considerably.[4] Changing notions of colonial governance coupled with the recognition that famines threatened the expansion of the colonial state fuelled British interventions into famine in India.[5] Famines caused mass migration, resistance to tax collectors and crime that posed new problems to the colonial administration. In addition, the loss of tax revenues during and in the wake of famines provided a strong economic incentive to mitigate collective starvation.[6] "The humanitarian discovery of hunger", as James Vernon has demonstrated, created additional pressure on the colonial administrators to reconsider their stance towards famine.[7] The influence of philanthropists and humanitarians gained considerable momentum towards the end of the nineteenth century, when developments pertaining to the print media and the improvement in communication facilitated the mobilisation of empathy for the famine-afflicted in distant parts of the world.[8] Finally, the growing interest in the relationship between meteorology and food scarcity formed the basis for the colonial state's intensifying response to famines.[9]

From the mid-nineteenth century onwards, the colonial answer to famines relied on a three-pronged approach. The state employed parts of the famine-stricken population capable of hard manual labour on public works. It provided doles of cash or grain and meals free of charge to the elderly, infirm and young children (so called "gratuitous relief"), and advanced loans to cultivators.[10] Until the end of the nineteenth century, the priority was to alleviate famines through labour, for which "famine workers" received a subsistence wage. Under the influence of the free market paradigm, the provision of labour was favoured to circumvent even more direct interventions in the market, such as regulating food prices and importing grain into scarce regions.[11] Making labour a prerequisite for state assistance also followed utilitarian reasoning. Colonial administrators considered the employment of the famine-afflicted population to build railways, roads and irrigation tunnels useful for the advancement of infrastructure and communication networks and thus for the modernisation of India.[12] The expansion of India's railway networks even ostensibly prevented local food shortages from turning into famines by enabling trade between provinces; it also helped the consolidation of the colonial state by facilitating troop movements.[13] After all, it was consistent with colonial assumptions about Indian culture and society to enforce famine-affected populations to work. Believing that Indians were shirkers by nature, work was supposed to encourage industriousness and became a pillar of colonial attempts of moral reform.[14]

Prison Diets and the Search for Minimum Requirements

Whether assistance took the form of wages, food or cash doles, the size of rations was a central concern of the colonial administration. Efforts to determine the level of famine relief were embedded in larger questions about the nature and limits of the colonial state's responsibility to its citizens and fuelled colonialism's foray into the rule of bodies.[15] The colonial state's attempts at mitigating famine involved managing the population, categorising and controlling it, with the regulation of the starving population's food consumption being an essential part of the process.[16] Like poor relief in Britain, colonial famine relief in India was based on the firm belief that government support needed to be unattractive. British colonial officers in India endeavoured to keep famine relief to a minimum in order to discourage those purportedly undeserving of government assistance from seeking relief, to protect the integrity of the market and avoid high government expenditure.[17] Since colonial famine relief only aimed at providing the worst affected parts of society with enough food to merely survive famines, knowledge of the food requirements of labouring populations was essential to set the level of relief. This knowledge had been generated primarily in the context of prisons.[18]

Poor and monotonous meals were a pillar of the disciplinary regime of prisons and a part of the punishment of inmates.[19] The revolt of prisoners against such practice and high mortality rates among them led to a rethinking of dietary regimes in Indian prisons. Medical opinion further fanned this change.[20] Surgeon Major William Robert Cornish (1828–97) studied prison diets in Madras in the 1860s. He became part of a group of physicians who advocated to abandon the punitive nature of diets and to preserve the health of inmates through proper nutrition. He discouraged critics who claimed that dietary improvement was too costly, explaining that a healthy body was the prerequisite for labour productivity and prevented future medical expenses.[21] Cornish substantiated his claims with studies of diets in and out of prisons, and estimates of food requirements of prisoners and workers.[22] In so doing, he applied a method that was common at the time. To study dietary requirements, physicians simply observed the diets and health of different societal groups to determine dietary standards.[23]

In comparing the diets of prisoners and the labouring population of Madras, Cornish noted that prisoners ate poorer than workers outside of penal institutions. (Whereas the daily diets of labourers in Madras commonly included 32 to 40 ounces of cereal and 2 to 3 ounces of pulses and meat, fish and vegetables each, prisoners received between 24 and 28 ounces of grain and little supplementary foods.[24]) This imbalance, while keeping with the disciplinary regime of prisons, endangered the health of inmates, according to Cornish. Under the influence of chemical food science, the methods for determining adequate diets were already becoming more sophisticated at the time of Cornish's study. From the 1840s onwards, chemists and physiologists emphasised the balance of carboniferous and nitrogenous elements of meals. This also reflected in the studies of prison diets in India. *The Sanitary Condition and Discipline of Indian Jails* of 1860 noted that "dietaries ought never to be estimated by the rough weight of their constituents, without distinct reference to the real nutriment in these, as determined by physiological and chemical inquiry."[25] With regard to prison diets, medical officers in India acknowledged that in addition to size, the composition of diets was also crucial for adequate nutrition.[26]

Such medical knowledge on food requirements was of no concern to Lieutenant Governor of Bengal Richard Temple (1826–1902) who was sent to Madras in 1877 to gain back control over escalating famine conditions in the province.[27] The failure of the South-West monsoon in southern India had signalled famine in the Madras Presidency in mid-1876. Drought started to diminish harvests in the South and the grain surplus yielded in other parts of India was exported to England instead of being used to alleviate food scarcity in Madras. Famine conditions spread unhindered across Mysore, the Bombay Deccan into the North Western Provinces and culminated in the worst famine India had seen in decades.[28] Before Temple's arrival in Madras in 1877, the Viceroy of India, Lord Robert Lytton (1831–91), had instructed him to apply fiscal stringency and prudence. Following the example of Bombay, where administrators had introduced a cut in the size of famine rations, Temple restricted the maximum amount of food that people labouring on governmental famine relief works could purchase daily to 16 ounces (1 pound) of grain. The Temple Wage, as it became known, was insufficient by any contemporary standard.[29] Cornish, being the Sanitary Commissioner of Madras, pointed out that the amount of food allocated to famine labourers was 8 ounces short of prisoners' diets and even lower than the minimum requirements Temple himself had determined in Bengal three years earlier.[30] Although Cornish and other medical officers urged that rations be increased and supplemented with meat or fish and vegetables, in particular protein-rich pulses, the Temple Wage remained in place for three months.[31] Temple's management of famine in Madras significantly contributed to the exorbitant mortality during the crisis that claimed the lives of six to ten million people.[32] By limiting famine rations of workers to one pound of grain, Temple wilfully ignored scientific opinion. This was not a one-off decision or administrative glitch. Temple set the tone for decades to come. Although the Indian famine codes that were drafted and reformed between 1883 and 1901 (and subsequently revised on a provincial level) made concessions to medical opinion, they prescribed a wage system in which the earnings of labourers still fell at, and even below, the Temple wage. This somewhat unacknowledged afterlife of the Temple wage is the subject of the following discussion which demonstrates that the gap between contemporary dietary standards and famine relief became built into the administrative response to famine.

1.2 The Indian Famine Codes and the Administration of Food During Famine

The Famine Codes

In the aftermath of the South Indian Famine, the colonial government tasked the Famine Commission of 1878–80, also known as the Strachey Commission, to formulate a set of general principles on state-sponsored famine relief.[33] The document became the basis of the Provisional Famine Code of 1883, which in turn served as a blueprint for the provincial famine codes that were gradually adopted across British India in the following years. The government of British India introduced further amendments to the famine codes in 1892 and 1893 and two additional famine commissions were summoned in 1898 and in 1901 to survey the relief responses of provincial governments. They provided recommendations for the famine codes' adjustment. Famine commissions commonly reflected on methods already used across India to achieve a more unified response. Their reports reveal that differences continued to prevail in the administration of famine relief in British India. This was not least because the famine codes were lengthy documents and established a complex administrative system that local administrators had trouble following in detail. After the last colonial (all-) India Famine Commission of 1901, provincial governments continued to introduce changes to the famine codes by assembling additional committees in the aftermaths of famines. In the following discussion, I review the famine codes' system for allocating food and, where available, use additional sources to trace how aid practice deviated from the famine codes.

Debates on the size and quality of relief rations in general, and of wages in particular, surfaced in the report of the Strachey Commission and gained further traction in the 1890s, when they became part of the reports of subsequent famine commissions. The Strachey Commission suggested a (slightly) higher standard in future famines. Deviating from the Temple wage, it recommended a daily ration of 1.25 pounds to 1.5 pounds of grain or flour for famine workers.[34] The Provisional Famine Code of 1883 also added a portion of pulses, vegetables, oil or ghee (clarified butter), salt and other condiments.[35] The increase of rations, though minimal, was a concession to the demands of Cornish and other medical officers. However, the diet affordable to famine workers continued to violate (what was then considered) basic food requirements, both in terms of its nutritional composition and size.[36] This violation of nutritional standards in colonial famine relief was in consonance with the famine codes' underlying principles. To prevent public works from absorbing labour rather than providing a last resort, labourers needed to earn less (and eat worse) on famine relief works than outside governmental institutions. Studies on diets and food requirements assisted in establishing a benchmark against which the level of famine relief was to be set. To fix the exact level of famine

Table 1. Ration Sizes according to the Provisional Famine Code of 1883
Based on a table provided in the report of the Indian Famine Commission, 1898, 256.

Descriptions of ration.		For a man.		For a woman.		For children.
		lb	oz.	lb	oz.	
Full	Flour of the common grain used in the country, or cleaned rice	1	8	1	4	¾, ½ and ¼ according to age and requirements.
	Pulse	0	4	0	4	
	Salt	0	½	0	½	
	Ghi or oil	0	1	0	½	
	Condiments and vegetables	0	1	0	1	
Minimum	Flour of the common grain used in the country, or cleaned rice	1	0	0	14	¾, ½ and ¼ according to age and requirements.
	Pulse	0	2	0	2	
	Salt	0	⅓	0	⅓	
	Ghi or oil	0	⅓	0	⅓	
	Condiments and vegetables	0	½	0	½	
Penal	Flour of the common grain used in the country, or cleaned rice	0	14	0	12	Not stated.
	Pulse	0	1	0	1	
	Salt	0	¼	0	¼	

rations—the minimum that workers needed to sustain themselves—, the famine codes adopted a simple intake-output equation. Rations aimed to compensate the exact amount of energy lost through physical activity. The benchmark for ration sizes was the “ordinary working male”—an approach that was not only used in the colonial administration of famine relief. In the context of institutional feeding programmes in South Asia and beyond, nutritionists and administrators used the “average man” or “average worker” to estimate food needs. Although nutritionists knew that the approach fell short of calculating individual needs, the bureaucratic requirements of institutional feeding ostensibly rendered such simplification necessary.[37] The method neglected physical differences among men, but the use of the male body as a benchmark particularly disadvantaged women and children. Their diets were deducted from the male average.[38] Consequently, the famine codes allocated “a little less” food than the “average working male” to women; children received a mere fraction (see table 1).[39] Although the way in which the gender gap was enacted changed during the revision of the famine codes, it was not abolished. The Indian Famine Commission of 1901 not only endorsed unequal pay, it further altered the wage system, widening the gender gap and even reducing women's absolute earnings.[40] To better understand how much famine workers were entitled to consume and why this system often prevented them from even accessing the rations prescribed, we need to turn to the wage system of the famine codes.

Cash for Work: The Wage System

> The principle upon which the Government of India has framed the scale of wages embodied in the Code is that the wage should be the lowest amount sufficient to maintain health under given circumstances. *While the duty of the Government is to save life, it is not bound to maintain the labouring community at its normal level of comfort.* To do so would be unjust to other sections of the community, besides prolonging the period for which the labouring population would cling to relief works.[41]
>
> Provisional Famine Code, 1883

Populations employed at public works during famine were given a cash payment to enable them to buy food at the nearest market or government-controlled shops. Thus, while the famine codes detailed dietary components (see table 1), what famine workers purchased with their cash payments was not controlled. Nutritional recommendations had therefore only a limited influence; the demand for an increase of green vegetables for instance could work in favour of higher wages, but would not guarantee consumption of the particular food item.

Since 1883, the colonial management of public works during famines rested on three wage units: the full, minimum and penal wage (see table 1). In addition, the Provisional Famine Code defined three categories of labourers (A, B and C) and distinguished them by expected productivity. Wages, hence, were tied to colonial estimates of labour productivity which meant that unrealistic expectations on the one hand and underperformance on the other hand diminished wages. This system was in consonance with the idea that workers did only need to consume (and were only entitled to) the amount of energy wasted through work. Consequently, the first generation of the famine codes (1883–1898) categorised all "able-bodied" men and women as class A and B workers. Class A consisted of labourers "accustomed" to the work demanded of them who were entitled to the full wage. Class B labourers were unaccustomed to the work. Class C labourers on the other hand included men and women who were unsuited for hard manual labour but fit to carry out "light employment." To reduce the earnings in accordance with (expected) productivity, the maximum wage of Class B and C labourers amounted to 75 per cent of the earnings of Class A. The first generation of Indian famine codes worked with a bottom line of wages, i.e. the minimum wage, which was set at 16 ounces of grain for men and 14 ounces of grain for women. Hence, men's wages ranged from the full wage of 24 ounces to the minimum wage of 16 ounces, while women earned a maximum of 20 ounces and a minimum of 14. Two further restrictions ensured that workers were never to eat more than the energy lost through physical labour. First, what was considered the refusal to work resulted in the penal ration that was even less than the minimum wage.[42] Second, wages were adjusted in accordance with performance. Underperformance resulted in a percentual cut of wages.[43]

Contemporaries in the 1890s noted that administrators often applied the penal wage and fined workers excessively even when the reasons for a labourer's unproductivity were beyond their control, for instance when weather conditions prevented work.[44] This was also noted by the second and third famine commissions.[45] The threat of going without a day's earnings intensified after the minimal wage was abolished in response to mounting opposition against this safeguard in 1901. Opponents argued that a bottom line for wages bred idleness among the workers and the Indian Famine Commission of 1901 concluded that "the evidence of the demoralizing effect of the minimum wage is overwhelming."[46] It established a system of payment by results (that was already followed in different provinces across India at the time) that entailed a maximum limit to daily earnings but no minimum.[47] Hence, no check prevented wages from falling to and even below the much-criticised Temple wage.[48]

Red tape prevented workers from receiving their subsistence wage in more ways than one. The colonial government introduced the "grain equivalent" in 1893 which meant to simplify the administration of wages and resulted in a decoupling of wages from the market price of food commodities.[49] Prior to the grain equivalent, famine administrators used the market price of every item of the famine ration to set wages. Now they only needed the grain price to do the same. The grain equivalent was computed by multiplying the weight of the grain ration with a predetermined factor; the cash payment in turn was calculated on the basis of the grain equivalent and the grain price. The wage thus calculated intended to cover the costs of all items of the ration.[50] This meant that officials started to determine wages independent of the actual prize of pulses and other supplementary food items. Now, whether workers were able to buy their food ration depended on the administrators' use of the actual grain price and presupposed that the prices for supplementary foodstuffs (pulses, condiments, oil and vegetables) did not fluctuate. The gap between the grain price used as the basis of wages and the actual market prize of grain widened after additional reforms of the grain equivalent were made in 1901. Prior to the reform, colonial administrators were instructed to consult the nearest market price of grain to calculate wages. Now the grain price was set for entire districts. The standardisation of district wages intended to prevent the migration of workers within a district to public works that offered better rates, but it carried the risk that grain prices on local markets exceeded fixed district-level estimates, thereby resulting in a cut of famine rations.[51] Thus, with the introduction of the grain equivalent and its later reform, wages could crunch to an extent that famine rations fell far below estimated food requirements. When the costs for food items rose above general estimates, workers failed to afford the entire ration. If prices dropped, the labourer was left with a little margin in excess. While in theory the system could lead to benefits for famine workers, even slight reductions of the meagre wages amplified suffering and hardship, risking the health and lives of the famine-afflicted in government-care. Contemporaries identified additional flaws. Although the famine codes stipulated that payments needed to be made daily at best, but at least fortnightly, the lack of available cashiers resulted in delayed payment. Workers were therefore forced to borrow money to buy food and the interest payments reduced their purchasing power and thus the food they consumed.[52] In other words, the wage system of the famine codes did not guarantee that workers consumed the diets they were entitled to on paper.

"Gratuitous Relief": Aid in Poorhouses and Relief Kitchens

Apart from employing the famine-affected population, the colonial state provided gratuitous aid during famine periods. Gratuitous relief took the form of meals, doles of cash and uncooked food that was given free of charge or at subsidised prices. In addition to people who were too weak to labour on public works, the Provisional Famine Code of 1883 listed "idiots and lunatics", "cripples", "blind persons" and parents taking care of a sick infant as suitable recipients.[53] In light of the frailty of these sections of society, testing their needs through the demand of hard manual labour was considered inappropriate. However, the fear that relief was given to people undeserving of state assistance remained omnipresent. The second and third Indian famine commissions emphasised the careful inspection of villages in order to assure that relief did only reach the needy. When famine was imminent but had not yet been officially declared, district officers were to prepare lists of people, who needed to apply for gratuitous relief during famine. An inquiry into "the applicants' whole circumstances" was recommended to avoid too much relief being given.[54] Gratuitous aid was provided through kitchens built in proximity to public works to assist dependants of famine workers; a mix of cooked food, cash and grain doles was handed out in villages. The most common form of administering gratuitous relief in India in the second half of the nineteenth century, however, was the poorhouse.[55] People travelling to relief works at times sought temporary refuge in poorhouses; the bulk of people kept at such institutions, however, was admitted without consent and kept there by force.[56] Poorhouses were used to control "paupers" and "wanderers" and although they were meant to shelter only those too frail to labour, the colonial administration still used the institution to promote industriousness. Inmates able to perform "light work" were demanded to fetch water, sweep, grind grain, repair fences or even made to do stone-breaking—remunerated tasks attributed to inmates according to a colonial understanding of gender and caste.[57] Poorhouses rendered inquiries into indigence obsolete, because the institution itself was considered a deterrent. A means to deter the "undeserving poor" from seeking admission to poorhouses and relief kitchens was to serve pre-cooked meals, a practice which gradually replaced the prior use of grain and cash doles. It was in the 1860s, as Sanjay Sharma notes, that the practice of using prepared meals "stepped out of colonial prisons and became the defining feature of poorhouses."[58] The provision of prepared meals was adapted from the management of prisons and applied first in poorhouses before it was also used in villages and on public works. Whereas the introduction of messing to prisons in the 1840s had resulted in riots of captives in Bengal and was eventually abandoned, it was retained as part of colonial famine relief.[59] In line with the logic of deterrence that was already in use in other areas of colonial famine relief, the food provided in poorhouses and relief kitchens was intentionally plain. On top of that, it was rarely sufficient. The Famine Commission of 1878–80 recommended a ration size of one pound of grain for men and reduced rations for women and children. This was overhauled in 1898, when the second famine commission recommended an increase of gratuitous rations for non-working adults—men and women alike.[60] The third commission, however, re-introduced a gender differentiation, which led to a decrease of gratuitous rations for women, who were now eligible to less than non-working men.[61]

The high mortality among inmates of poorhouses during the famine of 1896–97 prompted the Famine Commission of 1898 to instruct medical officers to exert greater flexibility in administering food in these institutions.[62] Although the "special dietary needs" of recipients of gratuitous relief were now considered, rations remained poor even in the early twentieth century. Thus, the Bombay Famine Relief Code of 1912 differentiated three diets: an ordinary diet (consisting of rice and pulses), a milk diet, and a *conji* (a form of thin milk gruel) diet.[63] The Famine Code of Bihar and Orissa of 1930 recommended the provision of pulses and *chapatis* (flat bread), and for those unable to digest the bread, boiled rice and *dal* (lentils). The weakest received a soup made out of flour, water and salt.[64] In parts, the poverty of diets was called for by the physical condition of inmates that prevented them from properly digesting food; in many cases, poorhouses had turned into infirmaries. However, poor and tasteless diets were also part of the disciplinary regime of colonial famine relief.

The provision of meals instead of doles of cash or grain intended to facilitate the identification of needs through another mechanism. Since caste regulated the consumption of food in India in complex ways that were not always transparent to the state, colonial officials assumed that the provision of meals required those seeking help in government institutions to risk violating caste norms. In doing so, colonial officers commonly overlooked other reasons for the widespread reluctance to accept government relief that ranged from the insufficiency of rations and the fear of being kept in poorhouses involuntarily, to the overall punitive measures employed by the colonial state to administer relief in these institutions.[65] Colonial administrators' frequent mentioning of "caste prejudice" and the exceptions they made to preserve caste discrimination by allotting doles of grain and cash to the "respectable classes", fuelled a gradual rethinking of the use of cooked food. The Indian Famine Commission of 1898, taking notice of this practice, sanctioned such exceptions "in order to guard against the possibility of deserving persons, who from caste or other prejudice cannot accept cooked food, being excluded from all relief although they really require it."[66] The famine codes began to make room for the assistance of "respectable persons"— Indians of high caste and class who were not expected to seek relief at government-institutions— as well as *pardanashin* women (women observing *purdah*, living in seclusion), through doles of cash and uncooked food. In poorhouses, however, where meals were still served, Indian cooks and overseers were now employed to observe caste rules to soften opposition against colonial famine relief— a practice that had already been in use before the famine codes adopted it.[67]

The colonial state started to give greater emphasis to the diets of children during famines in the 1890s and began to stress the importance of kitchens to provide relief to them. Although kitchens were considered more expensive than providing parents with cash or grain doles, the Indian Famine Commissions of 1898 and 1901 recommended the use of kitchens to alleviate starvation among children. Instead of giving relief in the form of additional allowances to parents, the direct provision of food to children was championed, because it allowed greater control of children's food consumption. As Nadja Durbach observed, this change of practice was linked to the British "strategy of colonial development" that wished Indian children to grow into healthy adults to maintain the colony's economic productivity and guarantee future resource extraction. The colonial state emphasised the alleged failure of Indian parenthood in general and of Indian motherhood in particular and staged itself as children's guardian.[68] As we shall see in the following section, the colonial claim of mitigating starvation among children more effectively through providing them with meals rather than channelling relief through their parents had detrimental effects. As nutritional experts in Madras would soon point out, colonial relief did not prevent malnutrition among children nor irreversible damage to their health.

1.3 Famine Relief and Nutritional Reform in the Early Twentieth Century

The flaws of the Indian famine codes, which had contributed to high mortality in the 1890s, carried over into the twentieth century. Yet local administrators and provincial governments recurrently succeeded throughout the next decades to contain scarcities and prevent high-mortality during famines. The following example, although not generalisable, points to the importance of the decision of local administrators to deviate from the famine codes and, as a result, to grant higher wages and larger rations. The increasing gap between the system of famine relief outlined in the famine codes and actual aid practice provides part of the explanation why famines proved less deadly in early twentieth-century India. Nutritional science and medical opinion, although resulting in little adjustment of the famine codes, nevertheless influenced relief practice. It was in particular the amplification of nutritional opinion through Indian social workers and political activists which helped modify aid practice.

The partial failure of the monsoon in 1907 led to drought and poor harvests in parts of the United Provinces (of Agra and Oudh). "Men who had known the course of Indian famines since the seventies told me that unless rain came within a fortnight that year's famine would be worse than any that India had suffered", wrote the British correspondent Henry Woodd Nevinson (1856–1941).[69] Famine conditions soon reached a troubling dimension. With thirty million people out of the total population of forty-eight million afflicted by famine, contemporaries were undoubtedly reminded of the famines of the last century.[70] In 1907/8, the Lieutenant Governor of the United Provinces, John Hewett (1854–1941), oversaw the anti-famine policies of the provincial government. Under the guidance of Hewett, the government reacted promptly and managed to keep famine mortality low. Hewett's success resulted from major deviations from the United Provinces Famine Code that governed the administrative response to famines in the province. Instead of adhering to the chronology of administrative activities that the code prescribed, Hewett acknowledged that the sharp rise in grain prices already impeded people's access to food and ordered the use of gratuitous relief at a time when "test works" still probed the existence of famine conditions.[71] Apart from doling out aid gratuitously at an early stage of the famine, the government allowed gratuitous relief to exceed usual limitations. Whereas the Indian Famine Commission of 1901 had advised to cap gratuitous relief at 42 per cent of the total of people on government assistance, the average adopted in 1907/8 was 54 per cent. In the district of Mirzapur, where people living in the secluded hill tracts and hinterlands suffered severely, the number rose to 67 per cent.[72] Apart from Hewett's decision to widen the circle of recipients of gratuitous relief, the government also provided larger rations. As "the scale of the diet prescribed by the Famine Code was discovered early in the year to be insufficient; a revised and more liberal scale was therefore prepared under medical advice", noted the official government report of 1908.[73] As government policies prevented that starvation weakened the population, only a handful of poorhouses were opened.[74]

The UP famine of 1907/8 illustrates changes in famine relief in India in the early twentieth century. Not only did the responses of provincial governments to famine change, but the Servants of India Society (SIS), which would become an important Indian provider of famine relief in the following decades, made its debut by sending three of its members to Mirzapur during the famine.[75] The founder of the SIS, Gopal Krishna Gokhale (1866–1915), had started to devise his plans to launch an organisation devoted to social service in the aftermath of the famines of the 1890s, and eventually inaugurated the SIS in Poona (today's Pune) in 1905.[76] The importance of the assistance provided by the SIS cannot be established in numbers as its relief efforts were marginal in comparison to that of the colonial state. The SIS remained a small organisation. Exercising a strict selection of applicants, it would grow from four members in 1905 to nineteen members in 1917.[77] Despite the small size of the organisation, and its limited resources at hand to relieve famine, it provided crucial impulses for the reform of colonial famine relief, substituted it and held colonial officers accountable when they failed to mitigate hunger.

In response to the famine in UP, volunteers of the SIS opened their headquarters in the centre of Mirzapur in July 1908. After that, they began to tour villages to inspect the situation and compiled lists of those in need of assistance. They issued tickets to those identified as needy and after a month and a half, the volunteers ran four relief centres, which catered for an estimated 3,000 persons daily. The assistance consisted of doles of food, money and clothes and concentrated on widows, orphans and other children, who were currently without government assistance.[78] The volunteers also opened poorhouses, which intended to cater for the weakest. The famine relief of the SIS in general and the management of poorhouses in particular resembled the colonial system. This resemblance also showed in the enforcement of caste. The SIS assigned "lower classes" with sanitary work and with bringing fuel to the poorhouses, whereas Brahmin women were tasked with preparing food. Meals were provided according to caste-divisions and a mixing of groups was prevented.[79] Gopal Krishna Devadhar (1871–1935), a founding member of the SIS and the leader of the mission, enthused that the poorhouses created a "*real home* where people were made to breathe the atmosphere of real love and affection and where some efforts were made to give the inmates an idea of a better and higher life."[80] Rather than love and affection, however, the SIS exerted control to ensure that the famine-afflicted took up the expected routine. As Devadhar himself noted:

> Work in the Poor-House was in no way smooth and pleasant. People had to be trained to get up early and to be clean before going to their appointed work. Children had to be forcibly

> taken from their mothers to be sent to school [...] In fact for the first fortnight we had to be very strict, rigid and at times cruel.[81]

The relief work of the SIS during the year 1907/8 demonstrates that colonial anti-famine policies left an imprint on non-governmental famine relief in India, as Indian organisations copied and adapted the famine codes in their own relief efforts. By the turn of the twentieth century, colonial anti-famine policies in India were widely regarded as a standard of modern famine relief in- and outside the subcontinent.

Watchdogs and Critics: The Role of Indian Social Service Organisations

Members of the SIS were not uncritical of the rules and regulations enshrined in colonial anti-famine policies. They used the famine codes to hold local governments accountable, recommended changes of the famine codes and oversaw their reform. After the SIS had started to provide famine relief in 1907/8, mitigating famine became an important part of the organisation's commitment to social service. The expansion of this strand of activities also reflected in the growing geographical scope of its famine relief. Whereas volunteers of the SIS initially concentrated on relieving distress in UP, they soon travelled to other parts of India to alleviate famine.

Alarmed by the current food situation in Orissa in 1920, the SIS sent one of its members to the province. Amritlal Vithaldas Thakkar (1896–1951), a trained engineer who had joined the SIS in 1914 at the age of forty-five had received his training from Devadhar and became an ardent follower of Gandhi.[82] After gathering experience in famine relief from 1914–20, Thakkar took the lead in the famine relief operations of the SIS in different parts of India. Thakkar arrived in Orissa in May 1920 and spent the initial two weeks travelling across the province to assess the situation. After inspecting nineteen villages in total, he concluded that acute starvation was rampant in Orissa.[83] Unlike the timely response of the UP government to famine in 1907/8, the government of Bihar and Orissa ruled out the possibility that people in the province died of starvation and attributed the rise in excess mortality to diseases. Based on his inquiries, Thakkar estimated that at least 3,000 people had died of hunger in Orissa and accused local officers of concealing starvation.[84] The widely reputed Oriya social reformer and *swadeshi* activist Gopabandhu Das (1877–1928) likewise penned a report that incriminated the Bihar and Orissa government, charging it with "criminal neglect of duty" in the *Modern Review*.[85]

Indian social reformers, using the print media to amplify their criticism, held colonial administrators accountable if they failed to respond to famine swiftly and adequately. This was also the case in the Madras Presidency, where a drought called the SIS again to action less than a year after its volunteers had travelled to

Orissa. In 1921/2, parts of Madras were gravely affected by the serial failures of the south-west and north-east monsoons. Irrigation networks failed to prevent the drought from destroying harvests.[86] In 1921, R. Suryanarayana Rao, member of the SIS and the Social Service League of Madras, was deputed to the region to assess the situation and to oversee the relief work of Indian relief agencies. Although the government of Madras responded swiftly, declared famine in the districts Bellary and Anantapur and opened relief works in March of the same year, Rao argued that the participation of Indian organisations in the relief effort was crucial. Rao was a vocal critic of the Indian famine codes. According to Rao, the flaws of the famine code and thus the official response to famine rendered the relief work of Indian organisations "absolutely necessary to avoid suffering."[87] In 1921, Indian relief providers, including the SIS, operated cheap grain shops, distributed *conji* to children and milk to babies and offered medical comforts and clothing in parts of Madras.

The famine of 1921/2 marked the beginning of a prolonged period of agricultural crisis in the Madras Presidency. Madras was gravely affected by the global depression that led to a credit crisis in the province and rendered the agricultural population vulnerable to famine.[88] Between 1922 and the beginning of World War II famine visited parts of Madras five more times, in 1924, 1931–32, 1934–35, 1937–38 and 1938–39.[89] In this period, members of the SIS took part in the administration of relief and pushed for the reform of the Madras Famine Code.

Rao had offered his services to the committee that evaluated the relief operations on behalf of the government of Madras for the first time in 1921. He led an expedition of the committee members to the famine-affected areas and made concrete recommendations for the revision of the provincial famine code.[90] The *Servant of India*, the weekly of the SIS, summarised Rao's recommendations for its readers, helping to bring his demands to the attention of a wider audience.[91] Improving the wages of famine workers emerged as one of Rao's key concerns. He cited nutritional studies that outlined the minimum requirements of prisoners to substantiate his demands for a rise in the earnings of labourers on famine relief works. Comparing the diets in prisons and on public works during famine, Rao noted that not only were Indian prison diets grossly insufficient in terms of calories and vital nutrients, but famine workers received even less. Rao concluded that "a convict is treated better than a hard-working honest cooly"—an observation which had been made recurrently from the mid-nineteenth century, without leading to a major revision of the wage system of the famine codes.[92] Whereas jail diets had gradually improved in India, the nutritional content of famine relief remained static.[93] Medical officers had long advocated that famine labourers received the same (if not better) diets than prisoners, yet famine workers still lived on rations that were inferior in both quantity and quality. The rejection of the demand to align the diets of famine workers and prisoners was commonly justified with the different duration of state assistance for these two groups. Famine relief was only intended to tide over brief periods. In the event of famine, the temporary nature of state assistance legitimised diets that were known to cause malnutrition if given over a longer period of time. In other words, diets sustaining famine sufferers for weeks and months were not applicable to prisoners who served longer sentences.[94] As will be shown below, medical opinion in the interwar period demonstrated that in areas where populations recurrently fell back on government assistance, the health of famine-sufferers, in particular children, was severely affected by this policy.

The committee tasked to advise on the revision of the Madras Famine Code in 1921 was inclined to some of Rao's recommendations. The government of Madras however refused to implement them. Rao continued to lobby persistently for a general overhaul of the Madras Famine Code. After a decade and a half of studying the consequences of colonial anti-famine policies in Madras, Rao found a more conducive political environment in 1938. At the first elections under provincial autonomy in Madras in 1937, the Indian National Congress had won a landslide victory.[95] Rao joined the Madras Famine Code Revision Committee in 1938, when parts of Madras were still in the midst of a famine.[96] Alarmed by the scale of the famine that affected several districts at once, the committee evaluated the current response and reached the conclusion that a "bolder policy" was needed.[97] It offered far-reaching recommendations for the reform of the provincial famine code, such as the abolition of the grain equivalent and the increase of wages and rations.[98] It noted that women workers received an insufficient amount of calories and that children's rations lacked fat. It also pointed out that the allowances of vegetables were lower than in jail diets and thus generally insufficient.[99] These minute recommendations were overshadowed by the demand of a more fundamental change. The committee recommended that the preamble of the Madras Famine Code broadened the duties of the government in times of famine.

> It must also be remembered that while the main object of state intervention is to save life, it is non the less essential to maintain people in good health to prevent physical deterioration and dispiritedness among them so that they may be in a position to resume their ordinary pursuits with advantage to themselves and the State on the advent of better times.[100]

In a departure from the earlier wording, this amendment was intended to introduce a duty on the part of the government to preserve the health of starving populations rather than to prevent their (immediate) death. The committee's recommendations resulted in an overhaul of the famine code, which became known as "the liberalisation of the principles underlying famine relief."[101] Whereas the change to the preamble was adopted, it remains unclear to what extent the celebrated liberalisation of famine relief led to a substantial change of relief practice—in particular since the onset of World War II geared India's economy towards British war needs. The revision did not end Rao's efforts to improve the state's famine response either. In 1939, during another outbreak of famine in the Ceded Districts, Rao again travelled to the affected areas to document whether officials implemented the changes made to the Madras Famine Code in the previous year.[102]

The Rise of Nutritional Science in the Interwar Period

Provincial governments sought the advice of nutritionists with increasing regularity in the interwar period to evaluate anti-famine policies. Their influence however remained limited. In 1935, when Wallace R. Aykroyd (1899–1979) became the director of the Nutrition Research Laboratories in Coonoor, he advised the committee summoned to assess the recent famine and provide recommendations for the reform of the Madras Famine Code. Aykroyd and his colleague B.G. Krishnan had toured the famine-afflicted districts in Madras to study the health of children in labour camps and noted the gross Vitamin A-deficiency of children's diets.[103] In response to his findings, Aykroyd proposed the inclusion of a larger portion of green vegetables in famine rations to counter the prevalence of deficiency diseases, but his recommendations were not taken up by the government. Little had changed when three years later, in 1938, Reginald Passmore (1910–99), an expert on energy metabolism and a colleague of Aykroyd at Coonoor, took up the role as the official nutritional adviser of the Madras Famine Code Revision Committee.[104] Passmore reiterated Aykroyd's earlier findings on the deficiency of children's diets, which he attributed to "the continued use of the grain equivalent and inadequate supervision of the kitchen and stores."[105] Following on from Aykroyd's earlier research, Passmore studied the effects of gratuitous relief on children in the district of Bellary in Madras. His study found that 35 per cent of the 1,175 children examined suffered from Vitamin A-deficiency which was likely to result in their stunted growth and development.[106] The findings of Passmore stood out for two reasons. First, they threw light on the flawed nature of a colonial rhetoric that had presented the state as a surrogate parent. The fact that colonial anti-famine policies fed children in Bellary on diets that caused stunted growth reveal the failure of colonial *in loco parentis*.[107] Second, the nutritional expert noted that the official definition of famine in India as a short and exceptional situation was misleading. Given the frequency of famine in Bellary, children, on average, lived on famine relief three to four times for a period of six months before reaching adulthood. The provision of diets designed to prevent (immediate) death but not tailored to maintain health had serious consequences for children's development.[108] Despite this alarming finding, Passmore concluded that it was not possible to bring children's rations into line with nutritional ideals and proposed a diet that, although improved, continued to violate nutritional standards.

> It should be realized that the recommendation of the above diet schedule modified from the present Code cannot be a source of satisfaction to the nutrition worker. It contains no first-class protein, and children reared on it will not reach their full size, or acquire an optimum physique. The schedule represents no permanent standard or ideal; it should be taken as a purely temporary compromise with the economic conditions. The best that can be said for it is, that, if accepted and put into practice, a marked improvement on the present condition of children in the Bellary famine camps would result.[109]

This was not the new "bold policy" which the committee wished to adopt in 1938. In accordance with the organising principle of the famine codes that had long pledged "to maintain health" but not "the normal level of comfort", Passmore argued that relief rations of children needed to be poorer than children's diets in "normal times".[110] In other words, the general prevalence of malnutrition among children in the district of Bellary—"the normal level of comfort" in colonial parlance—mandated that children's diets fell below nutritional standards in times of famine. Although the advancement of knowledge on food requirements and malnutrition diseases allowed nutritionists and social reformers to present evidence to prove that colonial famine relief violated nutritional standards, such findings did not challenge the economic logic of colonial famine relief. Feeding children affected by famine on diets that complied with nutritional standards was considered beyond the capacity of the colonial state.

Beyond Famine: Debating Nutritional Health in the Interwar Period

The decision of nutritionists to study the effects of famine relief on the health of children was no coincidence. Children and mothers moved into the focus of nutritional scientists in the interwar period when the recognition of high infant mortality in India sparked a new wave of scientific inquiry.[111] At the time nutritionists were invited in Madras to consult on the colonial management of food during famine, famines were no longer considered the primary manifestation of starvation in India. Nutritional studies of the interwar period revealed a major flaw in the colonial response to starvation which neglected the general prevalence of mal- and undernutrition outside the context of famines. Nutritionists demonstrated the general state of malnourishment in colonial India, arguing that the focus on famine-induced starvation obscured micronutrient deficiencies. The poor nutritional health of India's population, leading health officers opined, rendered Indians prone to diseases and was no less lethal than outright famines.[112] High mortality rates among children and infants were traced to maternal malnutrition. The nutritionists' discovery of the long-lasting impact of mothers' poor health on the physical and mental development of children struck a chord with Indian social reformers and politicians who now strove to prevent the grim vision of a nation suffering under a debilitated population.[113] Poor maternal health challenged nationalist tropes of a strong and unwavering Mother India and led to efforts to improve the nutritional health of Indian women.[114] Whereas Indian social reformers and politicians flagged the need for nutritional education, the task to improve the diets of children and mothers became a tenet of the work of newly founded women's organisations.[115] Volunteers of the Seva Sadan (Mission to the Women of India) in Poona for instance began carrying out annual exhibitions of the Baby and Health Week Association that educated middle- and lower-class audiences about healthy diets and foods.[116] The Seva Sadan initially concentrated on promoting the education of abandoned high-caste Hindu wives and widows, but later engaged in the welfare of working-class women as well.[117] In 1927, the Theosophist and suffragist Margaret Cousins (1878–1954) founded the All-India Women's Conference (AIWC) in Poona, which soon after took up the task of improving the health of mothers and children. Nutritional education assumed a more prominent role in the programme of the AIWC especially after 1931, when it adopted a scheme to enlist public support for the improvement of child and maternal health.[118]

Although the colonial administration began to acknowledge that malnutrition constituted a major public health concern in India in the interwar period, it discouraged debate on its political and economic causes and constructed malnutrition as a medical problem. As Sheila Zurbrigg has convincingly argued, the recourse to a medical nutritional discourse allowed colonial administrators to deny the state's responsibility in producing and mitigating malnutrition in India by attributing imbalanced diets to ignorance and custom. Colonial rhetoric established Indian food habits, that were supposedly the result of "backward" culture, as the cause of malnutrition in India. It also denied that sections of the Indian society failed to fill their stomachs even in the absence of famine conditions.[119]

In terms of Indian food habits, health officers and nutritionists in India were particularly concerned over the "poor rice diet" that, in neglect of regional and local differences, was commonly associated with South India and Bengal. The rice diet that consisted largely of rice and pulses, was allegedly less nutritious than a diet that included wheat and meat. Claims about the nutritional inferiority of rice, especially in its milled and polished form, gained weight after David McCay published the results of his study of jail diets in 1912. The professor of physiology at Calcutta's Medical College provided a dietary explanation of the martial race theory that linked the consumption of rice to the allegedly poor physique of Bengalis and South Indians. He attributed the supposed superior physique of

Sikhs, Rajputs and Pathans in northern India to a diet that consisted of wheat and animal products and, thus, was richer in protein.[120] In 1926, Robert McCarrison conducted his widely cited rat experiment that assessed the effects of different diets on the health of rodents and drew conclusions about the result of such food consumption in humans. According to McCarrison, his findings validated the long-claimed inferiority of the Indian rice diet.[121] In 1937, researchers at Coonoor built on McCarrison's previous work, when they fed "typical Indian diets" to suckling rats in order to study the prevalence of mortality of their offspring. They found a corollary between the "poor rice diet" fed to lactating rats and the early death of their progeny.[122] Apart from investigating the prevalence of malnutrition, researchers also tested possible remedies. Aykroyd and his colleagues for instance stepped outside the laboratory in 1937 to test the suitability of protein supplements in improving the health of children. The experiment demonstrated that the health of children proved unaffected when they were fed soybean food supplements, but those who consumed milk powder not only gained weight but also proved healthier in general.[123] Skimmed milk powder had no nutritional advantage over fresh milk, but an economic one: as a by-product of the dairy industry, it was considerably cheaper than milk and thus was more suitable for the large institutional feeding programs that Aykroyd envisioned in India.[124] Milk was also much easier to store and to transport in its powdered form. A main caveat however remained. India did not produce sufficient quantities of milk powder for large-scale domestic use and would have to import substantial quantities. Although available at a low price outside India, an import duty of twenty per cent (from within the British Empire) and thirty per cent (from outside the British Empire) made imported milk powder expensive. Aykroyd therefore strongly argued for the end of import duties but failed to persuade the colonial regime.[125] In the following years, the world's leading nutritionists reiterated the importance of milk, either in its liquid or powdered form, in the human diet. This group included Aykroyd, who in 1940 again concluded that "milk is the best supplement to rice diets."[126] In India, the nutritional superiority of milk was not only flagged by British nutritionists. In 1936, Narasinh Narayan Godbole (1887–1984), Professor at the Benares Hindu University who had attained his doctoral degree in Berlin with a dissertation on the molecular composition of peanut oil, concentrated on establishing the advantages of consuming milk in comparison to meat.[127] His study, titled *Milk. The Most Perfect Food*, was the brainchild of the three-time president of the Indian National Congress, social reformer and leading figure of the Hindu Mahasabha Madan Mohan Malaviya (1861–1946).[128] Malaviya explained in the foreword of the book that the study was inspired by the recent surge of global interest in milk, which neglected India's traditional knowledge on the matter. Consequently, Malaviya had tasked Godbole with reminding the international scientific audience that the use of milk had long been advocated by Hindu medical practice and tradition.[129] Godbole, in fulfilment of his task to educate his readers about the older Indian knowledge on the advantages of milk, outlined in detail why Indian vegetarians should increase the intake of milk to consume more protein. As he merged ayurvedic and allopathic knowledge to advance his argument, he dwelt in length on the benefits of a strict Hindu vegetarian diet. Referring to the abstinence of Indian Hindu vegetarians from meat and eggs, he propagated the superiority of Hindus and Jains over European vegetarians, as well as over meat-eating British colonizers, Indian Muslims and Sikhs.[130] Thereby, Godbole turned long-held racial hierarchies (and the findings of McCay's earlier nutritional study) upside down and propagated a "vegetarian humanitarian ethic" that condemned meat-eaters for the death of living-beings.[131]

While Godbole's Hindu nationalist leanings undergirded his writing, the importance that nationalists awarded to the consumption of milk was much wider. One of India's most vociferous vegetarians, M.K. Gandhi, had recently begun to embrace the consumption of milk. Curiously, Gandhi strove in his lifetime to eliminate milk from his own diet. He had initially opposed milk consumption in his earlier experiments with veganism but returned to milk consumption after he had contracted dysentery in 1918. Gandhi decided to consume goat rather than cow milk, and at first only tentatively referred to his dependence on the white liquid, arguing that he needed milk to acquire the physical strength needed to carry out his political struggle. He later, however, turned into a vociferous advocate of milk—a change of mind, which was influenced by his frequent exchange with nutritionists who convinced him of the benefits of consuming adequate amounts of proteins. In the interwar period, Gandhi emphasised the importance of milk in the Indian diet, which was meant to supply Indians with the physical and mental strength needed to shake off colonial rule.[132]

An important stimulus for the growing interest in milk in India and beyond was a publication by Cicely Williams (1893–1992). In 1935, Williams, who worked as a colonial medical officer in the Gold Coast region, recognised a series of symptoms in her younger patients that were not recorded in the medical textbooks available to her.[133] She described her discovery of a severe form of protein-malnutrition among children and borrowed a term from the Ga language to label it: kwashiorkor. The term kwashiorkor translated into "disease of the deposed child", which resulted from early weaning and often caused the death of the child affected by it.[134] Williams noted the success of treating protein-malnutrition through the feeding of milk powder. In the aftermath of Williams's findings, new international nutritional standards attributed importance to milk consumption. In the 1930s, the League of Nations Health Organisation (LNHO) suggested "a daily intake up to one litre per day for pregnant and nursing women, as well as to provide an abundant supply for infants, children of all ages and adolescents."[135]

In 1935, shortly before Aykroyd took up his work as the director of the Coonoor Research Laboratories in India, the LNHO had tasked him and Etienne Burnet to oversee an inquiry of the nutritional health of people in the UK, France, the US, Denmark, Sweden, Norway, and the USSR. The results of the study informed the new dietary standards that the League of Nations and the International Labour Office published in 1935 and 1936.[136] The study also shaped dietary recommendations in India. In 1937, Aykroyd distributed the findings through the *Health Bulletin* of the Indian Nutrition Advisory Committee (NAC). *The Nutritive Value of Indian Foods and the Planning of Satisfactory Diets* was reprinted several times in the following two decades, with a fifth edition published in 1956.[137] When the Department of Nutrition of Bombay Government in 1959 published a diet manual for institutional feeding in hospitals, schools, prisons and factories, it still relied on Aykroyd's findings.[138]

In the context of new nutritional standards emerging in the interwar period, even more important than assessing milk consumption was the caloric measurement of diets. Informed by Wilbur O. Atwater's thermodynamic model that introduced the calorie to nutritional parlance, the LNHO defined a caloric minimum that in principle was applicable on a global level and allowed for the comparison of national diets in terms of their energy content.[139] Yet, it similarly introduced several qualifications that translated into a differential standard for "oriental" populations.[140] Reasons given for this deviation included the current low caloric intake of populations in the East as well as the agricultural basis of the countries' economies; peasants ostensibly burned less energy on the field than workers in factories.[141] The claim that Indians needed less calories than Europeans also found supporters among Indian researchers. In his monograph *Food Planning for Four Hundred Millions* (1938), Radhakamal Mukerjee (1898–1968), professor of economics and sociology at Lucknow University, argued that the small body height of Indians, coupled with their climate-induced low metabolism, translated in less calories being burnt by Indians than by people living in Europe and the United States.[142] Despite his embrace of differential caloric requirements, Mukerjee nevertheless acknowledged the general poor level of food consumption in India and recommended the improvement of diets.[143] The professor moreover stressed the impact of nutritional health on labour productivity and explained that Indian industrial workers compensated their lack of nutrient-intake with idleness and absenteeism.[144] In the aftermath of the Indian Industrial Commission (1916–18), the purportedly low productivity of the Indian industrial labour force moved into the focus of Indian social reformers, who emphasised the need for welfare measures to improve national efficiency.[145] Mukerjee noted that industrial work warrants a higher calorie-intake than agricultural work—a proposition that found wide support. Diwan Chaman Lall (1892–1973), a founding member of the All India Trade Union Congress, similarly demanded the increase of workers' food intake by remarking with resentment that industrial workers in India were forced to live on a "famine ration".[146] In *Coolie: The Story of Labour and Capital in India* (1932), Lall compared the famine codes' regulation of food consumption to prison diets and to the food intake of factory labourers in order to reveal the poor nutritional health of India's industrial classes.

Thus, the nutritional health of sections of Indian society who failed to fill their stomachs despite the absence of famine received greater attention during the interwar period. Yet, while the focus on mal- and undernutrition outside the context of famines partly replaced the earlier focus on the colonial administrative responses to famines, famine continued to be an important point of reference in debates on Indian public health.

1.4 Conclusion

The assumption that it was too costly and generally in conflict with economic principles to provide famine relief on a scale that met nutritional standards and medical opinions on healthy eating undergirded the Indian famine codes—the rules and guidelines that were meant to govern the colonial famine response. The famine codes, despite their reform and revision, consistently fixed food rations and wages below standard. This was to limit expenditure and prevent people from abandoning their regular occupations in favour of state assistance. In other words, people affected by famine had to eat worse than they normally did, because a higher level of relief would ostensibly entice people to seek state assistance who could do well without.

Medical officials such as William Robert Cornish pointed out in the nineteenth century a weakness of the economic argument against the use of dietetic standards in famine relief. He argued that improving the nutrition of people in state care made economic sense, because compromised health was a major obstacle to labour productivity. Members of the SIS, themselves involved in mitigating famine, similarly tapped into nutritional studies to advocate for improved famine relief. R. Suryanarayana Rao cited studies to reveal discrepancies in the administration of food in different colonial institutions, comparing prison diets and famine rations to demonstrate the poverty of diets administered during famines. Medical and nutritional research on dietary standards had no immediate impact on the famine codes, but it gradually shaped aid practice. As John Hewett's administration of relief in the United Provinces in 1907/8 revealed, colonial administrators, alerted by nutritionists to the inadequacy of aid, at times deviated from the famine codes and issued larger rations and higher wages than recommended. At other times, colonial famine relief remained inadequate from a nutritional perspective. Against the background of heightened interest in the health of mothers and children, nutritional studies carried out in Madras in the 1930s investigated the impact of colonial famine relief on children. Scientists associated with the Nutrition Research Laboratories in Coonoor revealed the detrimental effect of the relief rations on the development of children in the Bellary district of Madras. Food aid was fixed at a level that did not prevent stunted growth and life-long impairment. The low standard of famine relief in Bellary was once again explained with the economic principle of famine relief, which set children's food consumption below that in normal times. Surveys noting the generally poor nutritional health of children in Bellary thus legitimised the low standard of famine relief and conveniently absolved the state of its responsibility to maintain the health of children during famines. Advocates of this "economic approach" to famine relief, including the author the nutritionist Reginald Passmore, argued that trying to significantly improve the health of children risked throwing Madras into economic jeopardy. To many Indians, on the other hand, the insufficiency of famine relief measured against nutritional standards demonstrated the failure of the colonial state to protect the most vulnerable sections of the population. For famine-stricken populations in colonial India, especially for women and children, seeking government assistance continued to be a risky survival strategy in the twentieth century. The famine codes, working with a series of simplifications and shortcuts to facilitate the administration of food, made the average working man the basis of estimated productivity and food needs. This was at the expense of the needs of women and children whose rations often remained inadequate.

Colonial administrators were still celebrating the success of the colonial system of famine management when critics of colonial rule in India reframed the debate, drawing attention to the hunger that prevailed in the country outside the context of famine. Indian social reformers identified the nutritional needs of the population at large as a pressing concern in the interwar period and broadened their own activities to address them. Indian discourses on nutritional reform and education predated the interwar period. However, nutritional studies that generated public awareness of the prevalence of maternal and infant malnutrition in India in the 1920s and 1930s accelerated the involvement of Indian organisations in nutritional reform and education. Indian nutritionists and politicians advocated a change of food habits and diets to improve the nutritional health of Indians, as well as to strengthen them physically in their political fight against colonial rule. Mirroring this shift of the debate on hunger in India that had occurred in the interwar period, famine would again take a backseat in the public debate on nutrition in the 1940s and 1950s, which was more concerned with improving the nutritional health of the population in general. The prospect and realisation of political independence at this time discouraged efforts to improve famine relief in India, since independence supposedly removed political resistance against effective famine prevention. When India became independent in 1947, the state governments inherited the famine codes that were converted into scarcity manuals from the 1950s onwards.[147] Continuity outweighed the changes made to famine relief in the process. In Bombay, where the new scarcity manual replaced the state's famine code in 1954, the main difference was the abolishment of the differentiation of scarcity and famine. Whereas previously, the full mobilisation of counter-measures followed only after the declaration of famine, now scarcity sufficed to set them in motion. This change was not insignificant, but it was overshadowed by the perpetuation of the earlier focus on labour and by the retention of the wage system, which emerged as continuities from the colonial period.[148]

The famine codes (and now scarcity manuals) received less commentary after independence, but they were not removed from the public debate. Some of the later discussions bore a strong resemblance of earlier debates, which illustrate unresolved contradictions of state-sponsored famine relief. In 1975, the economic historian Morris David Morris (1921–2011) lamented somewhat surprisingly that "up to now famine policy has been a subject which scandalously little serious attention has been paid."[149] Morris penned his analysis in the aftermath of a severe drought in Maharashtra. For three consecutive years, beginning in 1970, fifty million people in Maharashtra were at risk from dwindling access to food. Through a combination of anti-famine policies, including the Public Distribution System, food-for-work programmes, and the ban of food exports from the state, the government contained the famine.[150] Despite the success, Morris deplored the excessive government investment in famine relief. He recommended scaling back early investments in famine protection to free up resources for development and strengthen the population's resilience to drought in the long run. Criticising hasty interventions, Morris suggested setting the administrative machinery in motion only when consumption fell "below some established level of daily calorie intake."[151] Minimum food requirements had been widely discussed in India for over a century by this time. Morris recommended reconsidering the application of international nutrition standards. Reminiscent of debates on differential nutritional standards in the 1930s which claimed that "oriental" people needed less food than Europeans and North Americans, Morris opined that India's historical record of starvation and chronic poverty proves that international standards were excessive for Indians. Their bodies had supposedly adapted to poor food intake.[152] Echoing the economic liberalism that undergirded colonial famine relief, Morris called for a "hard calculus" to achieve a balance between humanitarian concern and economic feasibility.[153] Applying international nutrition standards during famine "would result in more being allocated to people in a scarcity area than probably is available to them when there is no drought."[154] Not long after Morris's publication, the economist

David Seckler provided further support for the application of differential caloric standards. Relying solely on his own observations, he defined a subsistence wage for Indian workers that was lower than the minimum requirements stipulated by the majority of his contemporaries. He proposed the "small but healthy hypothesis" in 1980 which claimed the ability of the human body to adapt to poor food intake. In other words, Indians could live on a nutritional level below recommended international standards, because of their reduced body-height and their slow metabolism. The "small but healthy hypothesis" built on earlier demands for national rather than international nutrition standards which had emerged in the interwar period. As a hypothesis, Seckler's claim provoked debate within the scientific community. It however lost its innocence when translated into political advice. If Indians could live on poor diets, India's nutritional crisis was less severe than assumed and no longer needed to be a political priority.[155] Historicising the mobilisation of science to define needs and manage food, as attempted in this chapter, sheds light on the wide-reaching consequences of the scientific underpinnings of food aid and anti-famine policies.[156] Scientific arguments are easily instrumentalised by political actors to shake off responsibility, or when translated into policies, hypotheses become indisputable truth.

CHAPTER 2

Food Technology, Nutritional Science, and Indo-US Entanglements in the 1940s and 1950s

Abstract

The chapter traces the growing collaboration between Indian and American scientists, social reformers and philanthropists in the effort to meet India's food needs against the backdrop of famine in Bengal, World War II and Indian post-independence food policy. It examines South Asia as a site of knowledge production in nutrition and anchors South Asia in the global history of food aid. The chapter shows how the use of supplements, fortified and blended foods grew in importance in the 1940s and 1950s.

Keywords: Bengal Famine, food supplements, independence, US food aid

2.1 Nutritional Science and Famine in the 1940s

Food Supplements and Substitutes

Famine had not been absent in India in the early twentieth century, but the scale of starvation that shook the population of the subcontinent in the 1940s marked a return of the horror of bygone centuries (see Chapter 6 for a discussion of the Bengal famine). Marginal aid for Bengal was coming forward from Delhi and Whitehall in August 1943, when the Bengal government started to dole out cooked food in Calcutta (today's Kolkata) and to support the relief efforts of non-state agencies by allowing them to procure food supplies at a subsidised price. The general unavailability of food coupled with the illegal diversion of stocks and the adulteration of food doles in official kitchens resulted in the poor quantity and quality of the relief provided. Not only did relief start too late, it was also by no means sufficient. Indian organisations trying to aid the affected population took creative measures to compensate the shortage. In October 1943, the district branches of the AIWC in Bengal, which were amongst the first to respond to the famine and primarily aided women and children, reported on the unavailability of milk. It started using a texture that was made of soybean, barley, and shark oil to offer some nourishment.[1]

In early November 1943, the *Hindustan Times*, an Indian English-language daily from Delhi with ties to the INC and a readership in the millions, published a cartoon

Fig. 1. British Food Supplements for Bengal. The *Hindustan Times* November 7, 1943.

that mocked British food aid for Bengal (figure 1).[2] The cartoon captured the cynicism of the British gesture of supplying multi-vitamin pills, cod liver oil and halibut capsules after turning a blind eye to the suffering in Bengal for nearly a year.[3] With the British government prioritising the war effort, the Bengalis who had suffered and survived starvation had now reached a level of physical deterioration that required therapeutic feeding. Their digestive systems could no longer cope with regular food intake. The arrival of food supplements to Bengal however owed more to the preferences of governments and humanitarian organisations than to the needs of the affected population. Pills, capsules and milk powder were easy to store and transport and less bulky than other food items. This was important at a time when governments prioritised the transport of war supplies. In addition, the sending of food supplements allowed the government to appease the public in Britain, where calls for government intervention to mitigate the Bengal famine had become louder.[4] At the time of the cartoon's publication in the *Hindustan Times*, Britain had eventually started to send substantial relief to Bengal. The newly appointed Viceroy Lord Wavell rushed food into the province in November 1943. However, the aid came too late to prevent the deaths of an estimated three to five million people.[5]

The use of food supplements to alleviate starvation during World War II built on developments of the interwar period. Nutritionists had started to emphasise the nutritional and economic advantages of milk powder in the 1920s and 1930s before it turned into a central element of famine relief efforts in India and beyond. Powdered milk was a surplus product in the dairy industries and one of the world's leading producers was the United States. The growth of the US dairy industry coincided with the revised estimate of the nutritional value of milk powder in the US. In the interwar period, Americans regarded milk powder as a vital source of protein, calcium, phosphorous, and vitamin G.[6] The white powder also turned into a relief item after World War I, when the American Relief Administration (ARA) used it to alleviate famine in Soviet Russia in 1921. Canned dried milk brought its nutritional benefits to distant populations and also had a strong emotional appeal. As Bertrand M. Patenaude notes in his exploration of the ARA's relief work in Russia, "of all symbols of American beneficence—corn, kasha, white bread, cocoa—it was milk and its associations with maternal nurturing that seemed to strike the deepest emotional chord."[7] World War II intensified the use of milk powder which turned into the main ingredient in US American food aid, supplementing the diets of American troops and British citizens.[8] In Bengal, American powdered milk was applied to counter malnutrition in late 1943 and early 1944. Donations to US American missionary societies and the Quaker American Friends Service Committee (AFSC) advanced the distribution of milk in India. The American aid organisations forwarded the white powder to the AIWC, the SIS and the Indian Red Cross Society who used the milk in their relief campaigns.[9]

The experiences of relief agencies in Bengal substantiated the suitability of milk as food aid. In India, relief workers "were struck by the improvement in under-nourished destitute children which took place when they were given milk for a few weeks."[10] Further support for the food supplement was coming from the United States. A group of advocates of food aid for India assembled the India Famine Emergency Committee (IFEC) in 1946 to mobilise against the discriminatory nature of the US-led international food aid regime (see Chapter 7 for details). Apart from its political work, the IFEC contemplated the feasibility of providing material aid to India. Weighing the advantages and disadvantages of different forms of food aid against each other, it found inspiration in the work of the Cooperative for American Remittances to Europe (CARE) that was currently sending food packages to Europe.[11] The IFEC explored ways to launch a similar food package programme for India. While it deliberated over the content of the food parcels and debated which products to send, it sought advice from Wallace R. Aykroyd, who had then become the first nutritional expert of the FAO.[12] Aykroyd strongly recommended abandoning the idea of the food package programme altogether to concentrate instead on milk powder—a recommendation the committee eventually agreed to.[13] The IFEC, however, did not base its final decision only on Aykroyd's insights. The purchasing agent of the AFSC, H.H. Thompson, had similarly advised the IFEC to opt for milk powder. Drawing from the AFSC's experience in Europe, Russia, China and India, he considered individual food packages inefficient in terms of costs and instead recommended appealing to Americans to donate one dollar each for ten cans of evaporated milk sent to India.[14]

The AFSC was not the only American aid provider sending milk powder to India. Church World Service alone shipped 80,000 pounds to India in 1946.

Indeed, milk continued to occupy a central place in the relief efforts of missionary organisations in India well into the 1950s. By 1955, the National Christian Council in India would call milk powder "one of the most effective weapons in the war against undernourishment."[15] The organisation most strongly associated with milk powder, however, was the United Nation Children's Fund (UNICEF), which became the main purchaser of US surplus milk after 1946. UNICEF also began to engage in a milk conservation programme that sponsored the establishment of milk processing plants in various countries to boost local milk production. In 1955, UNICEF co-sponsored the establishment of a dairy plant in cooperation with a milk cooperative in Kaira (today Kheda, Gujarat), which would later become the famous Indian Amul brand.[16]

Starvation Science and the Medicalisation of Famine Relief

The now intensive use of milk powder to alleviate famine in and outside South Asia was only one of the changes that famine relief underwent in the 1940s. The use of food supplements and substitutes indicated a medicalisation of famine relief that favoured techno-medical solutions to starvation. In October 1943, the Nutrition Advisory Committee (NAC) of the Indian Research Fund Association (IRFA) began to criticise the insufficiency of relief rations in Bengal, pointing out that rations currently amounted to no more than 800 calories per person per day.[17] This was a third of the amount the Bengal Famine Code prescribed.[18] The NAC formed in 1936 to advise the Indian government on matters of nutritional reform and food supply. It was the result of growing international pressures on the British government that heightened in the aftermath of the findings of the LNHO. The NAC comprised of the Public Health Commissioner with the Government of India, the director of the Coonoor Nutrition Research Laboratories who at the time was Wallace Aykroyd, and a group of British and Indian nutritional, agricultural and economic advisors.[19] The evaluation of famine relief was not part of NAC's mandate, but in the last quarter of 1943 it began to deplore the insufficient caloric value and nutritional composition of the gruel doled out at government relief sites in Calcutta. It suggested that famine relief was to be based on "expert nutritional advice."[20] Apart from deploring the size of famine relief rations, the NAC wished to learn about possible avenues of treating the most famished sufferers of famine who were at that time collected in the streets of Calcutta and admitted to hospitals. To this end, the IRFA allotted a grant of 12,000 rupees to the All-India Institute of Hygiene and Public Health in Calcutta (AIIH&PH) to finance a clinical study of the best "scientific methods of treating famished people."[21] The results of the clinical study were summarised in a leaflet titled *Treatment and Management of Starving Sick Destitutes*.[22] It contained concrete instructions on how to treat starvation in Bengal and recommended that doctors classify famine patients into one of four categories, and specified treatment accordingly. For patients able to consume and digest food (group one), the leaflet recommended the intake of a gruel diet. The gruel was to consist of one and a half pounds of *dal*, six pounds of cereal mixture, and two and a half pounds of vegetables, to be cooked in water and seasoned with condiments and salt. The second group included patients who were unable to digest the gruel and hence needed nourishing fluid diets. The fluids provided were fruit juices, if available, but more often consisted of flour soaked and boiled in water, and enriched with sugar, salt, yeast, and shark liver oil (this would later become known as the Bengal Famine Mixture). For patients able to digest milk (group three), nutritional advisors recommended a supply of various milk preparations such as *ghol* (a yoghurt drink), *conji* or sweetened milk diluted with water. The size of rations was to be determined by the medical staff, who wrote the sort of diet and the number of calories to be consumed on the patients' cards, which were then placed around their necks. While medical officers needed to adjust the quantity of nourishment to the patient's condition, the leaflet nevertheless provided a general estimation. Thus, a fluid diet was supposed to contain 800 calories, a milk diet 1,200 calories and a gruel diet 1,900 calories.[23] Finally, the AIIH&PH advised treating patients found in a state of collapse (group four) through nasal feeding or intravenous injections of protein hydrolysates.[24] Whereas the administration of special diets to patients was grounded in earlier relief practices in India, the use of intravenous injections to insert nutrients into famished bodies was altogether new. To manufacture protein hydrolysates, the AIIH&PH used an enzyme to pre-digest meat and extract protein. The Indian researchers dissolved the extracted protein in a glucose solution, so that it might be used on patients.

Indian researchers tested the use of protein injections on patients in Calcutta's emergency hospitals in November 1943.[25] It was the first time that researchers in India had tried protein hydrolysates in humans. Only a few months earlier, physicians in the United States had announced some success in using protein injections on patients who were otherwise unable to consume or absorb protein.[26] In Bengal, researchers tried the protein injections on people in a state described as inanition collapse, which referred to a loss of eighty percent of their normal weight. These patients were unconscious and unable to take even small quantities of liquid food.[27] The fact that a quarter of the people brought to emergency hospitals in Calcutta were in a state of inanition was proof of the severe delay and insufficiency of relief measures.[28] Science had become a last resort, relied upon to prevent the death of people in an apparently hopeless physical condition. The fact that these people were believed to be beyond saving also seemed to justify the use of experimental drugs. By the end of 1943, the outcome of the use of protein hydrolysates in hospitals in Calcutta looked favourable: "I think it [protein hydrolysates] pulled round some of the patients who were, according to previous experiences, considered hopeless" reported B.N. Bhandari of the Indian Medical Corps.[29] After such positive evaluation, AIIH&PH recommended the use of protein injections for cases of inanition collapse throughout Bengal. It supplied medical staff in Bengal with protein hydrolysates and instructed them to document their use. In the following weeks, protein injections that were meant as a last remedy were applied far beyond their intended use. M.V. Chari served in the medical battalion of the Army in the Burma Reserve of Officers (A.B.R.O) and was delegated to work in Barisal in Bengal (today in Bangladesh) from November 1943 to January in 1944. Chari documented his use of hydrolysates on patients who were able to consume food.[30] One of his patients, a young girl of nine years, had been recovering on a salt free diet of rice and *dal*. Chari, however, sought to speed up her recovery by administering protein hydrolysates. The solution given to the girl however was contaminated, causing cyanosis and the quick death of the child.[31] As Chari pointed out in hindsight, maintaining sterile conditions and storing the protein hydrolysates safely remained challenges given the poor standards of the medical facilities at hand. Injecting hydrolysates therefore meant a much greater risk than conventional methods.

Chari's clinical study furthermore illustrates the wish of medical practitioners to contribute to the scientific discourse on nutrition. The impetus on the part of medical staff and nutrition scientists to advance their knowledge on starvation during the Bengal famine can be traced in leading Indian and British medical journals, which presented a variety of nutritional experiments and clinical studies in the aftermath of the famine.[32] The desire to advance the understanding of starvation is also visible in the letter of a nutritionist and physician named R.C. Bhattacharjee to Syamaprasad Mukherjee, the leader of the Hindu Mahasabha. In September 1943, the overtly ambitious graduate of the Universities of Calcutta, Leipzig and Paris approached Mukherjee for gaining access to the relief sites administered by the Hindu Mahasabha.[33] In an effort to convince Mukherjee of his plans to conduct research during the famine, he explained:

> From scientific as well as from social point of view this is a unique opportunity of making such a research. Such abnormal conditions will never prevail again and such a mass of experimental human objects in such abnormal malnutrition condition will scarcely be available to a scientist.[34]

Echoing the current trends in scientific discourse, Bhattacharjee was primarily concerned with exposing the insufficient protein intake and studying the effects of such a deficiency on the bodies of those affected by famine. He proposed to apply what were, by then, common scientific methods. These included closely monitoring people's pathology as well as chemically analysing foodstuffs before feeding and after consumption, the last through the careful collection and analysis of excretions. The designs of the proposed experiments illustrate an important point. Nutritional studies depended on the ability of scientists to exert tight control to fulfil the demands of modern science. Such control was facilitated by the involvement of the army in famine relief in Bengal. The Indian Army Medical Corps had been deployed to attend to the crisis in mid-November after the government recognised that Bengal's medical needs outstripped the capacity of the Indian Medical Service. With the arrival of the army, military hospitals, termed Relief Emergency Hospitals, were set up in Calcutta and in rural Bengal.[35] Poorhouses and relief camps had already given way to experiments in the reduction of food rations and the study of diets since the mid-nineteenth century. Military hospitals further facilitated nutritional and medical studies in 1943 by enabling control.[36]

The desire to advance knowledge on the appropriate treatment of starvation, which drove nutritionists in India to conduct research during the Bengal famine, also manifested in the United States.[37] Unlike in Bengal, where the clinical work of the AIIH&PH resulted from immediate and proximate needs, nutritional scientists in the United States were motivated by the needs of post-war Europe. In November 1944, the US American physiologist Ancel Keys (1904–2004) conducted a controlled human experiment at the University of Minnesota that came to be known as the Minnesota Starvation Project. The aim of the project was to refine the methods of humanitarian organisations that were about to assist in rehabilitating war-torn societies. The experiment involved thirty-six conscientious objectors, mostly Quakers, who volunteered to live on a diet of 1,800 calories while walking twenty-two miles daily over a period of six months. Keys and his colleagues monitored the men closely throughout this period to study the effects of starvation on their minds and bodies.[38] The men were offered psychological assistance and some of them used the initial weeks of the experiment to complete their university degrees. Though the experiment left a lasting impression on the participants, none of them suffered any permanent physical impairment. The voluntary framework chosen by Keys had been truly unique, and his ability to limit the effect of the study on participants owed much to the socio-political context in which it was conducted. The United States had been the best-fed nation during the entire war and was in a position to carefully select volunteers to starve for science and to treat them with extreme care.[39]

Research in the US and India contributed to a body of knowledge on the treatment of starvation that shaped the aid to post-war Europe. The findings of Keys resulted in a manual for US relief personnel working in Europe. The tests carried out by the AIIH&PH led to the decision of the British Medical Research Council (MRC) to produce protein hydrolysates in Britain and their use in Europe after the war by British aid workers. Following a Dutch request, protein hydrolysates became a means to alleviate starvation in the northwest Netherlands too. It was not long after that further studies in Europe revealed the weakness of the new remedy.

A delegation of British medical officers under the lead of Janet Vaughan (1899–1993) was sent to Europe to conduct a study of protein hydrolysates as a treatment of extreme cases of inanition in early 1945. The delegation's journey coincided with the liberation of the concentration camp of Bergen-Belsen in April 1945, resulting in the diversion of the British research team which began to provide medical relief for the former inmates of the concentration camp.[40] At the outset, British nutritionists observed the use of protein hydrolysates with great elation. The euphoria was misplaced. Vaughan reported that the use of protein hydrolysates in the former concentration camp proved disastrous. The tubes and needles required to puncture veins and administer the protein infusion reminded patients of instruments of torture and murder. British medical staff approaching patients with protein injections in their hands found that their patients panicked and collapsed.[41] Even patients who eventually received protein hydrolysates either recovered slowly or did not recover at all. Vaughan concluded that milk flavoured with tea or coffee was decidedly more successful in treating patients than the injections.[42] Protein hydrolysates were not the only treatment that had travelled from Bengal to Belsen. The recipe for the fluid gruel prepared in Bengal, which came to be known as the Bengal Famine Mixture, was turned into a canned, ready-to-supply product used in Belsen.[43] This transfer proved to be a failure too. Unaccustomed to the taste and texture of the Bengal Famine Mixture, inmates rejected the gruel, which also caused them stomach pain and diarrhoea.[44] The problems inherent in applying an overtly technocratic approach to remedy starvation, which entailed a shift from feeding to treating patients, became apparent in Bengal and Belsen in 1944 and 1945.

2.2 Indo-US Cooperation in Food Science: The Rise and Demise of Indian Multi-Purpose Food

The Age of Food Technology

In November 1943, the *Amrita Bazar Patrika*, a leading Indian newspaper that published from Calcutta in English and Bengali and had a readership all over the country, offered its view on the interplay of food and science in India. With scathing irony, it commented on the prospect of modern science trumping people's cultural and emotional attachments to food in India and ending all sensual pleasures of eating.

> The trend of modern scientific research and innovations in the common man's dietary will fill all gourmets and connoisseurs of dainty dishes with dismay. No more may he be called upon to preside over the table where heaps of caviar or *pilau* vie in attractive power with

> the fragrant curry that brings water to one's mouth and fills one with an anticipatory exhilaration of spirit. No more those lovely balls of nectar called *sandesh* or that miracle of the confectioner's art, *rasgolla* (of Bagh-bazar, for preference), the soothing *dahi*, the cooling ice cream of the tonic *sherbet*. Science will have none of these, at least for usefulness in the way of nutrition.[45]

Given the still disastrous food situation in Bengal at the time, the description of culinary delights seems misplaced. But the focus of the *Amrita Bazar Patrika's* editorial was not on the current famine conditions in Bengal, but on the nutritional health of the Indian society at large. At the heart of it was an assertion that was gaining support of Indian political and social elites. "Our national diet is basically wrong", stated the author of the above lines, urging the future independent government of India to draw from science to improve Indians' "physical stamina" and "resistance to diseases."[46] Criticism of Indian diets had a longer gestation, yet they were fanned in the 1940s when the prospect of gaining independence from colonial rule increased the importance of nourishing Indian bodies to enable India to stand as a nation. In the early years of independence, food planning took precedence. Indian state planners invested in agricultural reform and encouraged a change of food habits that was aimed to scale down the consumption of scarce commodities. The Indian government strove to increase agricultural production and to balance out scarcities through the import of grain.[47] Meanwhile food science offered the prospect of changing food consumption in India and improve Indians' nutritional health through altering their food habits.

In late 1950, two months before famine conditions in Bihar and Madras forced the Prime Minister Jawaharlal Nehru to request US food aid, Nehru welcomed the official inauguration of the Central Food Technological Research Institute (CFTRI) in Mysore. Preceding this appointment, unfavourable weather conditions had led to floods in the east and droughts in the south of India. Widespread crop failure affected people in Hyderabad, Kashmir, and Uttar Pradesh, but the poor harvests hit Bihar and Madras the hardest. It was the worst food crisis since India had attained independence in 1947 and it caused political turmoil in the country as Nehru and his cabinet strove to understate the famine (see chapter 7 for details). In late 1950, India officially requested a loan from the US to purchase its food grain.[48] With the inauguration of the CFTRI, Nehru, in the meantime, had tasked India's leading researchers to provide another solution to India's pressing food problem.[49]

The formation of the CFTRI, housed in the former residency of a member of the Mysore royal family on a 130-acre property, points to the role that food science and technology was expected to assume in national food planning. While in the first decade of independence agricultural experts laid out plans to raise harvest yields and politicians strove to increase food imports, Indian biochemists and food technologists were asked to improve the shelf-life of food products, identify new and unfamiliar food materials, and invent food supplements and food substitutes that balanced out dietary deficiencies.[50] The first director of the CFTRI, Vaidyanatha Subrahmanyan (1902–79), had received his training in biochemistry in India and Britain. He had set up a food technology section in the Department of Biochemistry at the Indian Institute of Science (IISc) in Bangalore in 1942.[51] The small group of scientists of the food technology section initially catered for military needs. This paralleled similar processes in the US, where concerns over the nutritional content and appeal of army rations drove innovations in food technology. Towards the end of the war, however, a grant from the Indian Council of Medical Research allowed Subrahmanyan to tailor his research to the needs of the civilian population. He became invested in the development of soy-milk that was modelled after a Chinese prototype and intended to remedy protein-malnutrition in children and mothers. Since soy was not grown in India, his experiments soon concentrated on replacing the former with other ingredients such as coconut, sesame, cottonseed, and groundnut.[52] His research was clearly in consonance with international nutritional research which gave priority to protein in those years. Adding protein to Indian meals became part of dietary reform plans that in particular targeted Indian "rice eaters". Whether Indians in general, and "rice eaters" in particular, were capable of adjusting diets was heatedly debated. In the colonial period, the assumption that Indians were unwilling to alter their diets contributed to the reproduction of stereotypes that depicted Indian culture and society as static. In the early years of independence, the deep-seated belief that "rice eaters" could hardly be accustomed to alternative foods undergirded experiments of Indian researchers with artificial rice products. Researchers were convinced that products that resembled rice would gain easier acceptance among "rice eaters" than other grains.[53] As part of this broader research agenda, India's food technologists tested a formula, called Multi-Purpose Food (MPF) which was advocated in India by a US non-profit called Meals for Millions (MFM).

Meals for Millions and the Development of Multi-Purpose Food in the United States

In mid-1946, Indian nationalist and New York-based entrepreneur JJ Singh (1897–1976) received a letter from "an American young man" who offered to share "American knowledge" on fortified and blended food with the Indian people.[54] The letter was accompanied by wrapping papers to show the wide selection of products available, and the advice that the IFEC, currently at work to mobilise political opinion in favour of America food aid for India, "send many tons of the powdered food and candy bars to the starving people of India."[55] Boosted by the need to improve army rations, research on effective meals converged with an increasing confidence in science and technology to try to solve what was understood to be a world-wide nutritional crisis. After the war, fortified and blended foods started to alter food consumption habits in the United States and beyond, also influencing the choice of food aid products.[56] This was at a time when the new confidence in science and technology drove the invention of new tools and technologies used in later development work.[57] It is therefore not surprising that, as the IFEC contemplated sending food aid to India, the organisation received offers and product samples from food companies that sought to advertise their solutions to India's nutritional crisis. The "American young man" who had written to Singh, however, was unaware that Indian institutes already conducted research on food supplements and substitutes and applied them across India. Meanwhile, Henry Borsook (1897–1984), Professor of Biochemistry at Caltech, was convinced that food science and technology could assist in feeding the world's hungry through transforming conventional food in new and more effective products.[58] Borsook invented the soy-based high protein food supplement that came to be known as MPF in 1945. The idea that undergirded the invention of MPF, however, had a longer gestation. Research on MPF was instigated by the restaurateur and do-gooder Clifford Clinton (1900–69) who had approached Borsook in 1944 with the proposal to develop a protein food supplement that was cheap, easy to transport and to store and with ingredients readily in supply in the US.[59] Clinton had been interested in the design of effective meals since the 1930s. His experience of the social consequences of the Great Depression had led him to open his first canteen in downtown LA which served meals at low cost but high in nutrition value and offered its customers to pay as much as they wanted or could. Clinton's experiments in composing efficient meals continued as he opened further restaurants over the next decade. His understanding of the demand for protein considerably deepened during the last years of World War II when meat was increasingly unavailable for private consumption in the US, and Clinton searched for protein alternatives that would guarantee his customers a steady supply of the vital nutrient. He also joined a group of restaurateurs who consulted with the government regarding the nutritional content and taste of army rations. This brought him in touch with Borsook.[60]

Clinton began advertising MPF in 1945. He adopted the slogan "3 cents buys a meal" arguing that two ounces of MPF (56 grams) compared to "a quarter pound of beef, a baked potato, a side dish of peas and a glass of milk."[61] MPF was sporadically distributed in the US in the following years and was, amidst others, used to improve the diets of Native American Navajo.[62] Clinton nevertheless envisioned MPF to create the greatest impact outside the US. Borsook and Clinton made the MPF formula available to researchers in other countries, hoping they would encourage the development of similar food products. In the meantime, they promoted MPF as a food relief item to be applied in Europe and beyond. Although MPF spoke to the nutritional and logistical demands of North American relief agencies, they showed little interest in the product. Humanitarian organisations could buy milk powder at subsidised rates or even receive it free of charge from the US government. Hence, they had little incentive to purchase the still unknown and untested MPF. Not even Borsook's claim that MPF contained a higher dose of vitamins and proteins than powdered milk convinced relief agencies to buy the supplement.[63] The reluctance of the American aid sector threatened to end Clinton's bold initiative abruptly. At this time, a meeting with Pearl S. Buck (1892–1973) and her assistant Florence Rose (1903–69) opened up new opportunities. Convinced that MPF could assist in relieving hunger and malnutrition in what came to be understood as the "Third World" in the post-war era, the two women assisted Clinton and assumed a pivotal role in attracting relief providers and sponsors.[64]

Multi-Purpose Food for India

Pearl S. Buck is well-known for her work as a women's rights activist, her support of Indian nationalism and of the American civil rights movement.[65] She also became an advocate of US American food aid for Asia during World War II, organising private relief for civilians in China and India. Buck publicly embraced Clinton's plan to apply MPF to combat global hunger against the background of her philanthropic work. But it was Florence Rose who worked full-time to put Clinton's plan into practice. Clinton founded the non-profit MFM in 1946 in order to promote the food supplement in the face of the lack of interest from North American relief agencies and Rose joined him as the organisation's secretary. Drawing from Clinton's church and missionary contacts, MFM swiftly began sending the product directly to relief workers in Europe, the Middle East, and Asia.[66] While American missionaries applied MPF sporadically in India, MFM won further supporters of the food product in the United States. The personal channels available for the dissemination of MPF in India were thus constantly expanded. John Haynes Holmes (1879–1964), a well-known pacifist and Gandhi disciple (and member of MFM) personally introduced MPF to the Mahatma in early 1948, who acknowledged the potential of the product. After Gandhi's murder, his son, Devdas Gandhi distributed MPF in accordance with his father's wishes to Indian research institutions.[67] Two years later, the physician, educator, and then Chief Minister of Bengal, B.C. Roy (1882–1962), visited MFM in LA, expressing a wish to use the food supplement in India. Following his visit, MPF became part of feeding programmes in Bengal.[68]

While Indian politicians and social reformers had already shown sporadic interest in MPF in the early years of independence, it was the famine in Bihar and Madras in 1950 that gave a considerable boost to the food supplement in India. Seizing the opportunity that famine provided to advertise MPF in India,

MFM assembled the United Emergency Committee on Famine Relief for India (henceforth United Emergency Committee) in January 1951. The United Emergency Committee collected donations in North America to send MPF and CARE packages to relief agencies in the subcontinent. From June onwards, its work was facilitated by the India Emergency Food Aid Act which granted India a loan to purchase North American grains after lengthy negotiations over the conditions of this aid. It also eased the work of American relief agencies in India.[69] While India guaranteed a duty free entry of relief goods and paid the cost of inland transportation, the US government granted free ocean transportation to organisations registered with the Advisory Committee on Voluntary Foreign Aid. The American Red Cross (ARC) and CARE had already started to use monetary donations to buy "direly needed high protein content foods" like dried eggs, powdered milk, and dried beans from the government owned Commodity Credit Corporation and, under the agreement, now received milk powder free of charge from the US Department of Agriculture.[70] Therefore, in 1951, American voluntary organisations including the AFSC, CARE, Church World Service, and ARC shipped tons of wheat, milk powder, multi-vitamin tablets to India. In addition, aid shipments entailed a remarkable quantity of MPF: 4.6 tons of the food supplement reached India through registered voluntary agencies in 1951—an amount doubled by direct shipments of the United Emergency Committee.[71] Apart from promoting the use of its own in-house product, the United Emergency Committee worked closely with CARE. It offered Americans the option to earmark their donations for CARE food packages, plough packages, cotton packages, or hand tool packages (which comprised of pitchforks, mattocks, weeding hoes and spades). This was the first time that CARE offered non-food aid to equip farmers with the adequate means to increase their harvests. The sending of the first "CARE plow" to India in 1951 had important historical precursors in the use of tools and technologies by American missionaries in the early twentieth century. However, CARE's response to the famine in Bihar and Madras marked the beginning of the organisation's experiments with manufacturing and providing ploughs. Based on the experience gathered in India, it began to provide ploughs in Latin America, Korea, and Greece.[72]

By the end of the famine of 1951, the United Emergency Committee listed twenty-six institutions in India that applied tons of MPF in their emergency feeding programmes.[73] Moreover, it had aroused the interest of the Indian Ministry of Agriculture and Food which distributed MPF in famine-afflicted areas of Madras and Bihar.[74] Once the famine had ended and the United Emergency Committee disassembled, its members continued to promote MPF, now envisioning the wider application of the food supplement in India. In compliance with the mission of MFM, they promoted research on MPF, hoping that a locally produced protein-food supplement would soon help improve the diets of children and mothers from

India's low-income classes. As a leading member of the National Council for Child Welfare and the AIWC, Dhanvanthi Rama Rau (1893–1987) was pivotal in the process.[75] She visited MFM in Los Angeles in 1953 to acknowledge the donation of 10,000 pounds of MPF to the AIWC that benefited the rural health and nutrition programme of the women's organisation.[76] In early June 1955, Rama Rau convened a meeting between Rose and representatives of the Indian government, the Indian Red Cross Society, the Rockefeller Foundation, the relief committee of the National Christian Council, CARE and delegates of the regional offices of the FAO in New Delhi. The creation of the non-profit Meals for Millions India (MFM India) was announced during the meeting and, given that research on Indian MPF had significantly advanced in the meantime, the organisation began advertising the food supplement in the country.

Making Multi-Purpose Food Indian

The first laboratory experiments carried out to invent Indian MPF followed current general trends of food science in India and hence initially envisioned Indian MPF to become an artificial rice product. In 1952–53, the biochemist R. Rajagopalan at the IISc experimented with a blend of peanut, sesame, wheat, and tapioca flour that was partially gelatinised, pressed into vermicelli, and finally broken to resemble rice. His experiments showed that the Indian version compared unfavourably to the "American" MPF formula, and further experiments needed to be conducted.[77] The formula that eventually came to be known as Indian MPF was developed three years later at the CFTRI. It was a blend of peanut (groundnut) and chickpea flour that was fortified with minerals and vitamins. Researchers of the CFTRI had adopted the idea of using MPF as a dietary supplement that could be added to meals and therefore did not require a change of food habits. Whereas chemical analysis of Indian MPF showed that the new formula was identical to the American counterpart in terms of its nutritional value, its ability to treat protein-malnutrition still had to be scientifically validated. A series of feeding trials and experiments were carried out in the following years and the results were published in Indian and international scientific journals.[78] According to Subrahmanyan and his colleagues at the CFTRI, the tests were successful. The experiments proved that Indian MPF benefited the health of children, who recovered faster from diseases, grew steadily and gained weight.[79] Later trials carried out in paediatric wards of hospitals and in schools further fed into the process of establishing MPF as a suitable remedy against infant protein-malnutrition in India.[80]

While researchers were still studying the health benefit of the food supplement, MFM India strove to popularise MPF by identifying suitable channels of distribution and by educating consumers on how to apply the product. The daily ration of approximately ten teaspoons of the MPF powder could be simply stirred in water and eaten straight to derive the intended health benefits. However, researchers believed that promoting MPF as an ingredient for daily use in Indian cooking routines would enhance its acceptance. The CFTRI consulted housewives to learn whether the product met their demands and the findings proved encouraging. Tests revealed that MPF could be added to *dal*, curries, vegetables, porridge and chutneys without affecting the taste. MPF also raised the nutritious content of Indian sweet dishes, such as *chikki, kal kal, seviya* and *barfi* and when ten to fifteen percent of the food supplement was added to the dough of *chapatis* while kneading, the food supplement even improved the consistency of the bread.[81] To promote such usage of MPF, the CFTRI published a series of recipes and organised public cooking demonstrations. The first exhibition of MPF took place at the Lady Irwin College in 1956, a well-known institution of women's education and home economics, which had been running education programmes in nutrition and diets for over a decade.[82] The choice of the institution suggests that MFM India advertised MPF among the middle-class—a strategy that might be explained through the institution's trust in the "trickle-down effect." Protein food supplements in and outside of India were often stigmatised as "poor people's food" (or in the case of MPF "cattle feed").[83] To counter such stigmatisation, members of the middle-classes were encouraged to consume the food product, thereby contributing to the acceptance of the food among low-income classes. This, however, rarely worked out as intended.[84] Thus, MFM India used other distribution channels as well. With the support of the central and state governments, public medical and health departments distributed MPF to hospitals, maternity and child welfare centres, community development projects and school lunch programmes where it benefitted small children as well as expectant and breast-feeding mothers directly.[85] MFM India considered government support as crucial. The Indian Minister of Agriculture Panjabrao Deshmukh (1952–1962) was president of MFM India and lobbied within the government to make MPF a part of India's national food plans.[86] To win the additional support of high-ranking politicians, MFM India also sent out MPF to Indian members of parliament followed by a demonstration of the food supplement at Rashtrapati Bhavan, the presidential residence, in 1956. The demonstration aroused the interest of Nehru who also agreed to sponsor the production of Nutro Biscuits (another MPF-related invention of the CFTRI) for their application in flood and famine-affected areas.[87] MFM in Los Angeles also gave crucial support for the production of Indian MPF through purchasing the food supplement and encouraging US relief agencies to do the same.[88] The presidents of both MFM foundations, Clinton and Deshmukh, approached Nehru again jointly in 1957, in the hope of obtaining permanent government support for the expansion of MPF production in India which until then had been limited to the capacity of a pilot plant.[89] Nehru allegedly testified his interest in MPF during the meeting. His statement "This is what India needs, this is what India wants; but we must do it in our own way" even made it into a promotional film that MFM showed to international audiences in 1962.[90] From 1958 onwards annual government grants allowed MFM India to distribute MPF to charitable organisations concerned with the health of mothers and children. In addition, charitable institutions now purchased the food supplement directly and regularly from the CFTRI. Beginning with the second grant of 1959–60, the distribution of MPF was further expanded. Apart from family planning clinics and pre-schools, the food supplement was now also applied in industrial canteens, collieries, labour colonies and the defence services to enhance labour productivity. Previously, tests of the generic MPF formula in Latin America had shown that the food supplement increased the productivity of labourers by ten percent, which seemingly fuelled the imagination of Indian national planners and industrialists in making the ideal productive citizen.[91]

While the demand for MPF in India was growing, increasing its production was a lengthy undertaking which suffered from frequent setbacks. The need for keeping the prize of Indian MPF low was a major difficulty. Moreover, negotiations with state governments were time consuming. When Subrahmanyan retired as the director of the CFTRI in 1963, the future of Indian MPF was still uncertain. Subrahmanyan hoped that Indian companies would start producing, distributing, and marketing MPF to save the food supplement from falling into oblivion. Such a private partnership had proven successful in the case of Nutro Biscuits. After the private biscuit company Britannia took over its production and marketing, the food product became increasingly popular.[92] However, before MPF could follow a similar path, the decision of the US government to make US produced Corn-Soy-Milk (CSM) a part of its food aid agreement with India in 1966 discouraged any future investments in MPF. Against the backdrop of emerging famine conditions in Bihar in 1966–67, the US government donated large amounts of CSM to relief agencies in India, thereby discouraging the latter from purchasing Indian MPF. In addition, the US Agency for International Development (USAID) sent two of its nutritional experts, Nevin S. Scrimshaw and Alan Berg, to India to oversee the development of a "local" formula of CSM. The outcome of what indeed were collaborative efforts of Scrimshaw, Berg and Indian researchers of the CFTRI was a mixture that came to be known as Balahar. Whereas MPF was made out of Indian staple foods, seventy percent of Balahar consisted of US cereals, whose import was covered by the Indo-US food aid agreement.[93] In light of such investments, the still unstable financial support of MPF dried up completely. While it was still applied alongside Balahar and CSM in Bihar in 1966–67, its production ceased in the famine's aftermath.[94]

Although interest in Indian MPF dropped sharply in the late 1960s, MFM in Los Angeles still advertised the MPF formula as a solution to the world's "protein crisis".

Indian researchers had been the first to mould the generic MPF formula into a local product which rendered Indian MPF an important test case for proponents of the formula. The "Indian pattern", a summary of all major steps in the production and popularisation process in India, was used as a blueprint to introduce MPF in other countries.[95] Even more important than the development of an altered formula were the comprehensive scientific studies that had been carried out in India to scrutinise the claims made by the inventors.[96] The publication of findings assured that Indian research on MPF contributed to a global body of knowledge on protein food supplements that undergirded the invention of later products in India and beyond. In addition, alongside the promotion of Indian MPF in India itself, Indian researchers and politicians had begun to promote the food supplement as a means against protein-malnutrition in the global south and assist other countries in developing similar products.[97] Deshmukh used the occasion of the World Agricultural Fair in Delhi in 1959 to introduce Indian MPF to an international audience and subsequently small portions of the food supplement were sent to Asian and African countries.[98]

2.3 Conclusion

In 1961, two years prior to his retirement as director of the CFTRI, Subrahmanyan wrote a letter to MFM in LA in which he strongly opposed the description of Indian MPF as "a copy" of the American formula.[99] MFM was quick to apologise for its rhetorical lapse and promised to stop referring to "original" and "copied" MPF formulas. The carelessness that undergirded the use of such labels in 1961, however, reverberates in the literature on post-war development in South Asia. Whereas the former brushes away decades of research on soy and groundnut that had predated the arrival of the generic MPF formula in India and the work of Indian researchers in inventing, testing and applying Indian MPF, the latter readily assumes US dominance over the production of knowledge and the means applied in development planning in India. Both characterisations miss the complexity of the processes that undergirded the production of development knowledge and the multi-directional entanglements it entailed. In the 1940s and 1950s, the desire to improve Indian nutrition within and outside the context of famine intensified the collaboration between Indian and US scientists, philanthropists, politicians and social reformers. Previous collaboration between Indian anticolonial activists and American anti-imperialists in the United States (detailed in chapter 5 of the book) provided the organisational structures for this collaborative endeavour that now stressed the importance of US food aid and scientific cooperation to boost India's national development. After the rise and demise of Indian MPF in India itself, the food supplement continued to inspire nutritional remedies outside South Asia. In

1963, Subrahmanyan proudly referred to the assistance the CFTRI had rendered to other countries "in the Eastern region, but also in Africa and in the Caribbean in their production programmes", further claiming that Indian research on MPF "became a model for other countries to follow."[100] In the same year, Subrahmanyan assumed his new post in Manila, where he assisted in the establishment of the food technology laboratory at the National Institute of Science and Technology as advisor to the FAO.[101] The history of Indian MPF thus continued outside the subcontinent and illustrates the role of India as a site in the production of knowledge of food and nutrition and concrete technologies that continued to shape development practice in the decades that followed. Yet, the rise of protein food products of which MPF was an early and important example, had a doubtful effect on the reduction of mal- and undernutrition worldwide. Doubts about the usefulness of protein food supplements in mitigating malnutrition accumulated in the 1970s. Studies revealed that protein products remained unpopular among those they intended to nourish.[102] They also seldom produced significant health benefits if consumed outside of institutional feeding programmes. Diets that lacked protein were commonly also deficient in calories and the insufficient intake of energy impeded the absorption of added protein. In other words, when people failed to fill their stomachs, the consumption of ready-made packaged protein bars, biscuits and powders seldom saved them from starvation.[103] Lastly, the focus on food supplements diverted resources and attention away from the structural causes of malnutrition—above all people's poverty.

Considering the longer history of famine relief and its relationship with nutrition science, the use of food supplements and substitutes to alleviate famine in India in the 1940s and 1950s might suggest a reassessment of this relationship. But famine relief continued to be out of sync with the nutritional needs of the population as political and economic pressures determined the timeliness and scope of relief. The logistical demands of warring nations and humanitarian organisations rendered milk powder, protein supplements, vitamin pills and halibut oil capsules preferred aid items during World War II. They consumed considerably less shipping space than bulky food items, such as grain. The popularity of food supplements and substitutes was also conditioned by a new trust in scientific and increasingly technocratic solutions to hunger. The aid that reached the starving population of Bengal belatedly in 1943 and 1944 illustrate this development. When the scope of suffering inflicted on Bengal became visible to a larger public outside the province only in August and September, demands for the use of all scientific means available to mitigate starvation emerged in India. Whereas substantial relief reached the province only slowly, the intravenous application of protein-hydrolysates was tested on those believed to be already lost. The new technique might have helped some patients in the hospitals of Calcutta where the procedure was used to rescue those suffering extreme body wastage. But as it was applied without sufficient supervision throughout Bengal, and later misappropriated to assist the recovery of Holocaust survivors in Bergen-Belsen, it produced harrowing results. This is another example of the malleability of scientific arguments, able to be geared conveniently to match political and economic demands.

PART II

From Famine Relief to Community Development: The American Missionary Movement in South Asia

CHAPTER 3

Worldly Needs and Religious Opportunities: The Famine Relief of American Missionaries in Bombay, 1870s–1920s

Abstract

Missionaries of the American Board of Commissioners for Foreign Missions (ABCFM) scaled up their famine relief work in the 1890s against the background of American imperialism and changes in humanitarian funding. Examining the work of missionaries in one regional hotspot of famine in India, Bombay, the chapter uniquely explores how famine prepared the ground of American expansion into British India.

Keywords: Missionaries, Marathi Mission, American humanitarianism, missionary famine relief

3.1 The Missionary Discovery of Famine in Bombay

The Making of Famine in Bombay

The Bombay Province (also Bombay Presidency) was an administrative unit in the west of British India. Although its boundaries shifted during the colonial era, for most of the period of British rule it stretched across Maharashtra to present-day Gujarat in India and Sindh in Pakistan. Ecological differences contributed to the particular geography of famines in Bombay. In contrast to the humid climate of the densely populated coastal strip, the eastern part of the Deccan plateau was an arid region, known as the famine belt. It covered parts of the six districts of Khandesh, Nasik, Ahmednagar, Poona, Satara and Sholapur. Peasants in the eastern Deccan largely grew millet (*jowar* and *bajra)* for subsistence. Agriculture depended on the September monsoon that provided the water for the *rabi* crop. The general paucity of rainfall during the rest of the year and the low-quality of soil foreclosed a summer (*kharif*) harvest. The failure of the September monsoon thus weighed heavy on the Deccan's agriculturalists.[1] Three times did droughts spill over into famines in Bombay: in 1876–78, 1896–97 and 1899–1902. Bombay's short-lived cotton boom in the 1860s had fuelled agricultural indebtedness and contributed to the vulnerability of the Deccan population to famine.[2] Even earlier, the colonisers' exuberant confidence in economic liberalism and the wish to extract resources drove the commercialisation of Indian agriculture. The government of Bombay had made advances to agriculturalists in the 1830s and had exempted cotton fields from land revenue to provide incentives for the cultivation of the cash crop. Not all regions, however, were suitable for cotton cultivation. Whereas it grew best in the southern districts of Belgaum and Dharwar, the soil in most other parts of the province was inhospitable to the plant. Peasants lured into cultivation in areas unsuitable to grow cotton suffered immediate losses. The rising demand for cotton during the American Civil War of 1861–65 convinced additional cultivators in Bombay to produce for export. Those who succeeded found themselves in crisis when American production resumed after the war. The global economic depression that set in in the early 1870s exacerbated rural indebtedness in Bombay. It led food prices to spiral out of control, resulting in considerable hardships for agriculturalists in the province.[3] Small landowners, tied to their land and the revenue system, commonly remained on their property during harsh seasons and strove to balance losses through agricultural wage labour or loans.[4] Many landless labourers on the other hand migrated to the provinces' urban centres for work in the industries. One such urban centre was Bombay city which was connected to the rural hinterland through circular migration that intensified during famines.[5]

The El Niño Southern Oscillation that governed the monsoon in Asia resulted in a widespread drought in India in 1876. The subsequent harvest failure drastically decreased the food stocks of Bombay, while grain from surplus regions was exported to Britain instead of being used to meet shortages in the country. In the course of the famine that unfolded across Bombay and Madras over the next three years, six to ten million Indians starved to death.[6] Governmental relief was insufficient and the extreme loss of life fuelled the institutionalisation of colonial anti-famine policies in the famine's aftermath. It was only until after the famine that the Indian Famine Code was drafted (see chapter 2 for a discussion). The existence of the Indian Famine Code, however, did little to prevent high mortality in the following famines. In 1896, the failure of the monsoon again destroyed crops in Bombay. Prices skyrocketed as a consequence of widespread scarcity. The drought developed into a devastating famine. Colonial officers discounted the warnings of Indian and international observers for too long and the relief that was eventually provided remained insufficient in many cases.[7] The province was still recovering from the previous disaster, when drought and then famine struck again in 1899. In parts of Bombay, famine had to be declared for three consecutive years, lasting until 1902.[8]

The Beginnings of the AMM's Famine Relief

The confrontation of its missionaries with the devastation wrought by famine in Bombay sparked the AMM's sporadic assistance of the famine-afflicted population in the 1870s. The South Indian Famine of 1876–78 hit Bombay's eastern Deccan hard.

It was here that the AMM had opened several mission stations in the early nineteenth century to expand its influence over Bombay's heartland. Its missionaries, confronted with widespread starvation in 1876, began to provide ad-hoc relief near its mission stations. In Sholapur, they distributed rice, bread, soup and *conjee* (a sort of rice porridge). With the financial help of other Christian missions in India, donors in England and Scotland, as well as Indian and British philanthropists, the AMM fed 300 to 400 people daily from July to mid-September 1877. It also gave medical assistance to people who were brought to the mission station.[9] The city and district of Sholapur was home to the largest share of the province's weavers and had started to supply western India with cotton and woollen cloth in the 1850s.[10] In 1877 Bombay's industrialists opened a first mill in Sholapur. The famine gave an incentive to this business venture. The hunger resulted in an abundance of cheap labour as peasants sought temporary employment in the mills to manage harsh seasons. Subsequent famines further fanned the mill industry of Sholapur.[11] During famines, those taking work in the mills were often accompanied by dependants. This required the provision of famine relief to family members who camped near the mills. The AMM, which had first begun helping workers and their dependents in Sholapur in 1877, maintained this focus during subsequent famines, but it also provided aid in other parts of the province.

After the AMM added famine relief as a new set of activities to its missionary portfolio in 1876–78, it considerably expanded its famine-related activities in the 1890s. The growth of the AMM's relief work showed amongst others in the rise of the number of women and children who were taken into the mission and supported through donations.

Harvests after Drought? Famine and the Missionary Zeal

Considering that the goal of the missionary movement was to win "heathens" to Christianity, the connection between famine relief and the missionary zeal deserves attention. This requires a brief retrospection of the beginnings of the Marathi Mission. After the arrival of the first missionaries of the American Board of Commissioners for Foreign Mission (ABCFM) to Bombay in 1813, missionaries soon discovered how difficult it was to meet the expectations of the American Board in Boston. The AMM hardly attracted new members to its Christian community. Twenty years later, the AMM still counted no more than fifteen converts. Although it found adherents among members of the rural low-caste communities (the Mahars and Mangs), the number of converts did not increase significantly in the following decades.[12] This was attributed to the failure of traditional evangelistic methods which led missionaries to intensify the building of schools. Missionary schools sought to foster the missionaries' connections to Indian communities while attempting to introduce Christianity through education.[13] The conflation of educational and evangelistic work was contested by Indian communities who began protesting the conversions of pupils and teachers in the schools of the AMM and other mission societies in the late 1830s.[14] This protest, however, did not hinder the AMM in becoming an important provider of education in Bombay in the following decades. Conversion numbers in the meantime remained low. To a certain extent, the AMM got in its own way. The difficulty to engage (the right kind of) Indians owed partially to its approach to conversion. Unlike other missionaries in South Asia, American protestants demanded converts to abandon caste in its entirety and were at pains to evaluate the incentives of applicants, renouncing them in case they failed to provide sufficient evidence to prove their sincere wish to become Christian.[15]

This approach to conversion was not abandoned during famines that rather escalated the efforts of missionaries to scrutinise the intentions of its "inquirers." H.G. Bissell, AMM missionary in Sirur (also Shirur or Ghodnadi), wrote in 1878 that missionaries "found it necessary to be very careful in receiving candidates for baptism, testing their motives sometimes by weeks of probation and close scrutiny of their conduct."[16] The AMM's view on conversion also showed in its relationship to other mission societies. The aggressive approach of the British Church Missionary Society (CMS) and the Society for the Propagation of the Gospel (SPG) enabled these missions to convert large groups of Indians of low-caste communities during the South Indian Famine. Between September 1877 and February 1878, the SPG gained 16,000 new members.[17] The AMM rejected such a practice. Its missionaries further lamented the encroachment of the SPG into its territory and the fact that this rivalling mission society apparently converted inquirers of the AMM who, in the protestant estimation, were not ready to receive baptism.[18]

The missionaries of the AMM still sought to evangelise. Yet, instead of seeking to raise the number of conversions through the hasty performance of baptisms during famines, missionaries in Bombay wished to gradually convert those they had admitted to the mission. Reports indicate that the AMM increasingly capitalised on the concentration and confinement of famine-afflicted populations. The famines drove people out of the villages into urban areas and the punitive environments of aid camps and poorhouses.[19] Missionaries of the AMM went to relief camps and addressed people as they migrated in search for relief. Pastors preached to the people and handed out leaflets to declare famine a sign of God's demand for repentance.[20] Starvation, in this regard, played into the hands of missionaries because it made it easier for them to approach the population.

While the AMM seemed to concentrate on using its famine relief work to draw in those directly afflicted by starvation, it actually sought to address a wider Indian audience. Given that philanthropy was an important marker of social standing in British India in general and in Bombay in particular, missionary famine relief meant to boost the popularity of the AMM among Indian elites. Sporadic signs of such appreciation can be traced back to the early twentieth century. Gopal Krishna Devadhar (1871–1935) of the SIS used the occasion of the AMM's centennial celebration in 1913 to applaud the mission for its previous famine relief work.[21] Equally appreciative was the moderate INC politician Narayan Ganesh Chandarvarkar (1855–1923), a leading member of the Hindu reform organisation Prathana Samaj (Society of Liberal Religionists) in Bombay. He was invited to serve as a member of one of the distributing committees of US aid in Bombay.[22] Both men were English educated Hindu social reformers of high-caste, who had themselves been involved in famine relief activities.

Donations were another form to display goodwill and acknowledgement. Members of the influential Parsi community for instance gave to the AMM in times of famine. Parsis not only donated during famine. The money of the textile magnate Dinshaw Maneckji Petit (1823–1901) allowed the AMM to open the Sir D.M. Petit Industrial School in Sirur in 1891 that offered vocational training to Indians.[23] Although many Parsis defied the attempts of missionaries at converting members of their community, some showed sympathy for the philanthropic work of missionaries and gave in its support.[24] Zoroastrians in general and Parsis in Bombay in particular considered philanthropy a matter of social service and civic duty. Parsis had used philanthropy strategically to bolster their standing much earlier than American missionaries started to embark on a similar path.[25]

The discussion of Indian responses to the AMM's relief work in the nineteenth century, however, would not be complete without mentioning that it also fuelled opposition. The most vocal opponent was the Arya Samaj who interpreted the relief activities of missionaries—in particular the sheltering of women and children—as a deliberate attack on Hinduism and strove to mobilise nation-wide opposition against missionary famine relief. The Hindu socio-religious reform organisation that was founded in Bombay in 1875 now began to assist Hindu orphans and widows.[26] In providing relief, the Arya Samaj wished to demonstrate the capacity of Hindus to assist their co-religionists.[27] Missionaries of the AMM rejected the claim that Hinduism could serve the betterment of the people, calling Hindu customs a cause of rural impoverishment.[28]

3.2 Giving to India: US Funding of Missionary Famine Relief in Bombay in the 1890s

Although AMM missionaries held more complex views about the link between famine and conversion and were often sceptical about promoting baptisms during famines, fundraising required a simpler message.[29] In appeals to the American public, missionaries clung to the idea that the larger the American involvement in famine relief, the larger the turn-out of converts as this helped to solicit the support of Evangelicals in the United States.

The expansion of the AMM's famine relief work in the late 1890s was due to its success in tapping American donations, which occurred against a backdrop of changes in the way Christians in the United States gave money to missionary work. At the outset of the famine of 1896–97, the American Board in Boston noted with concern that regular church donations dwindled. The consequence was a cut of funding. The AMM was already suffering from a steep rise in prices in Bombay when the cut of allowances of American and Indian mission members aggravated its financial plight and caused an existential crisis.[30] Missionaries in Bombay reminded the Board of the special needs of Indian converts, arguing that conversion had cut off native Christians from their communities and rendered them dependent on the allowances of the AMM. The harsh reduction of appropriations, especially as it occurred during famine, threatened to drive converts away from the AMM, undoing the missionaries' hard-won success.[31] Missionaries pleaded with the American Board to reverse the reduction. Although the Board was sympathetic to the appeal, church donations further declined and the Board considered it impossible to raise the allowances to the previous level.[32] The future of the AMM looked bleak at this point, but it weathered the famine and even expanded its activities in Bombay in the coming years. The answer to this conundrum lay in a change in funding. American Christians who had previously supported the foreign missionary movement through church collections, had begun to give to special funds.[33] Such "special objects" were instituted to cover missionaries' philanthropic rather than outright evangelistic work. Touting famine relief and the assistance of "native helpers" in India as such "special objects", missionaries were able to fill their funding gap. Contrary to the recent fears of the AMM, Robert Hume enthused in 1898 that the previous year "was unquestionably the most successful year in the history of the Marathi Mission."[34]

The Christian Herald of New York and the Alliance of American Imperialists and Evangelicals

Reverend Charles C. Creegan, ABCFM secretary of the US Middle District noted in 1897 that his congregation would rather donate to the *Christian Herald's* Indian famine relief fund than to the foreign mission funds.[35] The success of the *Christian Herald* to mobilise large funds in support of missionary famine relief in colonial South Asia in the late 1890s demonstrates the influence of evangelicalism on American humanitarianism in these years.[36] A leading religious journal in North America since the 1890s, the *Christian Herald* was run by Louis Klopsch (1852–1910) and the controversial Reverend Thomas De Witt Talmage (1832–1902). The *Christian Herald* undertook massive humanitarian fundraising campaigns in the 1890s that contributed to the popularity of the paper. Its editors gained experience in mobilising funds for the famine-afflicted population in Russia and Christians in Armenia before they started to raise money for famine relief in India in 1896–97. They renewed their efforts to collect money for India in 1899 and again solicited large sums in support of the famine relief missionaries provided. The *Christian Herald* now raised $640,000, topping the record sum of $400,000 it had solicited during the previous famine in India.[37] In addition to donating to the *Christian Herald*, the ABCFM's constituencies also gave to a famine relief fund of the *Congregationalist* and the *Advance* in Boston which collected a sum of $125,000 that benefitted the work of missionaries in India.[38]

The change in funding of missionary work echoed wider transformations. Although Americans had shown interest in alleviating social ills outside the confines of their own nation before the 1890s, the scope of such spending was unprecedented. Americans seemingly plunged with vigour into sharing their wealth with those seeking to alleviate famines, wars, and other disasters abroad. The American missionary movement gave crucial impulses to this discovery of humanitarianism in the United States. Missionaries made Americans aware of the world outside their own country and solicited support for their philanthropic work.[39] Of similar importance to the popularity of American-sponsored humanitarianism was the rise and transformation of US imperialism. US imperialism and humanitarianism became inextricably linked in these years. The fundraising of the *Christian Herald* serves as a case in point. The bulk of the money raised by the *Christian Herald* during both famines was invested to purchase grain surplus in Kansas that was shipped to India on the *City of Everett* in 1897 and on the *Quito* in 1900.[40] The material prosperity of the United States had enabled donations, while the will to gain access to markets that could absorb US agricultural surpluses advantaged the sending of grain.[41] US Congress voted in favour of bearing the costs of the aid shipments to India, hoping that it demonstrated the strength of the US agricultural sector and its aspirations to dominate the global grain market.[42]

US imperialism benefitted from the display of the purportedly moral, religious and cultural supremacy in which many American humanitarians firmly believed and that missionary famine relief in India allegedly displayed.[43] The editorial board of the *Christian Herald* made no secret of its efforts to use the aid as a sign of Christian America's moral and religious superiority. To spread this message widely, it sought broad media coverage of the aid shipments. It invited journalists, philanthropists, and politicians to bid farewell to the crew members in America as well as to greet them upon their arrival in India. When the ships landed at the docks of Bombay and Calcutta, the stars and stripes alongside the Christian cross were waving on deck; and when dockworkers unloaded the cargo, the names of the mission societies that had donated money surfaced in bold letters on the sacks of grain.[44] In 1899–1900, the *Christian Herald* even used its aid drive for India to lend more direct support to American imperial aspirations. As Heather Curtis has demonstrated, the decision of the editorial board of the *Christian Herald* to raise funds in 1899 was motivated by the wish to unite US evangelicals who had grown apart during the Philippine-American War.[45] The atrocities committed by US soldiers in the Philippines, brought to public attention by the Anti-Imperialist League, led religious leaders to speak out against US colonialism.[46] The *Christian Herald*, in an effort to defend US expansionist policies and to counteract criticism levelled against the sanctity of Christian America, emphasised the good Christian aid could achieve in India.[47]

American Business Philanthropy Reaches for India

The *Christian Herald* was still the largest contributor to Indian famine relief in the US in 1900, but with the formation of the Committee of One Hundred on India Famine Relief (*hereafter* Committee of One Hundred) it received a competitor.[48] The Committee of One Hundred united New York's leading business philanthropists, including the mining magnate William E. Dodge Jr. (1832–1903) and John D. Rockefeller Jr. (1874–1960) and assembled in the Chamber of Commerce in New York for the first time in May 1900. Its formation followed the initiative of a group of American missionaries who had gathered during the Ecumenical Missionary Conference in 1900. They demanded a "larger national movement" to aid the people afflicted by famines in India.[49] In an effort to represent such a national effort, the Committee of One Hundred opened additional fundraising bodies throughout North America in the weeks that followed its inception.

The Committee of One Hundred signalled the diversification of fundraising for India and a cleavage among American protestants. Rockefeller Jr. was brought up under the influence of evangelical Protestantism which motivated his and his father's philanthropic endeavours. Evangelical protestants like Klopsch and Charles Sheldon however sharply criticised New York's capitalists for enriching themselves and easing their conscience by contributing a fraction of their wealth to philanthropic purposes. Prior to the formation of the Committee of One Hundred, Sheldon decried that the dividends distributed to shareholders of the Rockefeller corporation could have saved millions of people in India. The *Christian Herald* continued to oppose Rockefeller in the early twentieth century, but eventually lost out to the financial potency of the Rockefeller Foundation.[50]

The Committee of One Hundred differed from the evangelistic outlook of the *Christian Herald*'s fundraising campaign. The Committee of One Hundred presented itself as a "citizen's movement" that intended "to show the sympathy felt by the citizens of New York for the victims of the famine in India."[51] Differences between the two committees surfaced as well in the disagreement about the composition of the committee in India that was in charge of distributing US funds and grain. The Committee of One Hundred rejected the idea of donating the money to the all-white and all-missionary committee that distributed the aid sent from the *Christian Herald*. Consequently, the Americo-India Famine Relief Committee was set up to administer the donations from the Committee of One Hundred. Its formal separation from the *Christian Herald* served to ensure that Indians perceived the Committee of One Hundred as "a civic agency" that provided aid irrespective of differences of "race or creed."[52]

The differences that in New York seemed fundamental, however, lost traction in Bombay. The Americo-India Famine Relief Committee merely counted Narayan Chandarvarkar as its only Indian member. And whereas it forwarded a part of its donations to a small group of Indian organisations, the bulk of its funds still benefited missionaries.[53] The formal differences between the committees further collapsed in light of an overlap of membership. The AMM missionary from Ahmednagar Robert A. Hume (1847–1929) had a leading position in both distributing committees, commuting between the city of Bombay and Ahmednagar to participate in their meetings.[54] As he occupied these positions, the Yale-trained missionary was turned into a heroic figure by the American press and missionary publications. The press seemingly enjoyed reproducing the narrative of the one man's battle against famine in the subcontinent. When the colonial administration awarded Hume for his famine-related work with the *Kaisar-i-hind* gold medal in 1901, it contributed to this narration.[55]

American Famine Relief and British Colonialism

Given that American missionaries relied on money that was sent directly from US donors, they had considerable latitude to design their "charitable work" to their own ideas. This, however, did not lead to a conflict with the basic principles of colonial famine relief. Similar conceptions of poor relief on the one hand, and the need to appease the colonial state on the other hand, account for the fact that American missionaries organised aid largely in accordance with colonial expectations and standards.[56]

The United States occupied a special place in the domestic, imperial and international efforts to relieve famine in South Asia in the 1890s. This is because the majority of donations that were mobilised across and beyond the British Empire (including India) by and large went to the Indian Famine Charitable Relief Fund.[57] Split into a series of semi-official committees on a district and provincial level that were centrally supervised from Calcutta, the Indian Famine Charitable Relief Fund was in charge of distributing the money in adherence with colonial ideas of legitimate charity. This included the identification of the "deserving poor" through a rigorous assessment of needs that was meant to prevent those purportedly "undeserving" from receiving assistance.[58] The colonial administration further wished to channel charity into particular areas of work such as the provision of 'comforts': milk, *ghee* (clarified butter) and vegetables to young children and the infirm, as well as garments and blankets. Charitable funds financed orphan care and the relief of "respectable persons" who the colonial state believed were inhibited by caste, class and social norms to join labour gangs.[59] Such money was also used to open relief works in villages for people who were unable to seek relief in distant government-run centres.[60]

The writings of Hume illustrate how American and British colonial famine relief converged. Confirming the basic principles of the Indian famine codes, he stressed the importance of a stringent need assessment and the employment of those afflicted by famine to prevent indolence.[61] The AMM also created relief camps that resembled British colonial institutions. Its "largest single relief camp" was set up in Sholapur, where famine sufferers conducted labour that corresponded with the sort of work demanded at official public works. At the peak of the famine, over 2,400 famine-sufferers were digging, breaking stones, and fetching lime and sand to earn their livelihood in the missionary relief camp. In adherence to official standards, the relief camp was situated within the compound of the AMM Sholapur station "on open ground just outside the yard" and was overseen by guards who maintained discipline.[62]

At times, American missionaries even occupied positions normally held by colonial administrators. At the request of the provincial government, Robert Ward of the Methodist Episcopal Church managed a poorhouse in Godhra in the Bombay Province (today in Gujarat) in 1900. In *Some Fruits of the Great Famine* Ward is depicted as a benevolent superintendent who "made it a point always to be accessible to the complaints of the people."[63] Despite the many convergences of British and American famine relief, American missionaries also intended their aid to be discernible as Christian American. The portrayal of Ward's management of the governmental poorhouse in missionary publications was thus infused with the idea that Americans were more sympathetic to the needs of Indians than British colonial officers. Additional anecdotal evidence points to the efforts of missionaries to distinguish themselves from the colonial authorities. When missionaries distributed doles of seed they attached a note that explained to peasants how the next harvest would testify "God's satisfaction with American charity."[64]

Meanwhile in the United States, some American missionaries defended the colonial state against mounting public criticism that emerged in 1900. American opposition to British imperialism grew against the backdrop of press reports of the British war against the Boers in South Africa that appeared alongside accounts of Indians dying of famine in India in 1900. The combination of war and famine diminished US sympathies for British imperialism and instead created US solidarity with Indians and the Boers.[65] Analogies between the fight of Boers in South Africa and the struggle of the forefathers of the United States for national liberation appeared in the public debate.[66] The *Cleveland Leader* called the disproportionate investment in the defence of the Empire in South Africa at the expense of famine relief in India "one of the sorriest spectacles which our poor human nature has presented in many years" and added that this was "especially disheartening in view of the fact that it is the work of the nation which claims to lead the van in human progress and stands for all that is best in civilization."[67] The *San Francisco Call*, which reproduced the statement, carried a cartoon that showed how civilization demanded John Bull to stop pouring money into its war machine to assist its famine-stricken colonial subjects. Protest against British war-spending even mounted in the heart of the empire itself. The *Illustrated Missionary News* from London noted in June 1900 that "war has slain its hundreds, but famine its tens of thousands."[68] While even some British missionaries were outspoken in their criticism of British war spending, American missionaries by and large remained silent during the controversy. When the Committee of One Hundred assembled for the first time, its members consulted the missionaries Henry C. Potter (1834–1908) and Justin E. Abbott (1853–1932) who had recently returned from India. Members of the committee raised concern over the "drain of wealth" from India to Britain and the insistence of the colonial administration on relief through labour during the public meeting.[69] Potter and Abbott discouraged such criticism, referring to the allegedly demoralising effect of gratuitous relief on the Indian society. They also defended the free market paradigm and praised the Indian Civil Service as "the best civil service in the world."[70]

3.3 The White Man's Shadow: Women, Indian Mission Members and the Human Legacy of Famine

Women and Indian Mission Members

The ubiquitous presence of men like Hume in the famine relief funds and committees in India and the United States overshadowed the contributions of women and Indian mission members (women and men) to the AMM's famine relief in India. Despite the many self-celebratory depictions of American aid that presented hagiographic accounts of male missionaries who seemingly single-handedly saved thousands of children and women from starvation, female missionaries and Indian mission members were at the forefront of missionary famine relief. Their stories are much more difficult to trace.

The first American women to join the protestant missionary movement in South Asia had accompanied their husbands. Harriet Atwood Newell, the earliest woman missionary of the ABCFM who set out to live and work in India never arrived in the country but died together with her new-born at sea in 1812. Succeeding generations celebrated her bravery and devotion to the missionary movement and her tragic fate curiously drew more women into missionary work.[71] Since women were denied ordination, they mainly became teachers and organisers of women's clubs and bible study groups.[72] The demand for additional female teachers and trained missionaries who could support the work of the mission by targeting Indian women in particular created new opportunities for single women in the 1860s and 1870s. After the inception of the Woman's Board of Missions in 1868 that financed single women and collaborated with the ABCFM to meet the demands of personnel of the foreign mission movement, American women ventured abroad in greater numbers to aid the mission movement.[73] The famines of the 1890s further emphasised the importance of the presence of women, while similarly reinforcing traditional gender roles. Women played a prominent role in overseeing the education of the female survivors and children as their gender purportedly cast them as their natural custodians. They were not alone in consolidating their place in the AMM in the famine years.

The AMM had begun to scale up the training of Indian members in the 1850s. In 1896 it counted 362 Indian workers, of whom 20 were ordained pastors, 23 preachers and 66 bible women.[74] The count included Sumantrao Karmarkar (1861–1912) and Gurubai Karmarkar (1862–1933).[75] The missionary couple were born as the children of Indian pastors in Ahmednagar and Belgaum. They had joined the AMM in the 1880s, but their frustration over gender and racial barriers that prevented them from pursuing higher degrees in medicine and theology in India had led to their decision to seek education in North America in 1888. They did so against the resistance of missionaries in Bombay and the American Board, who feared that their exposure to American culture would diminish their 'Indian-ness' and reduce their ability to earn the trust of Indian communities they wished to convert.[76] Managing to pursue their education in the US nevertheless, Gurubai Karmarkar became the second Indian woman to graduate from the Women's Medical College in Pennsylvania, following in the footsteps of Anandibai Joshee (1865–1887).[77] Sumantrao Karmarkar in the meantime, earned degrees in theology from Yale and Hartford. When they returned to India in 1893, they brought with them new contacts to American donors and changed ideas on missionary work. The fruits of such influences showed when the Karmarkars joined the Indian YMCA and YWCA. Their dual involvement left traces as some Y-methods made inroads in the more conservative Marathi Mission.[78] Sumantrao Karmarkar, who made a name for himself as "the premier evangelist of Bombay" at the turn of the twentieth century, also became known for his tent meetings and the use of a stereopticon he had brought with him from the US to attract Indian audiences.[79]

After their return to India in the 1890s, the Karmarkars contributed to the famine relief of the AMM. Indian members of the AMM visited relief camps and villages during the famines. They often camped near labour gangs, in which the famine-afflicted worked to earn a subsistence wage, to speak to the women who either laboured on the public works or had accompanied their family members. They also visited government relief works to assist Indian Christians in practicing their faith and to spread the gospel among non-Christian workers.[80] Female Indian mission members were going from house to house to identify women who were susceptible to the gospel and parents who were willing to give their children to the missionaries.[81] The Kamarkars took several children in their custody who had been rendered destitute by the famine. At the turn of the century, the AMM accommodated 3,000 Indian children, in boarding homes, orphanages and missionary families.[82]

Raising Children, Negotiating Differences

The photo below (figure 2) shows the Karmarkars and a group of children the couple adopted during the famines of the late 1890s.[83] Information on the rearing and lives of these children is scarce. One of the adopted girls, who was given the name Prithi Hannah after her conversion, died in 1900.[84] Another boy, named Vishvas Rao, had been taken in by the Karmarkars at the age of six, was raised by them and later followed his foster parents' example.[85] He went to America to study medicine in Pennsylvania and worked as a doctor at a hospital in Pittsburgh during the influenza pandemic of 1918–19, when he contracted the virus and died.[86] The ABCFM commemorated him tellingly as "the best product of India's life", while the Mayor of Pittsburgh and the Governor of Pennsylvania sent their condolences to his foster mother.[87] The adoption of children during famines was, to a certain extent, a continuation of earlier practice. From the early years of the AMM, the American Board and donors in the United States had encouraged the adoption of children in India who it regarded as being particularly receptive to the Christian message. In the nineteenth century, the AMM occasionally took the custody of children, who were either orphaned or placed in the care of missionaries by their parents.[88] Whether the children were to be raised in institutions or family homes, missionaries of the AMM hoped to turn them into members through their education and gradual conversion. At the same time, the missionaries were cautious about the extent of such a transformation. Since Indian members of the mission were to serve as intermediaries between the AMM and Indian communities, missionaries were anxious to limit the children's adaptation to their new American protestant environment.

Fig. 2. The Karmarkar Family: To the left and right of Gurubai Karmarkar, who is at the centre of the picture, are children the couple had adopted. The different status of these children within the composite family is rendered visible through the embroidered hats and dolls that distinguished them visually from the other children who although they lived with the missionaries, had not yet become part of the nuclear family. Source: "From the Report of Mrs. Gurubai Kamarkar, of Bombay," *Life and Light for Women XXIX*:3 (March 1899), 120.

In a 1911 article in the *India Alliance* of the India Mission of the Christian and Missionary Alliance (C&MA), an evangelical protestant mission society from New York that worked in Akola (Bombay), the author claimed to be the daughter of an American missionary couple who had taken in an Indian orphan in 1900.[89] Although it is not certain whether *My adopted sister* draws on the memories of the author, Ruth Andrews, it does shed light on how missionaries strove to preserve the children's 'Indian-ness'. After the adoption of the child, Andrews's mother considered it important "to keep her like her own people."[90] The Indian girl hence kept her name (Durie), ate separately from the family, consumed Indian food and wore Indian clothes. That the child tragically died only shortly after her adoption was meant to prove the importance of missionary work, since the child's timely conversion allowed her ascendance to heaven. Andrews reminded readers that "little Durie would not now be waiting for us in heaven, if we had not been able to take her in and provide for her."[91]

Stories like that of Durie were published to solicit money for orphan care in India. Missionaries in India depended on the generosity of American donors to finance the education of the children. Orphan care had become a part of the

ABCFM's work only recently, namely since the Armenian massacre of 1894–96 had pushed missionaries to open orphanages in Armenia.[92] Emily C. Wheeler had worked as a missionary in Armenia before she began channelling her energies into popularising child sponsorship campaigns that raised money for children in Armenia and India.[93] Substantial assistance was also coming from the *Christian Herald*, whose readers subscribed to annual contributions of $15 per child to ensure that the children taken in by missionaries would become useful Christians.[94]

Training a New Generation of Indian Christians

Child sponsorship remained a popular cause of American humanitarianism in the early twentieth century. Missionaries were concerned, however, about the unreliability of such funding, as these donations could cease at any time. Not long after the famine ended, the AMM lamented its dependence on American money which interfered with the Mission's goal to build a strong, self-reliant Indian Christian community. The wish to gain independence from American donations gave impetus to the AMM's vocational training that aimed to turn the children and women admitted to the Mission during the famines into self-supporting members.[95] These efforts similarly grew in light of the dissatisfaction of missionaries "with the result of a purely literary education" and the perception that it needed manual labour to change boys "from dullards into stirring active workers."[96]

Education had long made up a large proportion of the activities of the AMM and trades such as woodwork and industrial drawing had been taught in the Sir D.M. Petit Industrial School in Sirur since 1891. By 1900, the school had two model farms and departments of carpentry, blacksmithing and aloe fibre.[97] Vocational training was also offered at widow homes that identified sewing as the most adequate trade to be taught to the women. In 1900, Hume wished to expand the AMM's industrial training in Ahmednagar to include the children taken in during the recent famines and to this end applied for support from the American Board. Deploring that missionaries were not trained well-enough to oversee vocational training, Hume suggested bringing American university graduates to Ahmednagar as technical experts. The response of the American Board was lukewarm. While the board allowed Hume to proceed with his plan, it offered no financial support. Hume made another attempt, now soliciting the support of the American philanthropist Henry Phipps.[98] He eventually secured the seed money from the Americo-India Famine Relief Committee that until recently had administered the funds raised by New York's business philanthropists to support American aid efforts.[99] In his letters to sponsors, Hume explained that he wished to model education in Ahmednagar after the Tuskegee Institute in Alabama that under the lead of Booker T. Washington sought to revolutionise the education of African-Americans. Hume's writings exhibit the idea that African-Americans and Indians shared similar qualities that rendered the transfer of educational methods between the United States and South Asia possible. Such racial thinking also undergirded the use of the Tuskegee model in other parts of the world.[100] Hume enthused that testing the Tuskegee model in Ahmednagar above all other places was ideal, because its people "are among the best races of India."[101]

Hume brought David Carroll Churchill, a young MIT graduate, and J.B. Knight, a former student at the Massachusetts Agricultural College at Amherst to Bombay. The two men arrived in India in August 1901. Following an inquiry into local conditions, Knight opened a model farm to teach forty famine boys "the principles of scientific farming with their practical application."[102] He chose a small group of orphans, whose prior English-education supposedly made them more susceptible to American expertise. The training entailed a daily schedule of four hours of work in the field, one hour of livestock and poultry keeping, and three hours of school education. In addition, Knight oversaw the training of a group of boys and men, who he aimed to turn into "agricultural leaders."[103] To this end, their training also entailed daily lectures in chemistry, zoology and botany.[104] When funding ceased a year later, Knight's model farm in Ahmednagar came to an end.[105] The government of Bombay had initially advanced two grants to secure the continuation of the agricultural experiment station. Instead of continuing to fund the missionary farm, however, the government encouraged Knight to continue his work on a governmental model farm to which he not only brought his expertise in early 1902, but also his former students. He later became a Professor of Agriculture in Poona and continued his agricultural research.[106]

Churchill, in the meantime, had started to work in the Sir D.M. Petit School for Industrial Training. Similar to the model farm, the days of students were split between literary education and artisanal work. Workshops taught rug-making, wood-carving and the making of brass, copper and silver-ware. That such industrial education had a more wide-reaching agenda which aimed at the transformation of the students' character and physique surfaced in a report of the AMM in 1902. Its author confidently declared that "to those [...] who enquire what we make in these workshops [...] I always answer, *Men.*"[107]

Churchill identified one particular handicraft which he believed would benefit from the skillset he had acquired at MIT. Dissatisfied with the available hand-looms in Bombay, Churchill developed a semi-automatic loom to raise productivity.[108] He began to lead a weaver's workshop for boys in Ahmednagar, who were employed in a nearby rug factory after their training.[109] The prospect of finding employment drew additional children into Churchill's weaving workshop.[110] His loom initially won prices across India and was widely celebrated. Yet a decade later its inventor grew dissatisfied with his own accomplishments. The advancement of Churchill's invention resulted in high costs of maintenance of the looms and made repairs unaffordable to Indian weavers. Failing to reconcile the wish for technical development with the needs of Indian weavers, Churchill eventually left India in 1917.[111]

3.4 American Missionary Humanitarianism in the Wake of World War I

War, Famine and Disease in Bombay, 1918–21

After one and a half decades of relative stability, World War I wrought economic dislocation in India as the country's economy was geared towards Britain's military needs. Apart from soldiers and auxiliary forces, India supplied leather, hides, clothing, and ammunition to Britain, as well as railway tracks, locomotives and cattle to Mesopotamia. In addition, food was exported from India to supply troops in the Near East and civilians in the metropole.[112] The transport of war products and the movement of troops furthermore diminished available tonnage and created scarcities in India.[113] In the Bombay Presidency, where the monsoon failed twice in three years, drought intensified pressure on the food market. Famine struck Bombay's population in 1918–19 and again in 1920–21.[114] Although the end of the war brought relief to Bombay's food market, parts of the population remained vulnerable to famine.

The prices of food, cloth, kerosene and other basic commodities had risen sharply during the war.[115] The war-induced inflation struck wage labourers hard as their salaries were not keeping pace with the surge of living costs. Whereas labourers suffered from the steep rise of costs, Bombay's mill- and landowners reaped enormous profits during the war as they benefitted from the high demand of cotton and the rise of prices.[116] Their wealth empowered them to engage in philanthropy. Bombay's industrialists and merchants formed a relief committee that assisted the efforts of the government to mitigate the hunger of the population in Bombay. The Gujarati merchant Purshottamdas Thakurdas (1879–1961), also known as Bombay's cotton king, oversaw the work of the Bombay Presidency Relief Fund that received large donations from Shapurji Broacha (1845–1920).[117] After the failure of the rains in mid-1918, the government of Bombay opened poorhouses and test-works in the affected districts. As the drought continued, parts of Panch Mahal, Poona and Satara and the total area of the Ahmednagar district were declared famine areas. Scarcity was much more widespread.[118] In the famine-afflicted districts, officials now started to employ men and women in so called village works, where they cleared tanks, repaired roads, built railway tracks and broke metal to earn food doles. Governmental relief continued for nearly a year. Only in August to October 1919 did the government wind up its relief camps and poorhouses.[119] In 1920, district officials were again mandated to declare famine in Ahmednagar while parts of Nasik, Poona, Sholapur, Satara, Bijapur and Belgaum suffered from scarcity.[120]

The timely provision of relief resulted in the containment of famine, but Bombay was beset by an even greater menace to public health in the meantime. A first wave of the influenza pandemic hit the province in June 1918. When it flared up again in September, hunger lowered the population's resilience to the virus.[121] Between 10 to 20 million lives were lost in India in 1918/19; in no other country did so many people die from the influenza.[122] Troops had carried the virus into Bombay after which the disease spread from the city of Bombay to the rural hinterland. The influenza was particularly virulent in the crowded *chawls* that accommodated many of the city's industrial labourers.[123] When the epidemic entered villages, it preyed upon a population struggling to fill their stomachs.[124] Hunger and poverty intensified migration into the city of Bombay. Colonial officials considered these "famine refugees" a threat to public health and decided to stop them on their way, to round them up and confine them in poorhouses in the suburbs.[125]

The Famine Response of the American Marathi Mission

The septuagenarian Hume was still working in Ahmednagar in 1918 when the food situation in the Bombay Presidency again drastically deteriorated. The AMM was not immune to the social and economic turmoil that unfolded in the Bombay Province. Missionaries died of influenza and the rise of prices sapped the AMM's resources.[126] The financial woes of the AMM again struck its Indian members severely. Missionaries noted that their allowances fell below the salaries paid at governmental relief camps.[127] To assist its Indian members as well as the population that lived in proximity to its mission stations, the AMM began to mobilise funds in late 1918. The *Congregationalist* in Boston published a first appeal to give money in support of Indian members of the AMM in December 1918, stressing that the famine afflicted "the very people who count most for the spread of Christianity in India."[128] The subsequent relief followed the conventional script. A committee was appointed first to oversee the collection of funds and to identify where and how missionary aid was most effective. In addition to supplements paid to increase the allowances of Indian mission workers, the AMM provided food and cash doles to the infirm. It assisted famine-afflicted men, women and children who were trying to reach governmental relief camps and opened small-scale relief works themselves.[129] Missionaries also helped through their educational institutions. The women's homes and boarding schools of the AMM noted an "unusual number of applications for admittance" as parents admitted their offspring into missionary institutions and women whose fathers, husbands and brothers had left in search of employment, sought shelter in the mission stations.[130]

Indian mission members were again at the forefront of the relief effort of the AMM, but continued to receive little recognition in the press reports that detailed the missionary aid efforts. The AMM had almost doubled its Indian membership since 1896, now counting 608 Indian members. Most of them worked as teachers (169 men and 135 women). In addition, 59 bible women, 48 unordained and 44 ordained Indian pastors were active in the AMM.[131] Among them were Reverend Anand Sidoba Hiwale (1879–1922) and his wife Taibai Hiwale who had opened a shelter for famine children during the current famine in Sirur that accommodated approximately 200 children.[132] For this endeavour, the couple enlisted the support of the Parsi community. Converted into a boarding home for children, it continued its work after the famine years and became known as the Sir Ratan Tata Institution for Destitute Children.[133]

Anand Sidoba had been a student at the AMM's Byculla High School and the Ahmednagar Theological Seminar before he received further education in Maine, New England. He had graduated from the Bangor Theological Seminary in 1907 and Bowdoin College in 1909. When he returned to India in the same year he married his fiancé Taibai Patole and the couple started to work as missionaries of the AMM. Like the Karmarkars before him, A.S. Hiwale had established lasting ties to American supporters during his stay in the United States. After his return to India he received assistance from the YMCA and his alma mater in Maine. Bowdoin College prided itself on its Indian missionary. It reported on his work in the *Bowdoin Orient* and invited students to stereopticon lectures that featured Hiwale's work in India.[134] There were other similarities to the AMM's previous famine relief activities. The AMM reiterated earlier doubts concerning the performance of baptism during famines. "Many requests have been made for baptism, but many have been refused" noted the report of the AMM in 1919 that also recommended to "go very slowly."[135] Missionaries however did not cease to evangelise. Missionaries approached inmates of governmental famine relief camps and the women who passed through Ahmednagar on their way to and from the governmental famine works.[136]

Unlike in the previous famines, missionaries in India were now organised in the National Missionary Council of India which gave the AMM another instrument to mobilise the missionary aid response in these years. Robert A. Hume, among others, represented the AMM on the Bombay Representative Council of Missions, the provincial branch of the National Missionary Council. The Bombay Representative Council of Missions had early kept an eye on the deteriorating famine situation in the province. It made recurrent offers to supplement the governmental relief efforts in case famine was declared, however refraining from antagonising the colonial administration by acting on its own.[137] When famine was declared in parts of India in late 1918, the Bombay Representative Council of Missions set up the Special Committee on Famine Relief. Under the chairman Hume, the committee united missionaries of different nationalities and denominations who began to disperse funds to mission societies across India to assist missionary famine relief.[138] In line with the aim of the National Missionary Council to foster cooperation between mission societies across India, it now integrated the relief efforts of missionaries in the different Indian provinces in a national structure.[139]

US Missionary Fundraising for India

When in late 1918, missionaries in India began to solicit funds for famine relief, Americans were taking centre-stage in the relief efforts in Europe and the Near East. The havoc caused by the warring nations of World War I gave impetus to the formation of relief organisations in different regions of the world, but in light of the economic prosperity of the United States "the largest and most influential humanitarian organizations hailed from the USA."[140] Missionaries were not absent from this surge of American humanitarianism. Although the professionalisation of American philanthropy in the Progressive Era had seen the emergence of secular aid providers, faith-driven philanthropy or more traditional forms of charity did not lose their significance.[141] The relief work of the AMM in India in 1918–1921 is evidence of the continued importance of missionaries as providers of American foreign aid. While American humanitarians largely looked towards Europe, missionaries strove to raise awareness of the war-induced misery in other parts of the world, including South Asia.

Given that, instead of peace and progress, "Christian nations" had brought violence and devastation to the world, the foreign mission movement suffered an ideological crisis during World War I which fuelled a rethinking of the mission movement in the interwar period.[142] Yet there were also those who argued that the war had demonstrated Christian virtues and therefore saw new opportunities emerging on the horizon. Bishop Frank Warne, for instance, believed that "the war has given India a clearer conception of the true Christian spirit."[143] Warne referred to President Wilson's principle of political self-determination which he saw as a powerful demonstration of American Christianity. There were other ways in which Warne claimed World War I had created new openings for American missionaries in South Asia. Since missionaries had aided colonial troops who fought in various theatres of the war, Warne and other missionaries assumed that returning soldiers were more likely to embrace Christianity because of the help they had received.[144] The service of Indian soldiers in the imperial army had also supposedly eroded their attachment to caste which was held as a major obstacle of conversion in India. The Canadian mission scholar John Lovell Murray furthermore concluded that the exposure of Indian villagers to the outside world through army service bred appreciation of the Christian way of life.[145] Murray was an advocate of

Christian internationalism who considered missionary humanitarianism a means to demonstrate the relevance of the missionary movement after the war. As the author of *The Call of a World Task in War Time,* he lamented that attention given to the suffering of people in mission lands was fading.[146] "We have been solicitous for the hungry in Belgium and Poland during the present emergency. But more people have been suffering from the pangs of hunger in India than in Belgium and Poland combined."[147] Murray penned these words for the delegates of the conference of the Student Volunteer Movement (SVM) in Northfield in 1918. The SVM had been a source of personnel to the foreign missionary movement since 1886. By 1900 it had tipped the ratio of British and American missionaries on the world stage to the advantage of the latter.[148] By World War I, many of its members had already taken part in missionary philanthropic and humanitarian work.[149] The postmillennial advocates aimed to better living conditions to pave the way for the return of Christ and therefore devoted themselves to social work and reform.[150] In the interwar period, they provided nourishment to the claim that missionaries ought to invest in social work, including the mitigation of hunger. Aiming to mobilise aid for India after the war, missionaries of the AMM lauded the American willingness for "sacrifice and service" and hoped to extend it to India.[151] If we consider that most Americans perceived the needs of Europe as more urgent, the sum raised by the American Boards for famine relief in India in the year 1919 is testimony to the success of missionary fundraising. After the first reports of American missionaries in India reached the United States, the Methodist Centenary Department of War Emergency and Reconstruction promptly set aside $10,000 to support famine relief in India.[152] Throughout the year 1919, the American Boards raised $150,000 through religious and secular newspapers, church papers and their constituencies.[153] The *Christian Herald* similarly renewed its fundraising for Indian famine relief in 1919.[154] From January to November 1919, the *Christian Herald* that was now edited by G.H. Sandison, raised $65,000 for India. This was a fraction of the amount raised in the 1890s. The diminished support of the *Christian Herald's* fundraising efforts, that also showed in an additional campaign for China, illustrated the growing disagreement between American evangelicals. Many of its former supporters no longer believed that missionaries should engage in humanitarianism. Some even considered it a vital threat to the mission movement, because it diverted time and money away from outright evangelistic work.[155]

The general success to mobilise money in support of missionary famine relief in India in the wake of the war partly owed to the news coverage, some of which turned out to be exaggerated in hindsight. Readers in the United States who followed the reports of missionaries in India in the American religious and secular press gained the impression that the subsistence crisis currently unfolding in Bombay and beyond compared to famines of the 1890s. Whereas the official report of the government of Bombay noted in 1920 that no visible signs of bodily emaciation were to be detected in 1918–19, missionary reports suggest widespread starvation. In March 1919 Hume testified that Indian members of the AMM were "slowly starving to death."[156] Sam Higginbottom, writing from Allahabad in the United Provinces in May, noted that the province was going through "one of the most widespread famines of modern times."[157] At the height of the aid drive, newspapers reported about a figure of 32 million people that supposedly were at the brink of death in India—a number that the colonial administration sharply refuted.[158] Criticism of the British colonial government prompted an apology of the National Missionary Council in November 1919 when it stated its "regret and emphatic disapproval of the sensational and incorrect reports."[159] Although the National Missionary Council declared to scale back its appeals, it nevertheless stood by the missionaries' continued calls for foreign aid that was needed to assist famine-stricken populations in Bombay and other parts of India.[160] The colonial government had ended its famine relief in Bombay's Deccan in late 1919, yet the population remained at the brink of famine. In 1920, missionaries in Ahmednagar raised alarm at the renewed onset of famine conditions. "By the testimony of those who have long lived there" Hume wrote now, "the Ahmednagar district was never so hard hit!"[161] The missionary, however, used his reports to similarly appease the colonial government as he stated that "no government [was] able to create rain or to cause water to run uphill for itself."[162]

3.5 Conclusion

Tracing the involvement of American missionaries in alleviating famine in South Asia is part of this book's broader aim to break away from the spatial and temporal frameworks that dominate research on famine in South Asia. The beginning of American food aid in South Asia is commonly attributed to the decade succeeding World War II. The earlier missionary famine relief shows the longevity of American investment in South Asia in the field of food aid. Famines also prompted contact and exchange between American and Indian societies, albeit not on equal terms. In 1900, Robert A. Hume was convinced that the aid missionaries brought to Indians in the previous years had drastically altered the relationship of India and the United States. "Today multitudes of people in the United States of America who never before had any interest in India now have become interested in all that affects this country", Hume wrote enthusiastically.[163] According to Hume, missionaries helped generate American compassion for famine-stricken Indians and Indian gratitude for American assistance, resulting in a bond of affection between the two countries. This, he prophesied, was to serve as the foundation of future US involvement in South Asia. That this bond was never purely affective but equally material is evidenced by the movement of money, grain and people between India and the United States in these years.

To fan the willingness of Christian donors to engage in humanitarian work in South Asia, missionaries portrayed India as a country of evangelical opportunities. The missionaries of the AMM saw famine relief primarily as a means to gain ground in a mission field that proved inhospitable to its evangelicalism. The unprecedented sums to support the relief efforts of American missionaries helped the AMM to expand its activities in the province. The AMM had long struggled to raise the number of Indian converts, and the famines of the 1890s allowed it to make some progress in this regard. The migration of the famine-afflicted population to urban centres in search of labour and relief gave missionaries access to larger rural populations. The sheltering of women and children in the mission stations proved another way by which the AMM expanded its influence over Indian society. The efforts of missionaries to exploit famine also had limits and donor expectations were often not met. American donors sought formal distance from British colonialism, and therefore channelled money through committees controlled by US missionaries. Such efforts, however, were contradicted by the AMM's adherence to British colonial standards of famine relief and its cooperation with colonial administrators. And although many donors in the US viewed famine relief as a suitable channel to improve Indo-US relations, US-sponsored famine relief accelerated conflicts between Indians and missionaries in Bombay. The reliance on Indian mission members to implement relief further limited American control and efforts to stage missionary aid as uniquely American. Despite the many self-celebratory depictions of American aid that presented hagiographic accounts of male missionaries who seemingly single-handedly saved thousands of children and women from starvation, Indian mission members were at the forefront of missionary famine relief. Their stories are much more difficult to trace. When we do, they reveal that the change which the AMM aimed to bring about in the hearts and minds of famine-stricken Indians, primarily affected itself and its relation with South Asia.

CHAPTER 4

Promising Freedom from Famine: American Missionary Rural Reform, 1910s–1940s

Abstract

The urge of American missionaries to demonstrate their capacity of preventing rather than mitigating famine fuelled their growing involvement in agricultural training and rural reform in the early twentieth century. The chapter offers a new perspective on the link between famine prevention and rural development in colonial South Asia, discussing the contributions of American missionaries from the late nineteenth century to the 1940s.

Keywords: American missionaries, rural reconstruction, agricultural training

4.1 Latecomers to the Field of Famine Prevention

> The downtrodden masses of India's unreached multitudes lie like a beggar at our gate, full of sores and desiring to be fed with the crumbs that fall from our table. We can help them best, not by fitful famine relief in special times of distress, but by the prevention and provision which the gospel of Christ can give to India, with all its uplifting power.[1]
>
> —Sherwood Eddy (1912)

Behind this statement of Yale-educated Sherwood Eddy (1871–1963), world-travelling missionary of the Young Men's Christian Association (YMCA), was a shift in the responses of American missionaries to famines in India. In the first decades of the twentieth century, a vocal group of American protestants emphasised the significance of preventive action over short-term food and medical aid. In a sense, the turn of American missionaries to famine prevention was preordained. At the very least, it was a logical step. The increasing presence of American missionaries in rural areas and the provision of famine relief facilitated later experiments in rural uplift. This was not only the case in South Asia, where Robert A. Hume opened a model farm in the aftermath of the famine of 1899–1900 to train Indian children—many of whom had been taken in by missionaries during the famine—to become better farmers. This widening of missionary activities also showed in China, where American famine relief in the early 1920s subsequently fanned missionary rural reform, with agricultural education seen as key to the prevention of famine.[2] Not all model farms and agricultural training institutes that American missionaries ran in South Asia in the early twentieth century trace back to earlier missionary famine relief. Yet they similarly took poverty and hunger as a starting point to promote Christian rural reform.[3]

American missionaries were late in pointing to the importance of averting famine in colonial South Asia, where calls for its prevention had emerged in the previous century. Dadabhai Naoroji and Romesh Chandra Dutt, to name just two prominent Indian advocates of famine prevention, had studied the economic causes of subsistence crises in rural India in the late nineteenth century. They identified the impoverishment of rural populations and the deindustrialisation of India under colonial rule as causes of famines and called for the investment of taxes raised in India in the development of the country (see chapter 5 of the book for a discussion of Indian demands for famine prevention).[4] In the late nineteenth century, many in and outside of the colonial administration promoted the construction of railroads and irrigation networks in the name of famine prevention, although the effect of such infrastructural development on the reduction of food shortages was limited. Aggravated by the disagreement on the meaning of famine, famine prevention alluded to a spectrum of demands and measures that continued to expand throughout the twentieth century. On one end of the spectrum, populated by Indian nationalists and British socialists, famine prevention meant the eradication of all forms of hunger. At the other end of the spectrum, it was merely about containing mass starvation. Colonial administrators in general espoused this latter position which did not completely rule out the future occurrence of famine, nor address nutritional deficiency outside its temporal scope. Somewhat counterintuitively, efforts to prevent famine did not necessarily imply a refusal of the idea that famines were inevitable and natural phenomena. Demographic pressures, climate, and culture remained common scapegoats that were invoked to evade responsibility and depoliticise the causes of famine in India.[5] American protestants were represented across the spectrum, but a growing group began to embrace the claim of Sherwood Eddy that the "uplifting power" of Christianity enabled rural populations to durably overcome hunger and poverty, and with it, famine as well. It was no coincidence that such a claim was made by a prominent member of the YMCA.[6] The YMCA was tied to the Social Gospel movement that sought to improve living conditions and address social injustice to prepare for the Kingdom of God.[7] YMCA missionaries had recognised early that improving the living conditions of the communities they wished to convert was central to the success of their mission. Moreover, given its international organisational structure, the YMCA contributed greatly to internationalise rural social welfare in the early decades of the twentieth century—a topic I will return to later in this chapter.[8]

Given that famines in South Asia rarely made international headlines in the interwar period, one would expect the American public's interest in preventing famines in South Asia to wane. Yet famine control still drew support from the United States, because the interwar period provided an atmosphere that was conducive to appeals to counter hunger in South Asia. Nationalist mobilisation in India (and wider Asia) amplified the political ramifications of starvation and gave urgency to effective famine prevention. This further intensified after World War I, when Indian economic nationalist critique "broadened [...] into calls for self-determination."[9] Indian nationalists linked political independence more closely to the fight against hunger and poverty. In light of the Bolshevik revolution and the rise of Asian nationalisms, Americans considered rural poverty and hunger in (South) Asia a threat to global political stability. For liberal American protestants who sought to engage with, and benefit from, Indian nationalism, the motivation to engage in rural welfare (and famine prevention) followed a different logic. Since claiming the improvement of the conditions of India's rural populations through social service had become central to Indian nationalist mobilisation, missionaries had to demonstrate the contributions Christians were able to make to rural life.[10]

Americans back home realised that Indian nationalism was not a distant movement that was thousands of miles away from them. The exposure of the collaboration between Indian radicals and the German government during World War I and the persecution of Indian anticolonial activists in the United States affected American public opinion, as some North Americans feared being drawn into a conflict that was not theirs. Many blamed the prevalence of hunger and poverty as a key driver of both South Asian migration to the United States as well as the surge of Indian nationalism.[11]

Against this historical context, it seems conspicuous that when Sam Higginbottom, the founder of the Allahabad Agricultural Institute (AAI), visited the United States to solicit money in 1919, he described the prevalence of hunger in India vividly. An attendee of one of his talks noted that when "others [spoke] on famine conditions [...] all one sees is a thin man; when Sam describes him, one can count every rib in his body."[12] Prior to his journey to the United States, the missionary had reported on the current food scarcity in the United Provinces for the *Christian Herald* and explained how the introduction of American farming methods was the most adequate way to avoid future famines in India.[13] Two years later, he reiterated this claim in *The Gospel and the Plow*, arguing that "the rapid introduction of better farming [...] is the one sure way to rid India of the ever present nightmare, as well as the reality, of famine."[14]

Whereas famine relief had become a popular cause of humanitarianism in the mid-nineteenth century, rural development had yet to build its funding base. In light of the politicisation of famine described above, and since it was not far-fetched to claim that rural improvement helped to prevent future food crises, missionaries strategically invoked famine to attract funds.[15] This again was not only the case in South Asia. In the aftermath of famine in China in 1920–21, American missionaries applied for the surplus funds of the America Committee of the China Famine Fund to finance the curriculum revision at Nanking University. They pitched the need to include agricultural training at Nanking University by suggesting that it would serve famine prevention— "a term elastic enough to encompass anything" as Randall E. Stross noted in his study of the American agricultural reform movement in China.[16]

Despite the fact that American missionaries had started to attend to the material needs of rural people in the late nineteenth century, the attention that American missionaries gave to rural reform in the first decades of the twentieth century was novel. This was a global phenomenon from the outset. In the first decade of the twentieth century, missionaries in different regions of the world began offering agricultural education to promote rural reform.[17] It was in the interwar period that agricultural reform gained considerable momentum and rural reconstruction became a buzzword.[18] The most prominent examples of missionary rural reconstruction in South Asia of the interwar period were the AAI in the United Provinces and the YMCA-led Rural Demonstration Centre in the South Indian village of Martandam. The AAI was the brainchild of the aforementioned Manchester-born Sam Higginbottom who had gone to the United States for his studies and became an active member of the SVM at Princeton. He agreed to his first mission to India in 1903 and although he initially intended to leave after two years, Allahabad became his new domicile. The demand for English education in the Allahabad Christian College resulted in Higginbottom's involvement in teaching. Concurring with the growing popularity of the social gospel, Higginbottom began to envision an institution that applied American expertise and practical education to train Indians in modern agriculture. In preparation for this project, he temporarily left India in 1909 to study agriculture in Ohio and Princeton. He returned to the subcontinent in 1911 and founded the Allahabad Agricultural School (later AAI) in the following year. He led the AAI until 1944, when he retired to North America and was celebrated for his life-time battle against famine in South Asia.[19]

The Rural Demonstration Centre in Martandam was set up by Duane Spencer Hatch in 1924 and quickly became a lighthouse institution of rural reconstruction in the following years. Hatch, a Cornell-graduate, had gained practical experience in the university's extension work before he ventured out to India as a member of the YMCA. After serving with the war department of the Y in India and Mesopotamia, he was assigned to Martandam in 1921. The promotion of rural work within the Indian Y however traces back to the first Indian YMCA secretary, K.T. Paul.[20]

Despite discernible differences, the AAI and the Martandam Rural Demonstration Centre shared many commonalities that started to define American rural reform in India. Such was the focus on self-help (or its synonym, self-development), which under the guise of equity, voluntarism and democracy, imposed rules on those ostensibly in need of help.[21] Hatch and Higginbottom both looked to Booker T. Washington's vocational training in Tuskegee as a role-model, applying its methods to South Asia.[22] They experimented with the introduction of new seeds and livestock, and the promotion of village industries, engaged in physical education, emphasised hygiene and sought to stipulate moral reform. The aim was to incite a wide-reaching transformation of rural India and its inhabitants.

4.2 Gendering Rural Reform

The Chicago tabloid *Day Book* announced in 1912 that Elsie Leue from Cincinnati was expected to leave shortly for India. She was to teach American agriculture to Indian peasants to empower them to "raise crops large enough to feed the people." Leue was a member of the YWCA Cincinnati chapter and the first woman to ever complete the four-year bachelors in agriculture at Ohio State University. However, despite the *Day Book's* enthusiasm who pictured her as the "first farm missionary to the land of Hindoos", Leue decided to join the YWCA extension work in Ohio instead of becoming a missionary to India.[23] By the time the tabloid printed Leue's portrait, Higginbottom had received his first batch of students in Allahabad.

According to American missionaries in the first decades of the twentieth century, taming Indian agriculture was a man's job. Missionary agricultural training aimed to make boys into men and peasants into agricultural leaders. Women had a place in missionary rural reform, but they were encouraged to take up a different set of activities that allegedly fitted their abilities and role in society. Budget-making, infant welfare, gardening and poultry-raising were subjects of missionary education of rural women in the interwar period and aimed to promote modern domesticity; tilling the soil, on the other hand, was meant to be done by men. The gendering of agricultural life in American mission stations in India paralleled a similar differentiation in the US, where agricultural extension initially excluded women altogether. After it was realised that the exclusion of women was a major hindrance to the success of rural reform, agricultural extension started to be tailored to families, married couples and later, children. It still neglected women who worked the fields and depicted women primarily as homemakers.[24]

The gendered division of missionary labour and rural reform was evident in the AAI and surfaced in the shared life and work of the AAI's most prominent couple, Sam and Ethel Higginbottom. The lives of missionary couples who displayed protestant ideals of marital relationships and domesticity in the quotidian spaces of mission stations, offer glimpses on the gendering of missionary work. Jane Ethelind

Cody, a trained kindergarten teacher and aspiring missionary, had married Sam Higginbottom on the day of her arrival to India in 1904. In the initial years in Allahabad, Sam Higginbottom taught at the Allahabad Christian College and the couple became involved in missionary leper-care. They also oversaw a boarding home for students of a Christian high school, adjacent to the Allahabad Christian College. Ethel Higginbottom considered it her task to ensure that students were well-fed by overseeing the purchase and preparation of food. Sam Higginbottom, on the other hand, emphasised the children's physical education by playing tennis, football and cricket with the boys. Ethel Higginbottom furthermore started a Bible class and gave birth to the couple's first two children in 1905 and 1906. Motherhood prompted her to take informal medical training from a civil surgeon to prepare herself for the treatment of children's diseases. She also pushed her husband to start a dairy farm to ensure the children's milk-intake.[25] Ethel Higginbottom's role in the AAI began to change in the 1930s when she promoted and participated in female education. While prior to 1935, she had sporadically contributed articles to the *Allahabad Farmer*, the AAI's in-house magazine, from 1935 onwards the magazine regularly featured the writings of Ethel Higginbottom and other female educators at the AAI.[26] In the May-edition of 1935 she offered readers of the *Allahabad Farmer* a review of the educational work in rural Mexico, suggesting that similar schemes be implemented in India.[27] The same edition featured a new segment, the Rural Home Section, that was devoted to Indian rural women. Not long after, in late 1936, the AAI introduced a two-year course in house-keeping and home-building (the predecessor of the AAI's Home Science Department) which covered a broad range of subjects, reaching from (Christian) ethics, music and art to home furnishing and home industries.[28] The Home Science course of the AAI, led by Ethel Higginbottom, was designed for middle-class women. Applicants had to provide their matriculation certificate; only in exceptional cases did a proficiency in English suffice to enter the programme.[29] Another important condition for enrolment was the ability to pay substantial fees for tuition and lodging. Despite such limitations, the first twelve students had enrolled by the end of the year 1936.[30] The AAI did not have to search for suitable teachers of its new department, since wives of the AAI's American staff had brought degrees in home economics and medicine from American universities and institutes to India. External support was also forthcoming in Charlotte Wiser who was offering courses in food and nutrition. Wiser had recently acquired international fame for her monograph *Behind Mud Walls* (1930), an anthropological study of Indian villages.

Ethel Higginbottom's changing involvement at the AAI points to a larger trend. Missionary Home Science had important precursors in the vocational training mission institutions offered to women. The wish of the Anglo-American mission movement to expand its work with Indian women and children had brought new opportunities for women missionaries (American and Indian) to participate in social service activities. After women took up positions as teachers, nurses and doctors in boarding schools and widow homes and participated in famine relief and social welfare, they started to tap into the global Home Science movement in the early 1930s to expand the education of women.[31] Home Science boomed in India in the 1930s, where it drew strength from the simultaneous endorsement of Indian nationalist and feminist movements and also received visible stimulus from the United States. The American Ann Gilchrist Strong authored two well-used teaching manuals for Home Science in India in 1931 and 1945, and set up a Home Science Department at the University of Baroda.[32] After years of fundraising campaigns of the AIWC, the Lady Irwin College for Women opened its doors to students in New Delhi in 1932. It became a central post-secondary educational institution for women in India.[33] As the AAI demonstrates, American missionaries contributed to the Home Science movement in South Asia. While other institutions offered courses and study programs to women in urban settings, the AAI strove to educate rural women. It sought to integrate female education in modern domesticity into their otherwise male-centred approach to agrarian reform.

That the advancement of rural reform had suffered from the lack of emphasis on female education and the participation of women as rural workers began to filter through to male architects of missionary rural reconstruction in South Asia in the 1930s. Duane Spencer Hatch remarked with remorse in 1938 that women were underrepresented in rural reconstruction.

> It is a mistake, *though a natural one*, that those actually engaged in the first years of the rural reconstruction movement are at least ninety per cent men. Rural reconstruction workers should number fifty per cent women. We cannot possibly make creditable progress until we have with us not only women workers but the general interest, understanding and participation of the women of the rural areas. I regret that among all the 900 leaders who have come to us for training since 1926 only ten per cent of them are women.[34]

The YMCA had not followed in the footsteps of other institutions by offering classes in Home Science to female students. In response to the failure of the YMCA to attract Indian women, Hatch instead started to provide incentives and rewards to Indian YMCA workers and apprentices, who could demonstrate that their wives participated actively in the Y's mission.[35] The claim that wives were able to offer invaluable support to the advancement of YMCA's rural work promoted conjugality as a pillar of rural life. It might have been also the outcome of his own experience. As Harald Fischer-Tiné has pointed out, Emily Gilchriest Hatch participated greatly in "planning and implementing new schemes in rural development."[36] Emily Gilchriest Hatch had studied drama at Syracuse before she decided to accompany her husband to India. She later became Provincial State Commissioner for the Girl Guides and taught rural drama among other subjects in the Practical Training School in Rural Reconstruction.[37] She wrote a number of pedagogical plays, used to convey protestant modernity to South Indian audiences and her audience widened considerably after Oxford University Press published her plays in "Drama for the Village Teacher" and "Little Plays" (1932).[38]

4.3 (South) Asia and the Internationalisation of Agricultural Missions

In the first years after World War I, a small group of American protestants aimed to institutionalise the until then loosely bound network of agricultural missionaries working in different regions of the world. The purpose of this endeavour was to learn from regional experiences and to popularise agricultural education and rural uplift within the American mission movement. This meant standing up against fundamentalist criticism that wished to limit missionary activity to evangelism. Tying in with the rise of Christian internationalism in the interwar period, agricultural mission work became a domain of enhanced international collaboration that aimed to better international relations through rural improvement.[39] American missionaries working in India took part in this collaborative effort. The mobility of YMCA members including Indian secretaries, and a surge of publications on rural mission work in South Asia further contributed to the integration of India into a global dialogue on missionary rural uplift in the interwar period.

Under the auspices of the American YMCA and the President of the Massachusetts Agricultural College Kenyon Leech Butterfield (1868–1936), the World Agriculture Society was formed as an organisation striving to internationalise rural work in 1919. It grew out of a conference that Butterfield organised in French Burgundy. Its venue was the American Expedition Forces University in Beaune, established by the YMCA in 1919 to keep the soldiers in Europe busy while they waited to go home.[40] The YMCA had invited Butterfield to teach agriculture to the American soldiers to prepare them for their future as farmers in the United States. To prepare for creation of the World Agriculture Society, Butterfield assembled Canadian, Belgian, British, French experts, as well as the secretary of the Chinese delegation to the Peace Conference, at a four-day convention on the "World Problems in Agriculture and Country Life" in 1919. The workshops held during the conference covered topics such as the world's agricultural problems, scientific agriculture and rural life and welfare.

The food problem, that only recently had acquired the prefix "global", figured prominently in the writings of the World Agriculture Society.[41] To begin with, it considered itself "a voluntary fellowship of individuals and organizations who recognize the importance of seeking a solution of the *world's food problem*."[42] Its periodical *World Agriculture* featured the writing of agricultural experts who explained how future food shortages were to be prevented. Although American missionaries in the interwar period tirelessly emphasised the superiority of American farming methods over "local" knowledge and technology, this dominant narrative was punctured in *World Agriculture* which offered articles on China detailing how Chinese horticultural knowledge and seeds could improve American farming. George Weidman Groff (1884–1954), the director of Agricultural Work in the Canton Christian College and author of "Agricultural Reciprocity between America and China" (1911) described China as a "'gold-mine' for Western students of agriculture."[43] Other authors explained how the import of Chinese crops benefitted cultivation in the United States.[44] To further foster Chino-American collaboration and to promote "'better understanding' of world agriculture", the World Agriculture Society opened chapters in China and in the United States in the 1920s.[45] The World Agriculture Society belonged to Butterfield's efforts to push rural reconstruction onto the agenda of the American mission movement while also redirecting American attention to Asia.[46] Ian Tyrrell has argued that apart from China, India had been of particular importance to the work of Butterfield. Given his longstanding connections to the YMCA of India, he began to promote the transfer of lessons-learned from India to China.[47] Butterfield also helped the global popularity of Martandam in 1930, when he published the widely read *The Christian Mission in Rural India*. The book was the outcome of six months of travelling in India, a journey prompted by the request of the International Missionary Council (IMC).[48] Another keen admirer of Hatch's work in South India was John H. Reisner (1887–1965). The son of farmers had studied biology and agriculture in Yale and Cornell before becoming a missionary to China in 1914. He later turned dean of the College of Agriculture at Nanking University, where he oversaw the university's agricultural experiment station. In the 1920s, after witnessing famine in China, he advocated reforestation in China's flood prone areas to deter the destruction of crops that had preceded the recent and earlier famines.[49] It was in the 1930s that his interest in South Asia grew. Reisner visited India as the secretary of the Christian Rural Fellowship in 1939 to assess the state of rural reconstruction in the subcontinent. Reisner and Hatch had been corresponding prior to his visit to India and in preparation of his journey, Reisner had studied Hatch's *Up from Poverty in Rural India* (1932) and *Further Upward* (1938). His personal visit to Martandam further fanned his enthusiasm: "in my whole missionary experience I have never seen anything like that."[50]

Another stop on Reisner's journey through India was Allahabad where he visited the AAI. Sam Higginbottom's monograph, *The Gospel and the Plow* of 1921, had made him well-known beyond India. Higginbottom was a founding member of the International Association of Agricultural Missions (henceforth Agricultural Missions) that also named Reisner as one of its members. The first meeting of Agricultural Missions was organised by Benjamin H. Hunnicutt in 1920. Since 1907, Hunnicutt had been a missionary to Brazil, where he had opened an agricultural school in Lavras. By the 1930s, he was celebrated as the first agricultural missionary to South America and as "one of the outstanding leaders" of agricultural mission work.[51] Agricultural Missions were committed "to promote the interests of Christian agricultural work in all lands" and to this end commissioned studies, held training and assisted the exchange of missionaries involved in agricultural education.[52] Unlike the World Agriculture Society that seemed to have fallen into oblivion after Butterfield's death in the 1930s, Agricultural Missions proved more enduring.

4.4 Missionary Rural Uplift in the Aftermath of the Bengal Famine

At the end of World War II, the imminent independence of India caused another phase of introspection of the American mission movement. American missions looked to their future prospects in India with a mix of confidence and scepticism. Criticism of Christian proselytisation had grown in India in the 1940s and was prominently showcased in Gandhi's increasingly overt opposition of missionary conduct.[53] Responding to Gandhi's demand that missionaries limit their involvement to humanitarian work or leave India, William Hazen of the ABCFM concluded in 1943 that "it is probable that missions in future cannot expect as sympathetic treatment as in the past."[54] Hazen's prediction held the test of time to some extent as in 1956, the Niyogi Committee Report on Christian Missionary Activities recommended a ban on religious conversion. Although the Indian government did not follow the Committee's recommendations and conversion remained legal, missionary evangelical work in postcolonial India incited further controversy and outright violence.[55]

In 1946, the National Christian Council of India (NCCI) called for a conference to discuss the future development of Christian institutions in the subcontinent.[56] The conference resulted in a ten-year programme that aimed to cement the role of Christians as the alleged drivers of Indian development. It entailed a threefold approach: assisting Indian organisations that set out to increase economic performance and drive India's industrialisation; forming technical institutions that offered relevant practical education; and lastly, training village Christians in their respective villages. This last pillar of reform, summarised under the title "Self-Help Projects", planned to intensify the education and training of villagers to help them "to free themselves from the clutches of poverty by using their own resources."[57] The plan favoured those American missionary institutions that had proven their ability to shift from outright evangelicalism to rural welfare and reconstruction in the interwar period. Fittingly, the NCCI chose the AAI as its venue.

Although Higginbottom had returned to the United States in 1944, the AAI had survived the war and the partition crisis. Higginbottom was succeeded by John Goheen and later, by Arthur T. Mosher. In 1952, four years after Mosher joined the AAI as its director, the institute made a new foray into rural development. Financed by a Ford Foundation grant, it employed over a hundred rural extension workers to supplement the government's work in community development.[58] The Rural Demonstration Centre of the YMCA in Martandam was also still thriving in the aftermath of the departure of Spencer and Emily Hatch in 1940. In 1947, it inaugurated the YMCA Rural Welfare Workers Training Institute.[59] Thus, missionary rural uplift that commenced in the interwar period continued to influence postcolonial development in India. This continuity was also helped by the fact that missionaries started new projects on the eve of independence and in its aftermath.

Post-war missionary rural development in India received impetus from the Bengal famine, which (again) called attention to food insecurity in South Asia. In April 1945 John Fischer (1910–1978) recalled his memories of the Bengal famine for readers of the American *Harper's Magazine*.[60] The article was reprinted as a standalone booklet titled *India's Insoluble Hunger* in Bombay in 1947.[61] Fischer's writing seemed to be that of a journalist who, although shocked by starvation, death and indifference, witnessed the famine as a bystander. This was a misrepresentation. Although Fischer penned his words as the associate editor of *Harper's Magazine*, he had been a US diplomat at the time of the Bengal famine. As a senior representative of the US Board of Economic Warfare and as the special assistant to the President's Personal Representative in New Delhi, William Phillips, Fischer had overseen the gathering of economic intelligence.[62] He was amongst the first Americans to learn about the famine in Bengal and he was in a position to lobby for US food aid. Unlike Phillips who became known as a strong advocate of US food aid, Fischer did not seem to have been in favour of American assistance for Bengal.[63] Fischer, who left his government job to return to journalism after his arrival from India in 1944, spoke out against any kind of US involvement in South Asia. In *India's Insoluble Hunger*, he claimed to remember a conversation with an American general in India, who had told him about a nightmare that plagued him. "I dreamed that all the Englishmen quietly slipped out of this country during the night, and left us Americans holding the bag. Can you imagine anything worse?"[64] The deaths of millions of Indians in Bengal in the years 1943 and 1944 were, according to Fischer, inevitable.[65] He prognosed that India would see worse famines in the future. The chief culprit was "relentless fertility" that in combination with agricultural inefficiency provided a recipe for disaster.[66] There was no solution, he claimed. Fischer's nightmarish vision had a long genealogy. It traces back to Malthusian population theory of the eighteenth century which undergirded colonial governance in the nineteenth century. It reincarnated in the interwar period when under the influence of eugenics, population growth was framed as the "population problem."[67] Fears of Asian overpopulation gained further momentum after World War II, when another global war had devastated food economies. Elites now lashed out against poor sections of society and advocated fertility control.[68]

In 1945, Fischer's article provoked the response of the current executive secretary of Agricultural Missions, John H. Reisner. Reisner sent a copy and a questionnaire to sixty American missionaries in India to inquire about their views on Fischer's assessment. Agricultural Missions summarised the missionaries' responses in *The Future of Rural Christian Missions in India* (1945).[69] Somewhat unsurprisingly, the missionaries opined that contrary to Fischer's analysis, India was to be helped and although the problem was "terrible baffling," American missionaries had a duty to assist.[70]

In 1944–45, Reisner participated in the formulation of a new venture in the United Provinces (Uttar Pradesh after 1947) that would become known as India Village Service (IVS). The original idea to launch this missionary experiment in rural improvement is attributed to a New York-based Indian businessman named B.N. Gupta. Gupta had led India Famine and Medical Relief in New York that raised funds in the United States during the Bengal famine.[71] Agricultural Missions quickly took up the idea and commissioned William H. and Charlotte Wiser, who were known for their studies of Indian village life in Karimpur, to work out the details of IVS.[72] Gupta became part of the provisional American committee that steered IVS and that next to Reisner also included A.T. Mosher of the AAI.[73] IVS was based in Mahrera, 600 kilometres to the north west of Allahabad. The decision to work in Mahrera owed to the proximity to AAI, to which the IVS maintained close contact from the outset.

Like other contemporary rural development projects, IVS was an amalgam of various influences. It exhibited the continuation of a transnational exchange that had exerted its influence on the evolving body of knowledge on rural development since the interwar period.[74] IVS set out to function not so much as an institutional set-up but an educational programme, which aimed to establish a prototype to be copied and applied all over India and possibly beyond the subcontinent. It sought to train Indian village teachers, who were expected to act as intermediaries between experts and the villagers. The largely western expert "counsellors" offered knowledge and advice to the Indian staff, but never directly interfered in the village experiment. The Indian teachers on the other hand assessed the needs of the villagers and supervised their learning. Because of this, Indians who wished to become leaders of village progress were expected to exhibit empathy and a strong interest in understanding village life. While a college degree and training as a teacher were mandatory, agricultural knowledge and experience in farming was not only not necessary but rather viewed as a disadvantage. The adoption of this specific approach by the IVS traces back to Martandam. However, its underlying ideas had much wider breeding grounds. IVS also drew from the work of the Near East Foundation, which had run a rural reconstruction programme in Macedonia from 1928 to 1938.[75]

The recruitment of suited personnel was important to IVS. Despite its similarity to secular development initiatives, it strove to safeguard its protestant identity. According to the Wisers, IVS "[sought] to be frankly Christian in its inception, operation and spirit."[76] Indians who wished to become teachers of IVS needed to demonstrate that they were "good Christians" and sought their motivation in a Christian ideal of service. A further notable feature of IVS was its goal to employ an equal number of men and women teachers.[77] Although Indian women's organisations and missionaries long lamented the low number of women working in rural areas, women were outnumbered in Indian community development in the 1950s, with programmes tailored exclusively for men.[78] Vidyawati Singh, working as an IVS village teacher, identified a series of reasons for the underrepresentation of women in community development. According to the rural worker, especially young single women from cities were neither accepted by the villagers nor were they accustomed to living conditions in rural areas. Safety problems, alongside inadequate salaries and housing conditions rendered social work in rural India unattractive to women.[79] The need to integrate Indian women in community development was highlighted by Shanti Daniel, also a staff member of IVS, when she attended the National Seminar on Development Work Among Rural Women in 1956. Daniel presented a paper that sketched the women's programme of IVS. According to Daniel, IVS aimed at improving women's literacy and educating them in hygiene and cleanliness, childcare, nutrition, recreational activities as well as knitting, stitching and spinning. Women were also encouraged to organise village functions to celebrate Independence Day, Republic Day and *Gandhi Jayanti*—a task that demonstrated best how IVS tried to craft Indian citizens. As the women's programme of IVS demonstrates, missionary rural uplift cemented traditional gender norms rather than challenging them. When secular Indian community development began to give greater attention to women's education and communal participation after 1958, it took a similar course.[80] IVS thus demonstrates links between rural reconstruction and village uplift schemes of missionaries in the interwar period and secular rural development in the post-war era. In this context it is also noteworthy that the much better-known community development experiment of Albert Mayer in Etawah in 1948 that was expanded by the Nehru government into a nation-wide scheme in 1952, drew from IVS. The transfer of methods from Mahrera to Etawah can be illustrated by a training technique called Training Within Industry (TWI). Elaborated first in the United States, TWI was later used by IVS and adopted by Mayer.

TWI was developed by members of the War Manpower Commission in World War II. The commission was tasked to improve the training of women and labour migrants to compensate the drain of labour due to war service.[81] TWI was adopted by the American industrial sector in 1941, and by the Extension Farm Labour Program in 1943.[82] As the name indicates, TWI trained people on the job and aimed "to put knowledge to use in the shortest possible time."[83] TWI entailed the breaking down of a job into a series of basic steps that allowed unskilled labourers to carry out even technical and specialised jobs. After the war, Agricultural Missions, in cooperation with the U.S. Department of Agriculture, began to teach TWI to American missionaries. When the staff of the Department of Agriculture visited Mahrera in 1950, they were enthusiastic to see the results of their courses. "We go on every year after year with short courses for missionaries and never have an opportunity of studying the results in the Field. It is a joy for us to see our materials in use and to know that they are being of help."[84] Previously, William Wiser had taken a course offered by Reisner in New York and subsequently applied the method to Mahrera.[85] The reasons given by Wiser for choosing TWI as a pillar of IVS in India provide insight into contemporary development thinking that was not unique to IVS. Wiser believed that similar to the unlearned and predominantly black and female labourers in North America, Indian peasants could improve their skills through the right method of instruction. He was convinced that "one reason why villagers do not change their practices is because they do not fully understand."[86] TWI used "key point lessons"—a summary of the basic steps of a specific task—for instructional purposes. In the case of IVS, key point lessons were meant to teach Indian villagers what they needed to know to improve village life. The Wisers directed Indian teachers to write such instructions in cooperation with agricultural experts. In the process, the Indian teacher as the intermediary between villager and expert translated expert knowledge into a manageable form and content. The key point lessons thus generated proved very popular. While the AAI reproduced them in the *Allahabad Farmer*, William Wiser explained the method to the American agronomist Horace Holmes who took it to the Etawah project.[87]

TWI was just one of the methods that IVS developed to teach Indian villagers and that started to radiate globally. After the American funding committee pushed IVS to adopt "demonstration" as a teaching method, IVS began to employ dramas, games, and films to convey the lessons of self-help to the mostly illiterate villagers.[88] In IVS's own estimation, the *garhagraph*—an Indian adaptation of a flannelgraph—was particularly successful.[89] This was a story-telling device consisting of a painted cloth spun across a frame and a series of paper cut-outs placed on the canvas. Another source of missionary enthusiasm was an adaptation of the board game "Snakes and Ladders" intended to teach villagers about adequate diets and nutrition. When the token of a player landed on a space labelled "wrong food", he slid down a snake. If he landed on a space that depicted a "right food", he advanced via the ladder.[90] The main producer of such educational material was Gladys Rutherford, a missionary who worked for the IVS as a Public Health and Sanitation Counsellor. Her teaching material had a remarkable longevity. In addition to the IVS flannelgraph stories, Rutherford instigated the production of the IVS Jet Series, which remained the preferred means of instruction for the Peace Corps in the 1970s.[91]

4.5 Conclusion

The AMM missionary Robert A. Hume, as shown in the previous chapter, established a short-lived centre of agricultural reform in Ahmednagar 1901, so as to introduce American farming methods and technologies that aimed to modernise Indian agriculture. Hume's isolated experiments anticipated the arrival of a group of American missionaries in the 1910s and 1920s, who promised to free India from famine through Christian-inspired rural reform. Missionary famine relief anticipated the growing involvement of missionaries in rural reform in South Asia. The broadening of missionaries' famine-related activities was driven by developments in the United States and in India. On the one hand, the cause of famine prevention became increasingly popular among donors in the United States, influenced by the export of Indian radicalism to America and the anxiety of Asian nationalisms. On the other hand, the American mission movement's contribution to the fight against rural poverty and hunger became urgent in light of the significance Indian social reformers and politicians attached to poverty alleviation. The pressure on missionaries to demonstrate their ability to work for the benefit of India mandated their emphasis on famine prevention. In the 1910s and 1920s, American missionaries touted agricultural education and rural reform as a means to attend to the root causes of famine, knowing this would free resources. Famine thus remained an important frame that influenced the way American missionaries viewed and portrayed their growing involvement in India in the interwar period.

American missionaries saw and presented the prevention of famine in India still as a central reason for their involvement in South Asia in the 1940s and 1950s. The American mission movement, seeking to consolidate its role in postcolonial India, launched new initiatives of rural reform prior to Indian independence. IVS, a postwar missionary initiative in rural education, shows the continuity of the missionary involvement in rural development in South Asia. IVS was a product of transnational entanglements dating back to the interwar period. It drew on the YMCA's work in Martandam, India, and the Near East Foundation in Greece, and introduced training methods originally developed in the US extension work. Its close collaboration with the AAI and Albert Mayer's Etawah project allowed IVS to influence these well-known Indian institutions of rural reform, and exemplifies how the American missionary movement came to influence Indian Community Development.

Seeking to address a blind spot in the history of American missionary rural reform in India, the chapter examined the role of gender and protestant notions of conjugality and family. The efforts of missionaries to shape rural life through the reworking of gender relations exemplifies another continuity between missionary famine relief in the nineteenth century and rural reform in the twentieth. Missionaries contributed to the framing of agricultural education as a masculine endeavour that aimed to make men out of boys. Women, overwhelmingly associated with the domestic sphere, attracted the attention of American missionaries in the 1930s when home science acquired importance in South Asia and beyond. The change of activities of the AAI, that under the lead of Ethel Higginbottom and its female employees now taught Indian women in home science, attested to this development. The participation of women and the work they conducted in American missionary rural reform in South Asia expanded further in the 1940s. Linked to the growing prominence of Indian women in social service at the time, IVS made it a concern to employ an equal number of women and men as village teachers. Indian members of IVS, such Vidyawati Singh and Shanti Daniel, used their association with the organisation to further promote the contribution of women in rural development.

PART III

Anticolonial Famine Relief: Mobilising against Hunger and Colonialism

CHAPTER 5

Famine Amid Swadeshi and Swaraj, 1900s–1920s

Abstract

Documenting famines and organising relief for the afflicted population was part of Indian nationalist activity in the nineteenth and twentieth centuries, when promoting a vision of a future India free from famine became part of Indian political mobilisation. Tracing Indian nationalist activity in India, Britain and North America in particular, the chapter demonstrates the significance of famines in the transnational history of Indian nationalism in the early twentieth century.

Keywords: Indian nationalism, diaspora nationalism, anticolonialism, anti-imperialism

5.1 The Indian Famine Union in London: Advocating Indian Political Demands in the Colonial Metropole at the Turn of the Twentieth Century

The Indian Famine Union's Constitutive Meeting

Criticism of the colonial management of famine in India predated the twentieth century, but it was in the aftermath of the devastating Indian famines of the 1890s that the demands for the effective prevention of famine mounted in India and in Britain. In June 1901, London's Westminster Palace Hotel hosted the first assembly of the Indian Famine Union which brought together retired British members of the Indian Civil Service (ICS), parliamentarians, philanthropists and journalists. The hotel was a popular venue for politicians, diplomats and lobbyists due to its proximity to parliament and its luxury amenities. The nature of the gathering and the choice of its location made it difficult for the *Times of India* to believe that the Indian Famine Union would help the cause of famine prevention. It suspected that this new organisation intended to deflect state responsibility, making famine control a matter of British philanthropy instead.[1] This accusation proved unfounded, for the Indian Famine Union had no intention of relieving the colonial state of its duty to prevent another famine from ravaging India and killing millions of the Empire's subjects as a result.

Membership of the Indian Famine Union overlapped with that of the British Committee of the Indian National Congress in London, which advocated Indian political demands in Britain since 1889.[2] The impressive line-up of the first meeting of the Indian Famine Union included the journalists and authors Vaughan Nash (1861–1932) and William Digby (1849–1904), as well as well-known heavy weights of moderate Indian nationalism, most notably Dadabhai Naoroji (1825–1917) and Romesh Chandra Dutt (1848–1909).[3] The Indian Famine Union not only carefully selected its members but also chose the timing of the inaugural session deliberately. Under the lead of Antony MacDonnell, the third Indian Famine Commission was currently finalising its much-anticipated evaluation. British liberal William Wedderburn (1838–1918), long-time chairman of the British Committee, considered this an opportune moment to increase public pressure on the India Office to recognise the importance of revising its anti-famine policies.[4]

Many of the attendees contributed to Indian economic nationalism by analysing and criticising the impoverishment of India under colonial rule. They identified poverty as a central cause of famine in India, which was a radical proposition at the time.[5] Naoroji was a professor, journalist and social reformer by the time he began studying the impact of colonial rule on the Indian economy. Commencing his writing on the "drain of wealth" in the mid-1860s, Naoroji showed that the practice of financing exports and the colonial administration through money raised in India precipitated a flow of wealth out of India.[6] By calculating India's national and per capita income, he defined the Indian economy in national terms, and therefore facilitated the imagining of India as a spatially delimited community and entity.[7]

Some of the later members of the Union had already commented on the intermediate findings of the Indian Famine Commission in March 1901. William Digby, the author of *The Famine Campaign in Southern India* (1878) and *India for the Indians* (1885), demonstrated his disagreement with MacDonnell in an exchange of statements in the *Times of India*.[8] Digby particularly questioned MacDonnell's statements on the condition of the rural population and the extent of famine-mortality. "[The Indian peasant] is not the starving creature that some people seem to imagine," argued the head of the Indian Famine Commission. Digby, in return, recommended him to consult his forthcoming book to correct his statements.[9] His 700-page magnum opus *Prosperous India* (1901) used the government's own statistics and Naoroji's previous analysis to demonstrate how rural impoverishment bred famine in India.[10] The book was part of a series of related publications all released in the same year. These included, Naoroji's re-edited collection of earlier writings *Poverty and Un-British Rule in India* and Romesh Chandra Dutt's *Indian Famines: Their Causes and Prevention*.[11] Equally important were Dutt's letters to the Viceroy of India, George Nathaniel Curzon, published as a book in Britain in 1900 and pointing out the siphoning off of land taxes during famines.[12] Finally, the instigator of the Indian Famine Union himself, Wedderburn, had added to the list of publications on Indian famines. As a member of the ICS from 1860 to 1887, he had witnessed first-hand the suffering inflicted by famines in India. This drove him to study the conditions of the peasantry in India and to call for investment in rural development. Concurring with Indian writers who identified poverty as a central cause of the frequency and intensity of famines, he argued that the Indian peasantry ought to be in a position to save to compensate for seasonal shortages.[13] His 1897 pamphlet *Skeleton at the (Jubilee) Feast* condemned the preparation of the pompous ceremony of Victoria's Diamond Jubilee in a year of famine.[14] In the same publication, Wedderburn demanded an official enquiry into the economic condition of rural India—a demand that had grown out of the debates of the INC. It was this demand that the Indian Famine Union now tried to popularise among metropolitan audiences, arguing that such a study would help build a factual basis for the improvement of colonial famine prevention. Since the colonial state questioned the objectivity of previous studies of Indian and British authorship, the demand for an independent enquiry meant to force the colonial administration into a debate on the state and causes of poverty in India. As Digby described in 1901, a profound ideological disagreement had eclipsed any middle-ground between defendants and critics of colonial capitalism at that time.

> Two schools exist. One always referring to the increasing prosperity of the country and the people, and claiming unstinted praise for England as the creator of this prosperity; the other is incessantly dilating upon the rapidly-growing and now alarming impoverishment of both country and people [...] One is right; the other is wrong.[15]

Unsurprisingly, therefore, official reactions to the Indian Famine Union were negative. The responses of the Secretary of State for India, George Francis Hamilton, and the Indian Viceroy Curzon vacillated between ignoring the Indian Famine Union and discrediting its demands.[16]

The Indian National Congress and the Indian Famine Union

The response of the INC to the Indian Famine Union was naturally more benign: the 1901 Congress gave its formal blessing to the organisation, accepting it as a champion of Indian demands in London.[17] Despite these concerted efforts, the Indian Famine Union was unable to move the India Office into giving in to the request for an official investigation. When the INC met in Bombay in 1904, Hari Sitaram Dixit again brought up the Indian Famine Union to stress metropolitan support for an evaluation of rural poverty in India. Dixit called upon the colonial government to either accept the result of Indian studies or agree to an independent enquiry. Other speakers at the meeting found it difficult to accept that, given the frequency with which the colonial state convened commissions, it felt unable to create the one commission that Indians actually called for.[18] A year later, in 1905, the Indian

Famine Union further strengthened its ties to the INC, when it invited the members of an INC delegation to a meeting in Liverpool. In the midst of the British election campaign, when the imminent landslide victory of the British liberals was already anticipated, an INC delegation, consisting of the newly elected INC president Gopal Krishna Gokhale (1866–1915) and the Punjabi Congress politician Lala Lajpat Rai (1865–1928), travelled to London. The British Committee hoped that the visit would invigorate it, since the electoral campaign and the sickness of some of its leading members had slowed down its activities. Hopes were high that a liberal victory would bring political reform in India.[19]

In the resolution passed during the Liverpool meeting of October 1905, the Indian Famine Union expressed "its cordial sympathy with the aspirations of the people of India" and recognised constitutional reform as "the only effective way" to counter poverty and famine.[20] Although it declared Indian self-rule to be central to the reduction of hunger, the Indian Famine Union remained strictly within the confines of moderate demands that spoke of a gradual and indefinite process of reform. In hindsight, the fact that Gokhale had addressed the Indian Famine Union alone while Rai remained absent could be interpreted as a sign of dissent. The Indian Famine Union was of no appeal to the generation of extremists who formed a political counterweight to the old guard of the Congress. Rai, though he tried to unite the INC, was counted among the extremist faction that criticised the Congress for its conciliatory stance towards the colonial state.[21] Such criticism of Congress politics can be traced back to the 1890s, but more radical political forces in India gained ground in the wake of the decision of the colonial administration to partition the east Indian province Bengal along a Hindu-Muslim divide. The division of Bengal resulted in the first mass boycott of British goods, the *swadeshi* (of one's own country) movement.[22] Though the colonial state claimed that dividing Bengal constituted an administrative necessity, it was an attempt to secure the control of the province. Bengal was home to a growing group of critical Indian intellectuals and still the centre of the colonial administration. While moderates embraced swadeshi and welcomed the advocacy of self-help and sacrifice that would become central elements of Indian nationalism, they sought distance from the increasing militarisation of the movement. This split also reflected in a growing divide of the moderate and extremist faction of the INC.[23] The shift of Indian politics at this moment is often exemplified with Naoroji's own transformation. The Grand Old Man of Indian nationalism had long been convinced that Indians could win their greatest battles in the British Parliament. Naoroji had joined the House of Commons as the first Indian MP in 1892 and maintained his seat of Central Finsbury until 1895. After the defeat, Naoroji increasingly "radicalised"—a political trajectory that culminated in his call for *swaraj* (self-rule) at the Calcutta-Congress in 1906.[24] Shortly before, Naoroji had returned to India and was now elected Congress president for the third time in his life. During the Congress session, the INC declared its unanimous support of swadeshi.[25] Against the background of swadeshi, the idea that the political answer to famine lay primarily in demanding the colonial state to recognise the extent of poverty and bring about development lost ground. Swadeshi made the fight against poverty and famine the subject of Indian self-help.

5.2 "The duty which we owe to each other as members of a nation": Famine Relief as Indian Nationalist Mobilisation, c. 1890–1908

Social Service and the Transformation of Indian Famine Relief

The provision of aid during famine had a long genealogy in South Asia where different communities practiced philanthropy and alms-giving.[26] Indian famine relief was not invented in the twentieth century, but its organising principles shifted visibly between the 1890s and World War I.[27] The social service movement started to show in South Asia from the 1880s onwards and novel techniques of fundraising, volunteering and relief administration altered how Indian elites responded to famines in the early twentieth century.[28] Indian philanthropy intertwined with British, European and American debates on charity, self-help and national efficiency, and its own ways of organising social service and welfare activities in India emerged. As Carey A. Watt aptly pointed out, Indian social service became an exercise in nation-building, aimed at crafting citizens and creating a sense of national belonging.[29]

Though not limited to Hindu communities, Indian nationalist famine relief in the twentieth century was partly rooted in Hindu socio-religious reform that had gained pace in the previous century.[30] The Brahmo Samaj was one of India's oldest Hindu reform movements and was founded by Rammohun Roy in Bengal in 1828. An offshoot of the Brahmo Samaj, the Sadharan Brahmo Samaj, was established in 1878 due to a break of some of its progressive members with the Brahmo leader Keshab Chunder Sen. Social service became a pillar of the new group of Brahmos, who ventured out into the field of famine relief in the 1880s and 1890s.[31] Considerably larger than the Sadharan Brahmo Samaj was the Arya Samaj who started to engage in famine relief in the 1890s. The Arya Samaj was founded in Bombay in 1875 and Lahore in 1877, and was particularly successful in North India. It pursued a religious reform agenda. Its members called for a revival of a pure form of Hinduism by evading a series of social customs and practices that it deemed as alien elements to Hindu culture. Religious reform aimed to countervail the perceived numeric decline of Hindus in India and to strengthen the Hindu community. The cultural nationalism of the Arya Samaj, that linked Hinduism and nationalism, is considered an important precursor of later Hindutva ideology of the political right.[32]

Lajpat Rai joined the Arya Samaj in 1884 and quickly climbed its ranks. In 1897, alarmed by the doings of missionaries, he initiated the assistance of Hindu children during the famine. Rai interpreted the reception of Hindu children by the Christian missions as a failure of Hindus, because they had left the care of the children to "outsiders". In this context, he envisioned the rescue of Hindu children from starvation a test of Hindu national capacity.[33] Under the lead of the Arya Samaj, students of the Dayanand Anglo Vedic College assembled Hindu children in the famine-affected provinces in North India and placed them in orphanages run by the Samaj in Ferozepur, Punjab.[34] Rai also pushed the adoption of legal mechanisms preventing Christian missionaries from sheltering Hindu children, which influenced the decision of the third Indian Famine Commission to ban the removal of children from districts affected by famine.[35]

Lajpat Rai made famine relief a nationalist endeavour, but his statements and actions were ambiguous and even contradictory at times. On the one hand, he wished to promote a national philanthropic movement in which no Indian was left uncared for. While he travelled across India to solicit funds for the famine relief work of the Arya Samaj, he flagged the duty of Indians to assist their co-citizens and opined that India's more privileged committed a sin if they ignored the hunger of their fellow citizens. On the other hand, the Arya Samaj provided aid to Hindus, and excluded Muslims and Christians. Rai argued that each religious community was obliged to help their members in need, while he also opined that—given that Hindus constituted the numeric majority— Hindus were destined to take the lead in this wider national movement.[36]

Indian Famine Relief in the Aftermath of Swadeshi: Demonstrating Unity

When famine struck parts of the United Provinces in 1907, the response of the Brahmo and Arya Samaj was unprecedented in terms of the scope of their aid, the reach of their fundraising campaigns and the degree of professionalisation. This was partly the result of the colonial repression of swadeshi which helped the prominence of social service in India and strengthened its importance for nationalist mobilisation. After the ban of student participation in swadeshi activities in Bengal in 1907, Indian elites used social service to provide a new outlet for nationalist activity.[37] The Sadharan Brahmo Samaj initiated a central committee in Calcutta in 1907, which solicited funds from across India and oversaw the famine relief activities of Brahmos in the United Provinces. It appointed Abinash Chandra Mazumdar (1855–1925) from Lahore as the committee's chairman and under his lead, Brahmos began to dole out aid in Allahabad in February 1908.[38] While Brahmos were often considered pro-British, swadeshi altered their stance. Criticism of the colonial state became more frequent.[39] During the famine the Sadharan Brahmo Samaj exhibited an open critical attitude of governmental famine relief, which it deemed insufficient and of "humiliating character."[40]

A rechannelling of political activity into social service was visible in the work of the Arya Samaj in the aftermath of the colonial repression of rural protest in Punjab. In 1907, the colonial state considered the Arya Samaj a repository of radicalism. Rai, who was suspected of being behind the rural agitation, was deported to Burma.[41] When he returned, he emphasised the non-political nature of the Arya Samaj on the one hand, and launched a nation-wide campaign to assist the famine-stricken population of the United Provinces on the other.[42] Lajpat Rai established the Arya Samaj Famine Relief Fund, toured across North India to solicit support for the relief of the Arya Samaj and inspected the work carried out by the provincial government in the famine-affected areas. The rescue of orphans remained a central concern of the society during the famine, but it also expanded its aid to other affected parts of society.[43] In keeping with the use of social service as civic education, Rai emphasised that famine relief was an "object-lesson in self-help" and "most valuable training in public life."[44] Although the timely provision of famine relief hinged on the ability to recruit large numbers of volunteers quickly, Lajpat Rai took considerable time for the recruitment of volunteers, arguing that men (and exclusively men) had to be selected with great care.[45] The Arya Samaj offered social service training to women, but reduced the scope of their activities to the home. Social service at this time, including the provision of hunger relief, was predominantly perceived and constructed as a male endeavour.[46]

Rai's appeal to assist the famine-stricken population of the United Provinces also attracted a newcomer to the field of famine relief. The Servants of India Society (SIS) approached Rai and offered to contribute three of its members.[47] It was the first time SIS responded to famine and although both institutions were divided in their approach to social reform, the Servants believed that they had much to learn from the more experienced Samajis. The famines of the 1890s had instigated Gokhale's plans of launching an Indian social service organisation, but it took him until 1905 to inaugurate the SIS in Poona (today's Pune).[48] Although Hindu in character and outlook, the SIS understood itself as a non-communal organisation which set it apart from the Arya Samaj. Reflecting the moderate nationalism and liberalism that Gokhale professed, the SIS worked within the constitutional boundaries of the British Raj to make India and its citizens fit for self-rule.[49] Gokhale understood the organisation to constitute a training site, where future leading Indian citizens were to receive their education in service. This thinking also surfaced in the context of famine relief. As the SIS assisted the famine-stricken on various occasions during the first half of the twentieth century (a role that is discussed in chapter 1 of the book), its members often reflected on how the participation in relief work promoted civic virtues. Gopal Krishna Devadhar, a founding member of the SIS and one of the organisation's later famine specialists, claimed that the famine provided volunteers with "a very good opportunity to learn the first elements of real love for the country."[50] Yet despite the emphasis on the benefit of social work, the perceived need for a careful selection of volunteers limited the participation of Indians. This had already surfaced in the Arya Samaj's mobilisation of relief workers, and showed again in the general admission policies of the SIS. Future members of the SIS went through a careful and lengthy selection and training process, at the end of which they had to commit to a frugal lifestyle that ensured their dedication to the SIS and the nation.[51] The actual number of volunteers involved in relieving the distress of famine was thus small. In 1908, in addition to the forty volunteers of the Arya Samaj, three members of the SIS assisted the famine-stricken.[52] Although women remained conspicuously absent from the early famine relief efforts of the SIS, the organisation would later strive to increase the participation of women in social service. This was helped with the formation of the Seva Sadan (The Mission to the Women of India) in Poona in 1909. Women trained at the Seva Sadan later served as nurses at religious festivals (*melas*), in dispensaries and hospitals.[53] They also provided medical assistance during famines in the 1910s and 1920s and participated in nutritional education that sought to address maternal and infant malnutrition.[54] In 1919, in response to famine conditions in the Bombay Presidency, the SIS, in cooperation with the Bhagini Samaj in Bombay and the Seva Sadan in Poona, mobilised female volunteers to join Indian famine relief work. Volunteers of the Bhagini Samaj went to Kathiawar to assist middle-class women observing purdah, helping according to own estimates, 23,000 women in total.[55]

Although the overall contribution of Indian organisations to famine relief was admittedly small, Indian relief providers argued that it was important and meaningful: on the one hand, Indians provided aid to groups that received not enough or no attention in colonial institutions. On the other hand, famine relief was considered a duty and a right of Indians vis-à-vis their co-nationals. This was emphasised by Lajpat Rai who noted in 1908 that "nothing can relieve us altogether of the duty which we owe to each other as members of a nation and as fellow men."[56] Similarly, in a response to the alleged resistance of colonial district officials to Indian famine relief, Devadhar stated, "it is certainly the right of the people to help their own countrymen in times of calamity and this question must be fought out."[57] In the aftermath of the famine, Devadhar wished to attribute historical significance to the recent relief work of Indian elites. While he described the work of Indian organisations in great detail to the readers of the *Modern Review*, he argued that "it [had been] for the first time in the history of Indian Famines that a successful effort was made by the people at large to go to the rescue of their unfortunate brethren purely out of a patriotic spirit, in a rather systematic and steady way."[58] Despite the differences and at times rivalry between members of the Arya Samaj, Brahmo Samaj and the SIS, famine relief in 1907–8 became a demonstration in unity.

5.3 Echoes of Famine and Swadeshi in North America

"Gifts of Famine": Indian Famine and Anti-Asian Xenophobia in the United States and Canada

While Indian organisations were working to mitigate famine in the United Provinces in 1907/8, anti-Asian activists in the United States used word of another famine in India to legitimise xenophobia. Although South Asian labour migration to North America was limited, it was met with racist outbursts. In the western US-Canadian borderland violence erupted in 1907 when an angry white mob attacked a group of workers asleep in their barracks in Bellingham. In fear of further violence, Indians on the US West Coast moved towards Canada, hoping that their status as subjects of the British Empire would offer them protection. But their arrival in Vancouver was again met with violence.[59] In October 1907, the *International Wood-Worker*, the mouthpiece of the Amalgamated Woodworkers' International Union of America, strove to explain the recent violence against Indians. In *Gifts of Famine: Invasion of Sikhs from the Punjab*, the union claimed that Indian immigrants stole jobs and caused a drop in salaries, leaving the white working class in need for defence. It stoked fears, thereby risking further violence, by referring to the alleged unusual body height of Sikh men and their diseased bodies. The *International Wood-Worker* also reminded its readers how famines had devastated large parts of British India in the 1890s and predicted an "Asian invasion" would follow in the wake of renewed famines.[60] More specifically, the author referred to the arrival of Jat Sikhs from Punjab, a class and caste of former landowners, who had been dispossessed as a result of the restructuring of the land tax system in India under colonial rule and were consequently driven into wage labour and military service. The famines of the late 1890s had contributed to the impoverishment of agriculturalists in Punjab and fostered their migration.[61] The majority of those fleeing poverty and famine however did not travel to Canada and North America but to East and Southeast Asia. The poorest sections of society were forced into the indentured labour system that after the abolition of slavery supplied cheap labour to the British Empire.[62]

South Asian migration to Canada on the other hand had surged in the wake of imperial celebrations of the Diamond Jubilee of Victoria in 1897 and the coronation of Edward VII. in 1902, when Sikh battalions of the British-Indian Army who had taken part in the celebrations had crossed Canada on their way home. Some of them returned home with the wish to migrate. These South Asian immigrants began to earn their livelihood as labourers on farms, construction sites and within the timber industry on the US West Coast and in British Columbia. Following common patterns of overseas migration, the establishment of communal and family networks resulted in a second phase of migration. The number of South Asians in British Columbia and the United States rose in 1905 and now also included more Indians of rural background. Yet, if South Asians indeed planned to invade North America, as the *International Wood-Worker* claimed, they did so rather slowly. Between 1904 and 1908, merely 5000 Indians made their way into Canada, and 6,800 arrived in California between 1899 and 1914.[63]

Although the open display of racism was not unusual for the time, it is striking that US labour also directed its wrath against the British Empire. It claimed that colonial governance failed to remedy famines in India and charged the British with causing South Asian migration. To substantiate its claim, the *International Wood-Worker* used a historical analogy that equated the arrival of South Asians in the United States in the early twentieth century with the influx of Irish famine survivors of the *Gorta Mór*. This constituted a peculiar criticism of British imperialism that read the British failure of preventing famines as a threat to the wellbeing of the white working class in North America.[64]

The increase of tensions and outright violence in conjunction with the lobbying of white anti-immigration leagues led the Canadian and US governments to restrict the immigration of Indians (and other Asians).[65] Denying Indians entry into Canada was a slippery slope as Indians were British subjects and, as such, were entitled to travel across the British Empire. To halt Indian immigration nevertheless, Canada adopted the Continuous Journey Regulation in 1908. Immigrants now needed to arrive on a direct route from their homeland to Canada, but since there was no direct steamship connection between British India and Canada the provision effectively prohibited South Asian immigration. The new orders also made it mandatory that Indians owned at least 200 US dollars when they entered Canada—a sum that remained out of reach for Indian wage labourers.[66] Similarly, the US Immigration Service started to systematically deny Indians entry after 1909, claiming that they were likely to become a public charge. Restrictions to Indian immigration and the experience of everyday racism in North America made imperial citizenship a central issue of Indian nationalists who began to settle and travel in North America in the first decades of the twentieth century.[67]

"Victims of British Rule": Famine as a Driver of Political Radicalisation

In May 1908, the activities of Indian radicals on the US East Coast became the focus of *New York Times* investigative journalism. The author of *Aroused India Faces Mutiny and Invasion* reported the frequent visits of obscure-looking Indians at a lawyer's office in New York.[68] The article was a response to the advent of terrorism in Bengal. A month earlier, aiming to retaliate for the whipping of Indian students who had participated in swadeshi activities, two Indian men threw a bomb in a carriage. The bomb was supposed to kill the colonial official, Douglas Kingsford, who had ordered the whipping, but instead caused the death of two uninvolved women. The murder drew attention to Bengali revolutionary activities, marking the beginning of conflict between Indian revolutionaries and the colonial state that would intensify during World War I.[69] With Indian opposition against the British turning violent, the dubious Indian gentlemen who seemed to have appeared rather suddenly in the city of New York were seen as an indication that the United States was about to be involuntarily dragged into the conflict.[70]

Colonial repression of the swadeshi movement had further driven the migration of Indian activists, who espoused more radical political demands and methods. The year 1905 saw the settling of Indian radicals in London, where Shyamji Krishnavarma opened the India House that became a locus of radical anticolonial cooperation.[71] Following in the footsteps of Krishnavarma, Myron Phelps, an American lawyer of Irish origin, established an India House in New York in 1906/7. It was Phelps to whom the *New York Times* referred to in May 1908. The author cited letters Indians addressed to Phelps to illustrate the deep-seated Indian resentment against British rule. Famine, the *New York Times* argued, nourished anticolonialism, fuelling the political destabilisation of British India. Britain's alleged inability to contain famines in India furthermore threatened the American people, because they triggered nationalist uprising and migration to the United States. Referring explicitly to the *Free Hindusthan*, a publication edited by Taraknath Das (1884–1958), and its inflammatory writings, the *New York Times* claimed that Indians will bring terrorism to the United States.[72]

Taraknath Das had fled India to escape imprisonment for his involvement in political radicalism in 1905. After seeking refuge in Japan, he had continued to the US in 1906, where he claimed political asylum. In 1907, Das worked temporarily as an interpreter for the US immigration service in Vancouver and helped Indian migrant workers in asserting their right to remain in the United States. He then enrolled at Norwich University in Vermont, where he was soon expelled for his political activism. He found shelter in New York, and in late 1908 began printing *Free Hindusthan* in the office of the *Gaelic American*, the outlet of the Irish republican *Clan na Gael*. The influence of his journalistic venture showed in the copies of *Free Hindusthan* that were confiscated in Calcutta.[73] Das, as well as other Indians abroad, used their distance from the colonial authorities to circumvent censorship and publish more freely about colonial rule.[74] In the first edition of *Free Hindusthan* in April 1908, Das reported famine conditions in north-east India.[75] The edition carried a two-page report about the famine in which Das demanded self-rule, claiming that the end of colonialism was a necessary precondition for the prevention of future famine deaths.[76] Das cited the Irish-American economist Robert Ellis Thompson (1844–1924) to support the claim that only self-rule could durably free the country from hunger. Thompson had recently argued that political independence was a necessary requirement for the economic stability and prosperity of both Ireland and India.[77] On the cover of the November-edition of *Free Hindusthan* (figure 3) of the same year, two famine photographs alongside the heading "Victims of British Rule" were shown together with a notice that informed the readers about the paper's mission: "The Free Hindusthan advocates liberation of Hindustan that Millions may be saved from the starvation caused by legalised pillage of India by the British government."[78] The photographs showed two famished children; their bodies disfigured by hunger. Das neither mentioned the source of the photographs nor the year in which they were taken. They must have been familiar to readers, as the images, taken by American missionaries during the famine of 1899–1900, had circulated widely. They were reproduced several times in different contexts, amongst others in a booklet titled *Some Fruits of the Great India Famine*, which praised the success of converting a group of famine sufferers to Christianity. These earlier publications informed readers about the tragic death of the girls who both died shortly after the images were taken.[79] By using the images of the past famine without identifying them as such, Das intentionally blurred lines between the past and the present famine. In the accompanying text, Das remembered the previous role of Americans in mitigating famine in India, and explained that sending relief was not an adequate response. "We tell them again that no famine relief fund will ever be sufficient to stop the calamity until the people of Hindusthan get self-government."[80] The rejection of (foreign) charity was by now a common element of Indian nationalist writing in North America. Indian activists saw humanitarianism as a makeshift that did not address the real cause of hunger, colonial rule.[81] The Anglo-Indian William Charles Hopkinson, former police officer in India now acting as an inspector of Canadian Immigration, tasked to thwart Indian radicalism in Canada, suspected Das of exploiting famine to raise money to sustain revolutionary networks and placed him under permanent observation.[82]

Das' use of photography and his direct appeal to American audiences (in addition to Indian readers) distinguishes his writings on famine in 1908 from those of earlier Indian nationalists. But in terms of content and tone, it was also in continuity with publications of Marathi radicals, published over a decade earlier in India. Bal Gangadhar Tilak (1856–1920) was charged with sedition for his poem in 1896 that glorified the seventeenth-century ruler Shivaji and deplored the prevalence of hunger.[83] At the time he published the poem, Tilak had succeeded Mahadev Govind Ranade (1842–1901) as the leader of the Poona Sarvajanik Sabha (PSS). The PSS was a voluntary political organisation with significant influence in the Bombay Deccan

The Free Hindusthan

An Organ of Freedom, and of Political, Social and Religious Reform,

RESISTANCE TO TYRANNY IS SERVICE TO HUMANITY AND A NECESSITY OF CIVILIZATION.

"EVERY MAN IS FREE TO DO THAT WHICH HE WILLS, PROVIDED HE INFRINGES NOT THE EQUAL FREEDOM OF ANY OTHER MAN." - Herbert Spencer, "Principles of Ethics" Section 279.

"RESISTANCE TO AGGRESSION IS NOT SIMPLY JUSTIFIABLE BUT IMPERATIVE, NON-RESISTANCE HURTS BOTH ALTRUISM AND EGOISM." - The study of Sociology, Chap. 8.

VOL. 1. NEW YORK CITY, NOVEMBER-DECEMBER, 1908. NO. 8

THE FREE HINDUSTHAN FUND,

The object of the Free Hindusthau Fund is to collect money to defray the expense of publishing the said paper and to aid all our affairs of National importance, especially in SPREADING POPULAR AND SCIENTIFIC EDUCATION IN HINDUSTHAN. Any contribution to the fund will be highly appreciated.

TREASURER'S REPORT

Receipts :—Balance (from Oct.) $108.00. From Africa $7.00. From America a Comrade, $5.00. Other sources $1.00. Total $121.00. Expenditure :—Printing, Postage etc., $26.00—Balance, $95.00.

N.B.—The existence of the FREE HINDUSTAN and the progress of the cause advocated by it depend upon financial aid, so for the sake of humanity, please help the cause.

BLANK FORM FOR DONATION OR SUBSCRIPTION

Dated the..................190...

I herewith send the amount of..............................
as donation or subscription ot the Free Hindusthan fo
..........................months.

Name..

Address..

VICTIMS OF BRITISH RULE.

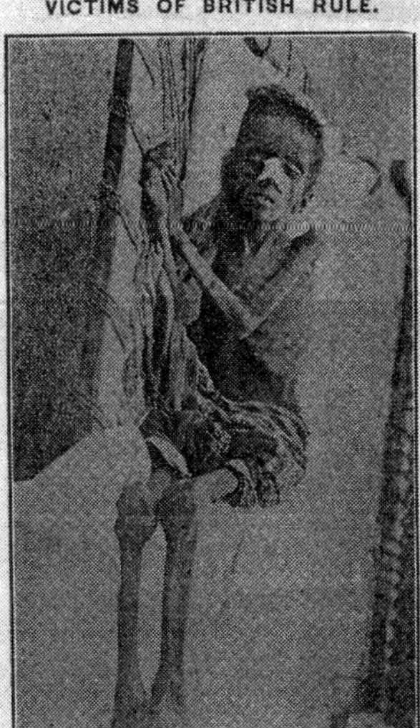

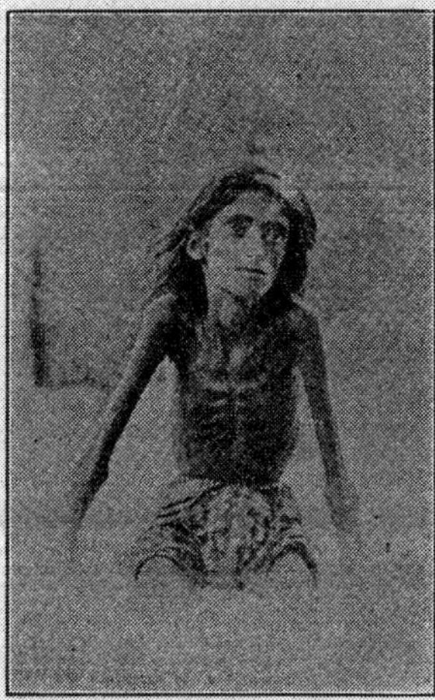

"THE FREE HINDUSTHAN"
Advocates Liberation of Hindusthan, that Millions may be Saved from the Starvation Caused by Legalised Pillage of India by The British Government.

FUNERAL OF THE MARTYE KANAI LAL DUTT.

Kanai Lal Dutt, one of the executioners of the informer Gossain, was hanged on November 10 at Alipore. There was a remarkable demonstration at the cremation of his remains at Kalighat.

The body was carried through the streets with the face uncovered, contrary to Hindu custom, so as to enable the thousands who pressed around the cortege to gaze upon the features of the "murdered Kanai," while purdah women followed in closed carriages. Hundreds of rupees were collected round the burning ghat in order to enrich the burning pyre, and there was a general rush to gather portions of the ashes, which were carried away in vases of silver and gold. Fragments and bones were preserved for despatch as relics to other towns. The uncle of the murderer delivered a funeral oration which was greeted with cries of "Victory to Kanai."

JOHANNESBURG, Sept. 17.—Mr. Harilal Gandhi, son of the leader of the Indian community, who was sentenced to a month's hard labor on the 18th of August, for refusing to obey the Magistrate's order and leave the Colony, was released to-day and quietly deported by the authorities to Natal.

Fig. 3. Victims of British Rule. Cover page of *Free Hindusthan* I: 8 (Nov 1908). Reproduced courtesy of the South Asian American Digital Archive.

region and a training ground for later INC politicians.[84] Ranade, who provided much of the intellectual foundation on which Naoroji was to build his drain theory, founded the PSS in April 1870.[85] In 1872, the PSS began to study the condition of the peasantry in the famine-belt. Its English-language publication, the *Quarterly Journal of the Poona Sarvajanik Sabha,* made the condition of Bombay's rural population a prime topic. In 1875, when peasants opposed revenue collectors and money-lenders in what became known as the Deccan Riots, the PSS was accused of being behind the movement. In the succeeding years of famine in South India, the PSS provided an important counternarrative to the official reports by giving detailed accounts of starvation and relief efforts; a practice it continued during the famines of 1896–97 and 1899–1900, when Tilak was in charge.[86]

5.4 Mobilising Against Famine and Colonialism in India, Canada and the United States, 1914–1920s

The Transnational Mobilisation of Indian Famine Relief

In 1914, the Congressman and member of the SIS, Hriday Nath Kunzru (1887–1987) was the secretary of the United Provinces Famine Relief Committee that under the presidency of the Vice Chancellor of the Allahabad University Sundar Lal (1857–1918) raised funds for non-governmental famine relief.[87] Several districts in the United Provinces went through another severe famine in 1914, the worst since 1907 as contemporaries claimed.[88] In addition to the SIS, the Arya Samaj and the Sadharan Brahmo Samaj were again leading Indian famine relief.[89] Intending to draw on famine for Indian nationalist mobilisation, Kunzru reiterated the importance of bringing Indians to aid their fellow citizens.[90] The fundraising of the SIS in 1914 also included South Asians in Canada and South Africa.[91] To mobilise the South Asian diaspora in Canada, Kunzru wrote to the editor of the *Hindustanee,* a publication of the United India League, to ask for its assistance.[92] "May I request you to draw the attention of our countrymen in Canada to the sufferings of the people of the United Provinces? I know they have grievances of their own, but I feel certain that they will nevertheless refuse help to their brethren."[93]

The United India League, as well as the Khalsa Diwan Society, were established in Vancouver and Seattle in 1907, committed to the welfare of Indian labourers in British Columbia and eager to challenge the anti-Asian immigration laws of Canada. In 1913, delegates of the organisations went to Britain and India to enlist support for the grievances of Indians in Canada. Although the delegation returned without concessions, it brought a group of Indians with them and, with the help of a lawyer, secured their right to enter Canada. Xenophobic fears rose in the aftermath of the court ruling, leading to further restrictions of South Asian migration.[94] But the legal victory now raised hopes among Indians who had been stranded in port cities along the route from British India to Canada since the Continuous Journey Regulation had come into force. In an attempt to oppose Canadian immigration policies, Gurdit Singh left Hong Kong on a chartered Japanese steamer in April 1914. The Komagata Maru entered port cities along the route to Canada, allowing Indians stranded on their way to Canada to complete their journey. The ship arrived in Victoria harbour in April, carrying on board 376 Indians, mostly Punjabi Sikhs, who hoped to disembark in Vancouver. Upon the arrival of the Komagata Maru, South Asians, however, were denied access to Canada. After months of dispute over the fate of the passengers, the ship and its passengers were forced to leave for India in late July 1914.[95]

The journey of the Komagata Maru was an important moment in the mobilisation of Indian transnational protest against imperialism. It followed in the wake of Gandhi's mobilisation of non-violent protest of Indian mine workers in South Africa against the state's repeated infringements on Indian citizen rights. Gandhi's experiment in *satyagraha*, or non-violent resistance, in South Africa was a stepping stone in his political career and received international attention.[96] The struggle of Indians in South Africa figured prominently in the publications of Indian organisations in Canada and the US which donated generously in support of the satyagraha. The experience of outright racism, manifesting in legal mechanisms to severely curtail the immigration of South Asians in white settler colonies, enabled Indians in British Columbia to relate to Indians in South Africa and to demonstrate larger structures of imperial discrimination.[97]

The activities of the United India League and the Khalsa Diwan Society made Kunzru certain that his appeal would fall on open ears.[98] And indeed, his appeal was taken up during a meeting of the United India League and the Khalsa Diwan Society. Umrao Singh, student of electrical engineering at Washington University, had previously taken part in Indian relief efforts during the Kangra earthquake in 1905. He now took the lead in urging the Sikh community to mobilise funds. Another man present, Raja Singh, cited the history of famines in colonial India and referred to Dutt's writings to locate the cause of famine in colonialism. Those present had already decided to raise funds when Nabhi Ram Joshi rose to speak at length on the ills of colonial capitalism. In the following weeks, the U.P. Famine Relief Fund Committee in Vancouver collected money and the names of subscribers and the sums forwarded to Kunzru in India were shared with the readers of the *Hindustanee*.[99] As the United India League reported on the progress of colonial and Indian famine relief in the following editions of the *Hindustanee*, it railed against the low pay of workers in government-run famine works and saw a nationwide swadeshi campaign an appropriate response to such imperial labour exploitation.[100]

Evidenced by its reference to Keir Hardie and Henry Mayers Hyndman, the United India League drew on socialist assessments of famine in India.[101]

At the time Sikh organisations in Canada were raising funds for famine relief, they had begun to pin their hopes on the Ghadar (Mutiny) movement. Aiming to take advantage of the mobilisation of Indian soldiers after the outbreak of World War I that left the colonial state vulnerable, the Ghadarites believed the time had come to drive the British out of India. In disagreement with the INC leadership who had commenced negotiations to advance the participation of Indians in key areas of state governance, the Ghadar movement strove for revolt. While North American anxiety had previously revolved around the arrival of South Asians, it was now their leaving that caused unease. Ghadar activists aimed to persuade Indian soldiers to turn against the colonisers and mobilised Indians across the globe, asking them to support the struggle against the colonial regime. Indians in North America began to board ships in 1914 to leave for India.[102] It was against the backdrop of efforts of Ghadar to mobilise Indians into a movement that aimed to overthrow colonial rule that its leaders also spoke about famine. Reiterating Naoroji's drain theory and referring to Digby's writings, Ram Chandra, leader of the Ghadar Party and editor of the widely circulating *Hindustan Ghadar*, accused the colonial state of obscuring famine in India. "Today India is in the midst of perpetual and endless famine […] and the officers of the government watch the terrific spectacle with the coolness of hardened villains."[103] Chandra lectured in California in July 1915 on the history of colonial India and circulated his analysis of famine also in a pamphlet.[104]

Lala Lajpat Rai and Opposition to Missionary Humanitarianism

In the midst of the Ghadarite efforts to launch an uprising in India, Lala Lajpat Rai arrived in the United States. He had originally planned a six-month tour of Europe, but the declaration of World War I diverted him to America. He remained there, intermittently, for five years from November 1914 and used his stay to educate Americans about India's political demands and to deepen his contacts with like-minded activists. The Ghadar Party found little favour with the senior nationalist, who rejected the violence Ghadar propagated.[105] Rai instead connected with the India Home Rule Movement that Annie Besant (1847–1933) brought into existence in September 1915 before Bal Gangadhar Tilak (1856–1920) took a hold of it.[106] Rai set up the Indian Home Rule League in New York in 1917 and started publishing *Young India*. Its editor was Jabez Thomas Sunderland (1842–1936), a Unitarian minister and long-time supporter of Indian nationalism. Sunderland had spent three months travelling in India from 1895 to 1896 to visit leaders of the Brahmo Samaj. In the aftermath of this journey, he became a vocal advocate of Indian nationalism and a critic of American imperialism. He became a member of the

Anti-Imperialist League and teamed up with Myron Phelps, who headed the Society for the Advancement of India in New York.[107] Sunderland also wrote on famine in India. In 1900, he presented a paper on *The Causes of Famine* that reiterated many of the claims Indian intellectuals advanced at the time. The paper was republished in 1904 as a stand-alone pamphlet and became well-cited by Indian nationalists in the US who used it as evidence of India's exploitation under colonial rule. Lajpat Rai for instance referred to it at length in his 1917 monograph *England's Debt to Britain.*[108]

Rai, like many others, believed that India's contribution to the British war effort during World War I would be reciprocated with political concessions after the war. President Wilson's Fourteen Points speech in January 1918 fuelled this hope, as nationalist movements interpreted the president's call for self-determination to equally include colonised countries. Such expectations of Indian nationalists were still high during the Peace Conference, even if they were ultimately not fulfilled. Rai, who had written numerous pamphlets and books during his stay in the United States, once again increased the tempo of his publications in 1919 when he wanted to draw American attention to the Indian demands.[109] First disillusionment set in with the publication of the Montagu-Chelmsford Report in March 1919 that fell behind nationalist demands and continued to exclude Indians from the most important areas of state administration. To make matters worse, the colonial state enacted the Rowlatt Acts in March 1919, which maintained martial law in peace-time and served to crack down on the nationalist movement. In response to the escalation of colonial repression, Mohandas K. Gandhi launched a nation-wide civil disobedience campaign. The violent killing of peaceful protesters in Jallianwala Bagh in April 1919 by British troops who had opened fire on the order of Colonel Reginald Dyer resulted in Gandhi's decision to call off the movement.[110] While the India Home Rule League provided a commentary to political events in India and the Peace Conference in Paris, it also published on starvation in India.[111] World War I caused economic dislocation in India, where prices of basic commodities skyrocketed.[112] Famine had to be declared in several districts across India, while scarcity was even more widespread. On top of famine, influenza spread in India, causing the death of an estimated 10 to 20 million people.[113]

The India Home Rule League was particularly vehement in its refusal of missionary fundraising efforts at the time. Contention arose after a Toronto-based fundraising committee consisting of Canadian and British philanthropists appealed to the public in the *Toronto Globe* in April 1919.[114] By its own estimation, the committee was driven by the wish to improve the strained relations between Canada and India. To feed Canadian empathy for the Indian victims of the famine, the organisers foregrounded imperial connections and suggested that both India's and Canada's affiliation with the British Empire was a source of commonality. The assertion of a commonality based on empire, which contrasted with the history of systematic discrimination against South Asians in Canada, infuriated the India Home Rule League. Putatively speaking on behalf of the people affected by the current famine, the League argued that Indians were willing to die in order to avail themselves of the help of the British-Canadian committee.

> India does not want charity [...]. We are sick of appeals for charity. We believe that charity debases both the giver and the receiver. Our hearts bleed for the miseries of our country. Yet, we do not desire to save the lives of our people by foreign charity. We have been reduced to this condition by the imperialistic, capitalistic exploitation of modern civilization. By offering us charity the British people add insult to injury. We are proud people and we will rather die and let our people die than ask for charity. Let the millions of India perish by hunger.[115]

The statement of the India Home Rule League provides the opportunity to reinforce some points made earlier in the chapter as well as to identify new elements of nationalist responses to famine. The rejection of British-Canadian aid was connected to a broader critique of foreign humanitarianism, which *Young India* argued obscured and perpetuated the root cause of famines in India, that is colonial capitalist exploitation. This not unfounded criticism of humanitarianism (or charity in the parlance of the time) had been part of Indian writing on famines since the late nineteenth century. Also evident in the statement is an escalation of language, both in terms of the emotional response to famine and charity ("we are sick"; "we bleed") and in terms of the suggested unity of Indians. Portraying the experience of famine as a collective experience helped to imagine the nation and its people as an entity, but negated that class, caste and gender (amongst other factors) were important determinants that split the Indian society into those who benefitted from colonialism and famine on the one hand, and those who were disproportionally affected by loss and death on the other. The collective experience of famine, expressed in the "we" here invoked, was a nationalist myth. Such assessments hark back to earlier Indian economic nationalist analysis of famine that, despite its strength of uniting critics of colonial capitalism, in so far as it reduced colonial exploitation to a standoff between Britain and India, tended to obscure other power asymmetries.[116]

The India Home Rule League's opposition to missionary fundraising continued in the following weeks and was fanned by the arrival of Robert A. Hume, missionary of the American Marathi Mission. Until his recent departure from India, Hume was the chairman of a relief committee in Bombay that collected and distributed money for famine relief on behalf of the National Christian Council of India. He now used his furlough to help the fundraising. *Young India* responded with scorn to the arrival of "one Dr. Hume" and reprinted excerpts from the missionary's appeal for funds to demonstrate the extent of the suffering that colonialism was causing in India.[117] Others joined in. According to the Bengali revolutionary M.N. Roy, the fact that even missionaries, "the most sworn and vociferous defenders of British imperialism" called attention to India's humanitarian needs, was evidence of the severity of the economic dislocation of the country.[118] Roy had recently prevented his arrest by fleeing from the United States to Mexico. From his hideout, he wrote a serial article entitled *Hunger and Revolution in India* that drew a link between imperialism, capitalism and Indian poverty.[119] He also noted that despite the many sacrifices India had made during the war, post-war rehabilitation concentrated on Europe and marginalised the plight of Indians. In this part of the analysis, the Indian radical found unexpected common ground with leading figures of the American missionary movement who currently flagged the needs of post-war India. Still, the purpose of Roy's writing could not be more different. "It becomes self-evident," Roy wrote in 1919 "that the liberation of India is more than a mere act of abstract justice, it signified a long step towards the redemption of the world from the jaws of the capitalist system." Against the backdrop of the Bolshevist Revolution, the Roy strove to make India an arena of revolutionary politics, that would put an end to colonialism and with it, famine.

Against heightened anticolonial resistance across the British Empire, the colonial administration became increasingly wary of the way Indian nationalists in North America capitalised on famine to draw attention to the exploitative nature of colonial governance.[120] When in 1922 reports about an Indian famine-relief fundraising event in North America reached colonial officials in Whitehall, the latter responded nervously. In early September 1922, the *Boston American* informed its readers of a three-hour entertainment programme at the Old South Building that collected funds for famine-stricken India. Whitehall identified the Friends of Freedom for India (FFI) as organisers of the event and asked the US government to intervene.[121] FFI was founded by Taraknath Das, Agnes Smedley, and Sailendranath Ghose in 1918. It assembled former Ghadarites and American sympathisers, among them the birth-control activist Margaret Sanger and the Chicago University professor Robert Morss Lovett. The size and membership of FFI pointed to a broadening alliance of American anti-imperialists and Indian anticolonialists that would considerably grow in the succeeding decades.[122] Instructed by Whitehall, the British consul, who acted in cooperation with US officials, tried to stop the aid drive for India. FFI, however, was unwavering. In a statement published in the American press, it responded furiously to the attempt to interfere in its activities and promised to send the collected funds directly to the INC.[123]

Gandhian Solutions to Famine

The outright refusal of missionary involvement and the emphasis of famine as a sign of imperial exploitation of India (although building on earlier Indian writings) arose within the context of the particular brand of Indian anticolonial nationalism which flourished outside the subcontinent in the World War I era. Gandhi's experiments in famine relief that commenced in the 1920s are addressed in the following to complement the review of Indian nationalist famine relief and to return the gaze to India.

After Gandhi's return to India in 1915, he ascended to become the leader of the Indian nationalist movement in the interwar period. Gandhi pushed for the reorganisation of the INC and expanded its support among the masses. His rise to prominence was helped by the erosion of the stalemate between the extremist and moderate faction in the aftermath of the death of two of its prominent representatives, Tilak in 1920 and Gokhale in 1915. In August 1920, Gandhi convinced the Congress to give up its conciliatory politics and to launch a national swadeshi programme that, in addition to the boycott of British goods, aimed at a complete renunciation of the British system of rule. In 1920, Gandhi confidently announced that swaraj could be achieved within a year.[124]

The swadeshi movement of 1920–22 overlapped with the mobilisation of famine relief for the people of the district of Puri in Orissa. In May 1920, Gandhi deputed Amritlal Thakkar of the SIS to Orissa to report about famine conditions and to oversee relief efforts. In July, he pressured him to remain in Orissa as long as famine prevailed.[125] Gandhi reprinted Thakkar's reports in *Navajivan* and *Young India* together with his commentary.[126] The relief work of the SIS was also covered by the India Home Rule League in New York, where Thakkar's reports were cited to accuse colonial officials of obscuring famine deaths.[127]

Lauding the donations from different strata of Indian society in 1920 as "a happy sign of increasing national consciousness and charity of heart," Gandhi strove to use the famine to stir nationalist feelings.[128] Gandhi's writings of famine were nested in swadeshi with an emphasis on self-restraint. He demanded Indians to forsake lavish meals until Orissa was fully relieved. "We would not have a feast in a family if a member was about to die of starvation. If India is one family, we should have the same feeling as we would have in a private family."[129] Gandhi's plan to arm India's peasants against famine became part of the promotion of *khadi* (handspun and hand-woven cloth) that aimed to reduce India's economic dependence on British goods and was a pillar of the attainment of full economic independence.[130] In 1920–22, Gandhi advertised the use of the *charka* (spinning wheel) to relieve and prevent famine. His followers implemented the first experiments with the introduction of spinning in villages affected by famine. Gandhi evaluated the experiment as "abundant proof of the spinning wheel being the greatest insurance against famine, and being also the best measure of relief."[131] Arguing that an increase of the agriculturalist's purchasing power was necessary to increase access to food, Gandhi recommended the introduction of *khadi* industries in villages affected by food scarcities. The idea that spinning represented a way to reduce famine stemmed from Gandhi's efforts of reviving village industries. It was also an expression of his renouncement of modern technology and emerged in opposition to the otherwise strong emphasis on the improvement of irrigation and transportation networks to prevent famines. Gandhi disagreed with contemporaries in- and outside the colonial administration who argued that the expansion of railway networks in India assisted in balancing out local scarcities. He outlined how the railway had led to the integration of remote Indian villages into the imperial market. Spinning and weaving, on the contrary, intended to free peasants from the enforced production of cash crops, and allowed them to exert full control over their labour. Gandhi was critical of employing the famine-affected in infrastructural work, believing that such forms of labour had no lasting impact on the peasant's resilience to famine.[132] The production of *khadi*, to the contrary, was a process that could address the problem of famine lastingly through multiple angles: the various stages of production, from the preparation of the raw cotton, and the spinning of the yarn, to the weaving of cloth, could create an income for a larger group of people. While *khadi* production was meant to achieve independence from foreign cotton trade, it also was intended to have a positive effect on the peasants' morale and restore their dignity.

Gandhi alternately used the term "famine relief" or "famine insurance" when he spoke about the use of the *charka*, despite the fact that the introduction of spinning provided no immediate help.[133] Through the creation of an alternative source of income, the introduction of spinning and weaving rather constituted a means to rehabilitate famine-affected localities and to strengthen the resilience of agricultural labourers against the rise of food prices. Gandhi noted in 1921, "I know, this is but the commencent [sic] of the transformation. But when it is complete, not a man nor woman having sound hands need beg or starve."[134]

In October 1921, after Gandhi's followers attacked a police station in Chauri Chaura, killing all of its occupants and police men, Gandhi called off the movement in February 1922 and in March was sentenced to imprisonment of six years.[135] The All-India Spinners' Association (AISA), tasked with the nation-wide promotion of spinning, continued earlier experiments in applying the charka in the aftermath of famines as well as other disasters in the 1920s. In 1925, AISA toured India with a lantern slide show that depicted the use of the charka in the flood-affected areas of North Bengal, where spinning was introduced to compensate for the loss of crops.[136] Apart from AISA, however, Indian social service organisations only sporadically took to spinning in their famine relief work. The SIS, although it did not use it in the aftermath of famine, nevertheless agreed that the promotion of spinning and weaving was a meaningful contribution of rural reconstruction—a field of engagement that would decisively grow in the 1920s and 1930s.[137]

5.5 Conclusion

In the 1901 session of the Indian National Congress Jagadindra Nath Roy (1868–1925) pointedly observed, "plague, we have been told, is a problem of Bacteriology, and famine of Meteorology, (Laughter), and as both these sciences happen unfortunately to be in their infancy, we shall have to wait long before any solution can be found."[138] Dissatisfied with the colonial administrators' explanations for the frequency of famines in India and the proposed remedies, Indian politicians and social reformers intensified their efforts to expose the poverty of India's rural population and called for the effective prevention of famines. Therefore, if the Bengal famine of 1942–44 "put the final nail in the coffin of colonial rule" as James Vernon has argued, Dadabhai Naoroji, Romesh Chandra Dutt and Mahadev Govind Ranade provided the casket and the death of millions of British subjects during famines in India in the late nineteenth century sunk the first nails into the wood.[139] Indian economic nationalism of the nineteenth century formed the bedrock of nationalist critique of the colonial handling of famine. Indian economists and politicians were among the first to demand that the colonial state addressed the root causes of famine, with poverty at its centre, rather than mitigating famine when it struck. While Indian assessments of the causes of famine in the nineteenth century have been well explored by earlier historians, this chapter offers a new perspective by demonstrating the importance of famine for Indian nationalist mobilisation in the first two decades of the twentieth century. It does so by examining publications of Indians in the United States and Canada and the mobilisation and provision of famine relief by Indian organisations. The historical context of such activities was the increasing transnational dynamic of Indian anticolonialism and the growth of the social service movement—both contributing to the diversity of Indian political activity in the early twentieth century.

The activities of the Indian Famine Union in London from 1901 to 1905 illustrates the widening of the support of Indian demands of poverty reduction and famine prevention in the colonial metropolis at the turn of the twentieth century. The partition of Bengal that set off the first mass boycott of British goods in Bengal and the boost of (South)Asian confidence after Japanese victory over Russia, fanned the outpouring of Indian radicals after 1905.[140] It was in the decade before World War I when Indian activists, many of whom were displaced in the wake of swadeshi, created organisations in North America (and elsewhere), to work for the rights of South Asian migrants and to agitate against colonial rule. Against the background of South Asian migration to the United States and Canada and the formation of Indian organisations in the early twentieth century, the demand to eradicate famine (and colonialism) in India was carried beyond the British Empire. The writings on famine of nationalists overseas tended to be more radical in their political analysis, indicating the widening of Indian political demands in the early twentieth century. Whereas early moderate demands aimed to alter the colonial governance of famine, after 1905, a growing group of Indian anticolonial activists made clear that colonialism had to end in order to permanently eradicate starvation on the subcontinent. Besides claiming that Indians collectively suffered from colonialism, nationalist estimations on famine tended to side-line issues of class, caste and gender which were (are) important determinants in deciding who was affected by loss, displacement and death during famines. Far from being the leveller that nationalists occasionally made them out to be, famines affected sections of Indian society with varying degrees of intensity.[141] With the diversification of Indian social and political movements in the interwar period, the political analysis of famine became more nuanced. This tentatively showed in the writings on famine published by those branded radicals and revolutionaries in the interwar period, which were infused with socialist criticism of global labour exploitation. It also showed in relief practice that from the interwar period onwards sought to respond more adequately to the needs of women through the inclusion of female aid workers.

The material aid provided by Indian relief organisations during famines and food scarcities, as a second form of nationalist mobilisation examined in the chapter, served to demonstrate political authority, claim legitimacy and craft citizens. It gained pace in the aftermath of the colonial efforts to supress swadeshi activism, when Indian organisations rechannelled political activism into social service. Though marginal compared to the resources that underpinned colonial famine relief, the appeals for assistance and aid across provincial, regional and "national" borders were intended to fuel the identification of Indians as co-nationals. Against the backdrop of the formation of the Indian diaspora and the global travel of Indian activists, Indians in the United States and Canada (among other places) began to participate in fundraising activities. The colonial state watched the growing entanglement of humanitarian and political agendas with nervousness. The fundraising efforts of organisations in North America that opposed British colonialism frequently provoked attempts of supporters of British colonialism to discredit such campaigns and refute the claims made. However, their ability to stop such initiatives outside of India remained limited.

CHAPTER 6

Famine Relief and Nationalist Politics on the Eve of Independence: The Bengal Famine of 1942–44

Abstract

The chapter reviews the history of the famine in Bengal and shows how political movements and organisations inside and outside Bengal worked together to provide aid. Their aid was underpinned by visions of India's post-independence political future, which were reflected in the mobilisation of the relief. The chapter examines the involvement of Hindu nationalists, Indian communists and Indian women's organisations, all of which drew on transregional and transnational networks to support their relief efforts.

Keywords: Bengal famine, Communism, Feminism, Hindu Mahasabha, AIWC

6.1 World War II and the Prevalence of Hunger in India

Although the Bengal Famine of 1942–44 remained unmatched in its scale and tragedy, it was part of a broader spectrum of hunger and poverty affecting India in the 1940s.[1] At the outbreak of World War II, parts of Punjab were already in the midst of a food crisis. Even though official relief was mobilised in December 1938, famine conditions were visible well into the year 1941 and produced alarming accounts of scurvy and rickets.[2] The people in Travancore and Cochin in southwest India were plagued by famine from 1941 to mid-1944. Food shortages left the poorer classes starving and malnourished.[3] The hunger drove about 15,000 Travancoreans up north to British Malabar, where they squatted in camps that remained largely unaided.[4] In the meantime, in Bijapur, a district situated in the interior of the Bombay province, the failure of seasonal rains in September and October 1942 had precipitated a food crisis.[5] Famine was officially declared in December of the same year, but the provincial government was consumed by the task of food rationing in the city of Bombay. In a country where the economy was geared to meet the needs of a nation at war, the provision of relief for rural populations, who had no direct relevance for the smooth functioning of the war machinery, turned into a secondary concern at best.[6] Famines continued to represent only the tip of the iceberg. In early October 1942, Sonia Tomara, who had come to the South-East Asian war theatre as the first female journalist to cover World

War II, offered the readers of the *New York Herald Tribune* a rare glimpse at the state of rural India. In her account of an unnamed village ten miles away from Delhi she wrote "I had indeed seen extremely poor villages in Russia, Poland, the Balkans, Italy and Spain, but hardly anything that could touch the distress here."[7] When British administrators in Whitehall read Tomara's account they were alarmed. In a continuation of well-rehearsed colonial responses to the American news coverage of hunger in India, officials undertook efforts to repudiate her report.[8] The poignancy of the article lay in its timing. When the colonial government declared war on the Axis Powers on behalf of India without previously consulting with Indian political leaders, the All India Congress Committee protested. Congress politicians withdrew from provincial governments and made the achievement of substantial political concessions a condition of its cooperation. The negotiations between the Congress and the British colonial state reached an impasse in 1942. In March, Sir Stafford Cripps arrived in India to renew efforts for a settlement, but failed to reach an agreement. In August, the Congress demanded the British to quit India and appealed for nationwide non-cooperation to enforce its claim. The colonial state responded promptly. In the following days and weeks, it banned the Congress and imprisoned over 60,000 Congress representatives. The suppression of the Quit India movement prompted widespread mass protests in return, which in light of a lack of leadership that could have moderated the masses turned increasingly violent.[9] Weakened by the war, the British colonial state, which was under attack from the world's liberals for its harsh reprisal of the Indian nationalist movement, wanted to prevent Tomara from inciting further agitation. Two months after her report from India, Tomara was writing again and once more, criticised the government for starving its colonial subjects.[10] She was now in the east of India, where she witnessed some of the events that marked the beginning of the Bengal Famine.

The Making of the Bengal Famine

Amidst the series of famines that affected Indian society during the war, the hunger in Bengal was by far the severest. The famine and the diseases following in its wake would take the lives of an estimated three to five million people. Historians have pinpointed the government's scorched earth campaign as one of the causes of famine conditions in Bengal. The fall of Burma in the spring of 1942 had turned Bengal's eastern districts and its coastal line into frontlines of World War II and had caused fear of a Japanese invasion.[11] In mid-1942, colonial authorities ordered the confiscation of boats and rice surpluses, which not only deterred Japanese invaders, but also removed the sources of income of Bengal's agriculturalists and fishermen. The first outcomes of this policy became visible in June 1942, when Indian Army recruiters noticed a sudden increase in the number of young men from the east of Bengal and commented on their miserable physical condition. Officers in charge had to create rest camps where the mal- and undernourished recruits were restored to health prior to their admission.[12] Meanwhile, with the cessation of rice imports from Burma in the spring of 1942, and later, the disruption of trade and communications networks, the balancing of local scarcities had become much more difficult.[13] The influx of large numbers of British and American soldiers into Bengal, who were to fight in the South East Asian war theatre, exerted further pressure on Bengal's food economy. To accommodate troops, agricultural land was cleared and trains that had previously transported rice across district and provincial borders now moved soldiers instead.[14] Adding a further layer of complexity to the unfolding of the famine, a cyclone ravaged the coastal districts of Bengal and Orissa in mid-October 1942. In Midnapore and 24-Parganas, the storm affected a population that was already strained by the government's scorched earth campaign. The relief response was severely impeded by the colonial efforts to quell political opposition.[15] Clashes between the provincial government and the Congress movement had become more frequent and increasingly violent in Midnapore since September. The District Magistrate, N.M. Khan, and the police officer in charge for breaking the Congress movement, aimed to exploit the crisis to gain back control over the district. They withheld information about the cyclone until early November 1942, and pursued and imprisoned volunteers of political parties, who had set out to attend to the calamity.[16] In December, Nripen Sen, an Indian relief worker who had been delegated by the Communist Party of India to visit the district, reported about how he was arrested upon his arrival and found himself in the company of other relief workers, whose aid efforts had been halted through the colonial intervention.[17] Measures taken to stop political parties from exploiting the crisis also impeded the work of volunteers of non-political relief providers. The Ramakrishna Mission reported how the ban of news prohibited the Mission from appealing for funds. Moreover, the implemented rationing schemes that had rendered petrol scarce and that had posed restrictions on the use of the railway for trade and travel obstructed the transportation of relief supplies into Midnapore. Volunteers of the Ramakrishna Mission had to rely on bullock carts to move food and medicine into the province.[18] In addition, the Sadharan Brahmo Samaj, who had set out from Calcutta, had been denied access to Midnapore. After lengthy negotiations, the Sadharan was eventually allowed to open its temporary headquarters on Ghoramara Island, situated in the subdivision in the Bay of Bengal in the 24-Parganas and not in Midnapore.[19]

Although in early November 1942, the ban of information eventually was removed and the official disaster response got on its way, the assistance remained inadequate.[20] In February 1943, a relief worker of the Sadharan Brahmo Samaj reported from Diamond Harbour about the still grim condition of the people affected by the cyclone. The deterioration of conditions in Midnapore did not escape Tomara, who criticised the government for responding insufficiently.[21] In response to mounting public criticism, the Secretary of State for India in Britain, L.S. Amery began to inquire with the Viceroy about the existence of a fundraising appeal of the Government of India in Delhi. When he was informed that no such appeal was made, he decided the time had come to demonstrate colonial benevolence. The result showed in January 1943 when Amery, in cooperation with the High Commissioner of India in Britain and the Mayor of London, began to raise funds in the metropole for the victims of the cyclone.[22] The appeal was published in the *Times* with letters of support of London's business class and was followed by a five-minute broadcast through the BBC. Mansion House also suggested using a film, but was unable to receive recordings from India. Given the lack of visual material, the organisers of the fund decided to stick to "colourful details" to solicit donations.[23]

The famine unfolded in Bengal in the following months against the backdrop of skyrocketing prices for basic commodities and hesitant and misguided market regulation that encouraged hoarding and led to a flourishing black market.[24] At the same time, the ability of journalists and private institutions to pressure the government into action became more difficult. Partly as a result, it would take until the end of the year to again shift the response of the colonial government from denial to benevolence. As conditions worsened in Bengal's rural areas in the first half of 1943, the presence of Subhas Chandra Bose in Japan, and later in Burma, made fears of a Japanese invasion more palpable. Bose had been at the forefront of a group of Indian nationalists, who had entered into an alliance with Germany and Japan to end British colonial rule in India. Before he had arrived in Japan, Bose had spent time in Germany, where in 1941 the *Sonderreferat Indien* had harnessed his presence to step up anti-British propaganda. Bose's appeals to Bengal's inhabitants to side with Germany and Japan were broadcasted into Bengal and pro-axis propaganda distributed through Bose's political party, the Forward Bloc. After his arrival in Japan in February 1943, Bose mobilised captured Indian soldiers into the Indian National Army. Convinced that Japan would use the evolving famine to launch its strike, the colonial government drew from the Defence of India Act to obscure the true extent of the famine in Bengal. Thus, in contrast to Bombay and other parts of India, famine was not officially declared in the east of India. Censorship, implemented by the Chief Press Adviser to the Government of India in collaboration with the Provincial and District Press Advisers, suppressed publications that might be of use to the enemy, thereby also significantly delaying the international media coverage of the famine in Bengal.[25] His task was facilitated by the war, which diminished the interest of the international press for events in India. In July 1943, the news blackout stopped to fulfil its main purpose. At this time, the *Sonderreferat Indien* in Germany, however, had already collected enough evidence to conclude that Bengal suffered from a famine. A month later, it recommended the intensification of propaganda linking the famine in Bengal to British imperialism.[26] In August, Bose broadcasted his offer to release Burmese rice for Bengal to ease the food situation in the province—an offer which the government found unacceptable and, hence, refused.[27] The colonial government instead continued their course of denial. It came under increased pressure after the English-language newspaper the *Statesman*, previously known to be loyal to the government, managed to run drastic images of the famine past the Press Adviser in late August. The rules applied to censor the press had been silent about images—a gap which the *Statesman* exploited.[28] While historians have concentrated on the role of the *Statesman* and its editor, Ian Stephens, in bringing the famine to international attention, they have overlooked how the publication unfolded a dynamic in Bengal itself.[29] Although the political landscape of Bengal was conflict-ridden, politicians united and used the publication of images of famine in the *Statesman* as ammunition against the press censor. In late August 1943, days after the images had appeared in the *Statesman*, the Press Adviser faced unprecedented opposition. He had just given the order to ban parts of a statement about the famine made by the Hindu Mahasabha leader Syamaprasad Mukherjee. H.N. Kunzru, now the president of SIS, immediately brought a motion before the Bengal Legislative Assembly that lamented the "lack of uniformity of censorship" and declared the censoring of Mukherjee's statement an injustice. Mukherjee, the current president of the All India Hindu Mahasabha was involved in the mobilisation of famine relief on multiple fronts. He was vociferous in his critique of the Bengal government, and as such, of the Muslim League ministry. Although they had been the main target of Mukherjee's criticism, not even representatives of the Muslim League welcomed the ban on Mukherjee's statement.[30] The parties agreed that the press censor was unjust. Only a few days after the famous publication of the *Statesman*, the *Amrita Bazar Patrika* ran a first series of images of the famine and contributed to rendering the crisis visible to a wider Indian audience.[31]

The drastic accounts of the famine provided in the national and international press in the last months of 1943 allowed humanitarians worldwide to put pressure on the colonial government. In October, the newly appointed Viceroy Archibald Wavell eventually called for the army to drive food into Calcutta and rural Bengal. Army convoys arrived in November and provided much needed food and medical relief. The aid drive however ended prematurely. At the end of 1943, the Viceroy and the Bengal government were too quick to announce the end of famine. The media response that unfolded in the last months of 1943 had facilitated the work of non-state relief providers in Bengal who now found greater receptiveness of their appeals for assistance and were receiving help from outside the province. The chronology of events seemingly provides historical evidence for the role of a free press in pressuring political elites to assist the starving.[32] However, at the time the news broke, efforts to mobilise relief from outside Bengal and outside India already had been permeating the colonial curtain of denial and censorship.

6.2 For Akhand Hindustan and Greater India: Famine Relief and the Political Aspirations of the Hindu Mahasabha

The Hindu right-wing party, the Hindu Mahasabha, was one of the largest non-state contributors of famine relief in Bengal in 1943.[33] Throughout the famine, Syamaprasad Mukherjee (1901–1953), the Mahasabha politician and later founder of the Bharatiya Jana Sangh (the predecessor of the BJP), was active in four famine relief committees. Mukherjee's prominent involvement in the relief activities reflected the political ambitions of the Hindu Mahasabha in general and Mukherjee in particular. Mukherjee and the Hindu Mahasabha played a key role in inciting communal conflict in Bengal and wider India, which would heighten in 1946, when in the aftermath of the failure of the British Cabinet Mission, Hindus and Muslims would kill each other in their thousands in Calcutta.[34] In 1943 famine relief became a field of public engagement that allowed the Hindu Mahasabha to advance Hindu communalism; in that regard, the Bengal famine of 1943 has been identified as an important prelude to the history of communal violence that followed.[35] At the same time, the politics that undergirded Mukherjee's relief work during the famine were more complex. Mukherjee employed different, and at times, countervailing secular and communal registers to also carve out a space for the Hindu Mahasabha on a national and international level. As the head of the Bengal Provincial Hindu Mahasabha Relief Committee, Mukherjee advanced the vision of a united Hindu India. Through his parallel leadership in the Relief Coordination Committee, he worked to broaden alliances with the Congress and non-communal institutions, with the aim of promoting the Hindu Mahasabha on an all-Indian political landscape; as the president of the Maha Bodhi Relief Committee he likewise demonstrated his efforts of cultivating pan-Asian solidarities between India, Ceylon and China.

The Political Context

The Hindu Mahasabha was formed in 1915 as the first of a series of Hindu nationalist organisations devoted to "defending" India against non-Hindu and particularly Muslim "intruders." Vinayak Damodar Savarkar (1883–1966) created a powerful ideological basis for the Hindu right-wing movement with the publication of his infamous *Hindutva: Who is a Hindu?* (1923).[36] In the 1930s, the Hindu Mahasabha accelerated its efforts to build a united India (*Akhand Hindustan*, later *Akhand*

Bharat) by inducing aboriginals and untouchables in Bengal to register as members of Hindu castes and by offering non-Hindus the chance to "convert" through ritualistic purification (known as *shuddhi*).[37] While the Hindu Mahasabha clung to restrictive caste hierarchies on the one hand, it allowed for a certain social mobility within the caste system to ensure that low-castes remained Hindu. The proselytising activities of Muslims and Christians sparked fears of a numeric decline of Hindus; such fears were often further stoked by British census activities. In preparation for the upcoming census of 1941, Mukherjee, the leader of the Bengal Provincial Hindu Mahasabha at the time, took active part in the caste consolidation and mobilisation practices.[38] In his role as party-leader, Mukherjee also strove to improve the relationship of the Mahasabha with the Congress movement in Bengal, which suffered from differences between the All India Hindu Mahasabha and the INC on a national level. Since the Government of India Act of 1935 enabled the election of Indian provincial governments, national and provincial politics frequently diverged and regional loyalties complicated national party politics.[39] This dynamic intensified during World War II when the Congress protested against the British declaration of war on behalf of India and the Hindu Mahasabha sided with the British. Savarkar, who was the Mahasabha's president at the time, saw India's participation in the war as a chance to oppose the Congress as well as an opportunity to militarise Hindus.[40] He ordered the formation of Hindu military associations and encouraged Hindus to join the British armed forces.[41] Over the last few months of 1942, however, the anti-Congress stance of the All India Hindu Mahasabha threatened to diminish support for its provincial counterpart in Bengal. In light of the colonial violence against the Quit India movement in Midnapore and the widespread sympathy it generated, the Mahasabha found it increasingly difficult to oppose the Congress in Bengal. Mukherjee, who was now part of a Progressive Coalition Ministry under the lead of the Bengali Muslim Fazlul Huq, chose to establish closer relations with the Congress-led movement.[42] In November 1942, after the government's violent reprisal against the revolt in Midnapore, Mukherjee resigned from his post as Finance Minister in an apparent act of solidarity with the Congress.[43] This, however, did not hinder either Mukherjee's or the party's continuing efforts to carve out a separate space for the Mahasabha as the protector of Bengali Hindu middle-class interests. Proximity to the Congress and toning down of the party's communal rhetoric were parts of the Mahasabha's claim to a political place at the national level.[44]

The Huq-Mukherjee ministry ruled Bengal from 1940 until first Mukherjee resigned in November 1942 and then Huq was forced to do the same in April 1943. The Huq-Mukherjee ministry had been unique in India for uniting the political representatives of communal parties and advocating inter-communal understanding without making a claim of secular politics. Thereby, it had been able to accommodate communitarian demands without exacerbating them and, for a brief moment in time, showed a way out of the spiral of communal violence.[45] This changed abruptly when the Governor of Bengal, Jack Herbert, used the Defence of India Act to install a new ministry under the lead of the Muslim League in Bengal. The Mahasabha now benefitted from the unpopularity of the new ministry, and Mukherjee was able to rally greater support for the party.[46] The Mahasabha framed its disapproval of the new ministry as nationalist and anticolonial (rather than simply as anti-Muslim).[47] The relief efforts coordinated under the auspices of Mukherjee and the Mahasabha in 1943 are evidence of this complex political manoeuvring of the party between promoting the Hindu communalism, appeasing the Congress movement, and criticising the Muslim-led government of Bengal.

Relief and Communalism

In July 1943, Mukherjee and the Marwari merchant and chairman of the Imperial Bank of India, Badridas Goenka (1883–1973), formed the Bengal Relief Committee (BRC). According to the BRC, its mission was to bring relief to Muslims, Christians and Hindus alike.[48] But despite its efforts to show that it cared about the needs of all Bengal, suspicion about its intentions soon cast doubt on its willingness to support non-Hindus. The Hindu Mahasabha and the Marwari community had been in frequent conflict with Bengal's Muslim population and their unity during the famine now inspired no confidence among Muslims. The Marwari community in Bengal originally came from Rajasthan and had migrated to Bengal in the late nineteenth century, attracted by the opportunities of imperial trade. Marwaris soon dominated the province's grain trade and made fortunes in the jute and textile industries. During World War II Marwari merchants profited immensely from the heightened demand for cloth and other war materials.[49] While many Marwari merchants grew wealthier during the war, their hoarding of and speculation in rice led to soaring food prices that contributed to famine. Many peasants who had chosen to cultivate a single cash crop (jute) rather than rice were dependent on the market to buy food and thus severely hit by the rising food prices. The great majority of them were Muslim peasants in the eastern areas of Bengal.[50] Allegations that Marwaris intentionally starved Muslims were made again when a group of Marwari industrialists in Calcutta supplied regular Hindu dockworkers with food rations while excluding Muslim contract labourers from the rationing schemes.[51] However, while some Marwaris aggravated the hardship of Bengal's population, and Muslims in particular, the Marwari community was also amongst the largest providers of famine relief. The Marwari Relief Society, created in 1916 in Calcutta to institutionalise Marwari flood and famine relief work, contributed generously to the BRC.[52]

Given the history of tensions and open conflict between Bengal's Muslims on the one side and Marwaris and the Hindu Mahasabha on the other side, it is not surprising that the Muslim League ministry soon accused the volunteers of the BRC of giving preferential treatment to Hindus. The Hindu Mahasabha in turn claimed that the Muslim League ministry deliberately discriminated against Hindus in official relief centres by providing cooked food that neglected the religious and caste prescriptions of Hindu relief seekers. Indian leaders outside Bengal fuelled the mutual suspicion. Savarkar spread rumours of enforced Islamic conversion of Hindus,and the Muslim League politician Mohammed Ali Jinnah sparked controversy by making a personal donation that was earmarked for the relief of Muslims to the Muslim Chamber of Commerce.[53] The All India Muslim League had refrained from opening its own relief centres to avoid communal tensions in Bengal, instead forwarding money to the Muslim Chamber of Commerce. The latter distributed its funds to Muslim organisations, which were instructed to give assistance without discrimination.[54] After Jinnah's donation was made public, the Muslim Chamber of Commerce argued that such demarcations did not matter because the number of Muslim relief seekers far outweighed the donations earmarked for their relief. By this time, the scandal had spread beyond Bengal and in response, Savarkar called upon Hindus to rescue their co-religionists.[55]

Although Mukherjee continued to emphasise the ability of the BRC to go beyond communal distinctions, he gave in to the extremist fraction of the Mahasabha and set up another committee that allowed Hindu donors to rescue their co-religionists exclusively.[56] The list of beneficiaries of the Bengal Provincial Hindu Mahasabha Relief Committee that largely included middle-class and high-caste Hindus, was in keeping with the ideological premises of the party. The Sabha offered relief to members of high-caste who felt inhibited by their caste and class sensibility to queue for food rations; teachers (*pandits*) at Sanskrit schools (*tols*) received particular mention.[57] Along with the Arya Pradeshik Pratinidhi Sabha, which united Arya Samajis in Punjab, Mukherjee appealed to those Bengali middle-class families unable to provide for their offspring to turn them over to the Samaj, thereby ensuring their upbringing and education in Punjab.[58] Their efforts led to the formation of the Arya Samaj Relief Society in early September, and Mukherjee oversaw the departure of a first batch of children soon after. On the morning of their journey, the Samajis selected 70 "peasant boys" aged six to twelve years. They were medically examined, provided with new clothes, "sumptuously fed", and thereafter entrusted to a Bengali *sannyasi* (Hindu religious ascetic) who was to oversee not only their journey but also their rearing in Punjab.[59] The rescue of children served to promote Hindutva and the transfer of them to Punjab showed the attempts of the Hindu Mahasabha to deepen interprovincial networks during the famine. Since the famine codes were not applied, organisations in Bengal appear to have used the lack of restrictions to remove children from the province, a course of action not limited to Hindu communal organisations. Even moderate institutions like the

AIWC appealed to families across India to adopt children from Bengal.[60] Not only children but others too were taken out of the province. The Hindu Mahasabha began to transfer widows to Punjab, and another organisation announced the large-scale rescue and transfer of Bengalis: the Khaksar Relief Organisation.

The Muslim *Khaksar* ("dust-like") movement, was founded by Inayatullah Khan (alias Allama Mashriqi) in Punjab in 1931.[61] It was one of several Hindu and Muslim paramilitary groups formed in the interwar period, and promoted discipline and fitness among members along with, by implication, the ability of Hindus and Muslims to defend themselves against the perceived "other." Although these groups identified themselves as social service organisations and occasionally assisted people affected by natural and man-made disasters, they spent the bulk of their time on training their members. Founding the Khaksar movement, Mashriqi demanded that Muslims militarise under his authoritarian leadership. Khaksars soon attracted attention as its members marched in khaki shorts through North India to display their strength in "mock wars." However, the Khaksars remained politically isolated. Not only did Hindu-dominated parties like the Mahasabha or the Congress oppose the organisation, even the Muslim League had no sympathies with Mashriqi. Although the Muslim League initially sought to collaborate with the Khaksars, who were occasionally tasked to protect party assemblies, the relationship soon deteriorated. Jinnah demanded that Khaksars renounced their affiliation with the movement prior to joining the Muslim League, which Mashriqi strongly resented. In addition, the latter was opposed to the Two-Nation theory, which asserted that Hindus and Muslims were destined to form two separate political entities. The Khaksars instead aimed to strengthen Muslims within a united India. In 1940, the colonial administration declared the Khaksar movement illegal and arrested Mashriqi.[62] However, Mashriqi and the Khaksars frequently dodged government restrictions. Tensions between the Muslim League and Khaksars escalated in 1943, when a Khaksar tried to assassinate Jinnah.[63]

In 1943, the *Amrita Bazar Patrika* noted that a Khaksar "information office" had opened in Calcutta, coordinating the influx of volunteers from other provinces.[64] The volunteers took up work under the Khaksar Relief Organisation, which opened 20 relief camps across Bengal to shelter survivors of the famine and nurture them until they were fit enough to be taken out of Bengal. In September 1943, Mashriqi gave orders about how Khaksars were to respond to the famine in Bengal that were both militarily precise and bizarre. He asked 600,000 followers in India to come forward to form groups of four.[65] These "Khaksar families" were told to contact Muslim representatives in Bengal, who in turn were to send one famine survivor to each of the volunteer groups.[66] The success was limited. By December 1943, Khaksars sheltered only 900 of the intended six *lakh* (600,000) Bengalis in Punjab, Sind and the United Provinces.[67] Despite the lack of success, the Hindu Mahasabha was outraged. It suspected Khaksars of using the famine as an excuse to round up and confine Hindus. While Khaksar representatives denied targeting Hindus deliberately, it did not repudiate the accusation of sending Hindus out of the province.[68]

In this controversy, Mukherjee and Mashriqi tried to appease colonial authorities by emphasising the non-discriminatory nature of their relief work. Even though the Famine Code was not implemented, charitable institutions still needed governmental approval. Mashriqi therefore clarified that "no refugee should be molested, or any Khaksar should interfere with his religion" and demanded that "Hindu and Muslim Khaksars should take part in the movement irrespective of caste or creed."[69] This emphasis on the non-communal character of the Khaksars' famine relief work was also connected to the wider ideological basis of the movement. Despite Mashriqi's efforts to militarise Muslims, the Khaksar movement remained open to non-Muslims in theory. Although this had not translated into an actual membership of Hindus in the movement, the Khaksars secured the support of a Hindu named Amar Nath Joshi in 1943.[70] Joshi's participation in the relief work of Khaksars allowed them to substantiate the claim to being a movement of much wider appeal. Joshi was tasked with overseeing the care of Hindus in their shelters.[71] Likewise adopting a conciliatory tone, Mukherjee declared that no Muslims were amongst the children sent to Punjab. He emphasised that the Hindu Mahasabha was "acting solely on humanitarian considerations without any distinction of caste, creed or colour" and offered to hand over ten Muslim children taken in by the Hindu Mahasabha out of necessity to Muslim organisations.[72] Mukherjee also repeatedly called on the Bengal government to stop Khaksars from moving Bengalis out of the province. He was not alone in this demand. Jinnah eventually joined in, and at the end of December of 1943, the government ordered Khaksars to end their relief activities. Those sheltered by the Khaksar Relief Organisation in Bengal were transferred to governmental poorhouses.[73] An outraged Mashriqi responded by announcing that he would gift Jinnah the corpses of starved Bengalis on his approaching birthday.[74]

Seeking Broader Alliances: The Hindu Mahasabha and the Relief Coordination Committee

Although the Bengal Provincial Hindu Mahasabha Relief Committee gave Mukherjee a platform from which he could please his right wing and Hindu middle-class voters, he also undertook efforts to project himself and the Sabha onto a larger political canvas. Aiming to benefit from the weakness of the Congress who given its persecution was largely absent from the relief efforts, Mukherjee demonstrated his leadership through the formation of an umbrella committee emphasising the Mahasabha's alliances with Congress workers and pro-Congress institutions.[75]

The Relief Coordination Committee led by Badridas Goenka was formed in September 1943. It united a broad range of relief bodies, and had the BRC (now operating on an unrivalled budget of 1.6 million rupees) as its largest contributor.[76] In addition to the natural allies of the Mahasabha, the umbrella committee included the Ramakrishna Mission and the AIWC, which although markedly "Hindu" in membership were renowned for their efforts to organise relief across religious distinctions.[77] This already pointed to Mukherjee's success in building broader political alliances, but his ability to work with the Congress showed further in the appointment of the committee's second vice-president (Mukherjee himself was the first vice-president). The Congress politician Bidhan Chandra Roy, or Dr. Bidhan as he was emphatically called, was a Bengal-born physician and educationist. He was the Vice-Chancellor of Calcutta University and the president of the Medical Council of India (1939–1945).[78] Roy was a close associate of Gandhi and was asked to attend to the Mahatma in February 1943. When Gandhi embarked on a 21-day fast during his imprisonment, Roy was called for medical assistance.[79] Roy was also a veteran Congress relief worker. He had led the INC's relief mission for the victims of the earthquake in Bihar in 1934 and organised the Bengal Civil Protection Medical Committee. The latter set out in May 1942 to provide medical assistance to stranded Indian refugees from Burma in Assam and Manipur, and to Midnapore in October 1942, where it worked side by side with the FAU to provide aid to the cyclone-afflicted province.[80] Showing that collaboration between the Mahasabha and the Congress preceded the famine, Mukherjee had sent members of the Sabha to assist Congress workers in doling out aid in Midnapore in 1942.[81]

At the time that Mukherjee helped to set up the Bengal Relief Coordination Committee, he became involved with the work of yet another famine relief body. He was already the president of the Maha Bodhi Society of India (1942–1953), when in September 1943, Mukherjee assumed the presidency of the newly formed Maha Bodhi Famine Relief Committee. In the remaining months of 1943, the Maha Bodhi Society of India opened orphanages and schools, provided cooked meals at a free kitchen, gave medical aid at a dispensary, and supplied milk to children and invalids. It also attended to the needs of *Bhikkus* (Buddhist mendicants).[82] The Maha Bodhi Society of India belonged to a transnational reformist Buddhist movement that had additional branches in South East Asia.[83] It was co-founded by the theosophist Henry Steel Olcott and the early Sri Lankan anticolonial nationalist Anagarika Dharmapala in 1891. Believing in a glorious Buddhist past in which Asia was united under the Buddha, the Society fostered pan-Asian ties. Philanthropic work supported its members in doing so. During the famine of 1896, the Maha Bodhi Society channelled funds from Japan, China, Siam and Burma to support its famine relief work in Bengal. In 1934 it assisted the victims of an earthquake in Bihar.[84] Its main work, however, was to promote the restoration of the sacred space of *Bodh Gaya* in North India, where the Buddha is believed to have attained enlightenment (the site was eventually restored in 1949).

Mukherjee's support for and work within the Society were connected with the Mahasabha's appropriation of the concept of Greater India that framed India as the nodal point of a united pan-Asian civilisation. While initially an academic and literary notion, the concept was transmuted into politics in the interwar period. In the 1930s, and especially under the lead of Savarkar, who assumed the leadership of the Hindu Mahasabha in 1937, the party appropriated the concept of Greater India, which now was deemed a non-Muslim entity. Arguing that Buddhism was essentially Hinduism, it advocated for the unity of these Asian religions and strengthened ties to Buddhist institutions across Asia.[85]

The Hindu Mahasabha was not alone to invoke Buddhism and its links to Hinduism to strengthen pan-Asianism. In the first years after independence, the Indian government under Jawaharlal Nehru would deploy Buddhism as a foreign policy tool, leading to a series of relic tours across Asia to nurture diplomatic contacts. At the same time, Buddhism played a special role in India's domestic politics after independence. In the mid-1950s, the Indian politician, social reformer and co-author of the Indian constitution Bhimrao Ambedkar (1891–1956) promoted the mass conversion of dalits to Buddhism to protest against caste discrimination.[86] In addition, Nehru frequently used Buddhist religious symbols to advocate for a secular vision of the Indian nation, thereby turning Buddhism into a means of cutting across communal distinctions. Clearly, this was not how the Hindu Mahasabha envisioned a united Buddhist Asia. Yet Mukherjee's role in the Maha Bodhi Society would lead to a strategic alliance between Nehru, Mukherjee, and the Maha Bodhi Society after independence. Although tensions between Mukherjee and Nehru would escalate after the assassination of Gandhi by a Hindu nationalist fanatic in 1948, Mukherjee would become the representative of the Nehruvian government for travel across South East Asia, carrying Buddhist relics to further pan-Asian internationalism.[87]

This role of Mukherjee as the unlikely pan-Asian liaison of the INC could already be glimpsed at in 1943 through the work of the Maha Bodhi Society Relief Committee. The latter mobilised funds through institutions in Ceylon and China. The largest contributor to the Maha Bodhi Relief Committee was the Chinese government that under Chiang Kai Shek forwarded its donations to the Maha Bodhi Society Relief Committee in Bengal through the Sino-Indian Cultural Association in Chungking.[88] The Sino-Indian Cultural Association (also Society) had been established in Nanjing in 1933, followed by an Indian branch in Santiniketan a year later. It was a part of the efforts made by the Guomindang and the INC to bring both political movements closer together.[89] In 1943, the Sino-Indian Cultural Association received money from a fund-raising committee in China, which worked under the auspices of Song Meiling, better known as Madame Chiang. The Committee assembled 40 delegates from the government, industrialists, and relief workers and appealed to the Chinese people to support the famine relief in Bengal as an expression of "growing Sino-Indian friendship."[90] At the inaugural meeting of the committee in China, Meiling announced that the Joint Board of Chinese Government Banks would advance 1.2 million dollars to forward the Chinese donation immediately. The Consul-General for China in India, C.J. Pao, joined the Maha Bodhi Famine Relief Committee in India as its vice-president.[91] That high-ranking Chinese politicians and diplomats contributed to famine relief in Bengal was symptomatic of the growing diplomatic and personal ties between the Chinese government and the Indian nationalist movement during World War II. The Chiang Kai Sheks visited India in March 1942, reciprocating Jawaharlal Nehru's 1939 visit to China.[92] The fact that the Chiang Kai Sheks channelled contributions through the Maha Bodhi Society adds a further layer to the efforts by the Chinese political elite to foster ties to India. Later that year, the Chinese government expanded its partnerships and channelled money to a much wider range of relief bodies in the province.[93]

6.3 Reconciling Nationalism and Communism: The Famine Relief of the Communist Party of India

In the aftermath of Germany's attack on Soviet Russia in 1941, the Communist Party of India (CPI) acceded to the international communist embrace of World War II as "the people's war." Communists worldwide were called upon to postpone the fight against imperialism until after the war. The policy complicated the CPI's political position as Indian communists now had to balance conflicting goals. On the one hand, the CPI wished to identify with a nationalist movement that aimed to shake off colonial rule. On the other hand, the party needed to demonstrate its solidarity with the Soviet Union, which demanded an endorsement of the allied war effort. The CPI was thus forced into the unfavourable position of opposing the Congress-led Quit India movement by having to swing into full war support. To convince Indian communists of the new party line, many of whom had doubts about supporting the British, the CPI stoked fears of an immediate Japanese fascist threat to the Indian nation. The emphasis on Japanese aggression enabled the CPI to argue that the sudden advocacy of war was in India's best interests, as the immediate threat on India's eastern front outweighed the need to confront the colonisers for the time being. The CPI's decision to support the British in the war led to the lifting of the ban on the party and its publications. The removal of the ban gave the CPI considerable liberty to describe the famine in Bengal. The new freedom, however, was limited as colonial authorities remained sceptical of the party and attempted to renew the censorship. The outcome of these multiple pressures showed in the party's rhetoric. The CPI instructed members to refrain from strikes and anticolonial protests and to support the war instead. It also called on peasants to step up India's agricultural production so as to sustain the urban labour force engaged in producing essential war materials. At the same time, the CPI touted its role as the primary protector of peasant interests and framed its social and political demands as evidence of its patriotism and allegiance to the Indian nation. The party's involvement in famine relief enabled communists to regain sympathies that had been lost by its political manoeuvring.[94]

Organising Famine Relief

In September 1943, the CPI and its General Secretary P.C. Joshi (1907–1980) formed the People's Relief Committee (PRC).[95] It was the second largest umbrella committee to coordinate the relief response of non-state providers of assistance in Bengal after the Relief Coordination Committee steered by Goenka, Mukherjee, and Roy. For Indian communists this was the first step towards subsuming all relief providers under a unified body that would provide relief in a truly "non-party, non-factional, non-communal manner."[96] At the outset, the PRC enlisted the support of a diverse set of representatives. It included Syed Nausher Ali (1891–1972), member of the Krishak Praja Party and speaker at the Bengal Legislative Assembly, Chaudhuri Moazzam Hossein (1905–1967), a graduate of Aligarh University who had joined the Muslim League in 1943, and the Congress supporter and anthropologist K.P. Chattopadhyay (1897–1963).[97] The ability to work across political boundaries, however, also had limits: The Hindu Mahasabha was absent from the committee. In its role as the protector of Bengal's capitalist class, the Mahasabha was a natural enemy of the CPI. The committee instead included mainly leftist and communist organisations, such as Trade and Worker's Unions, the Indian People's Theatre Association (IPTA), the Anti-Fascist Writers' and Artists' Association and the *Mahila Atma Raksha Samiti* (commonly translated as Women's Self-Defence League). Members of the CPI had already been working to mitigate the crisis prior to the formation of the Relief Coordination Committee through these independent, but communist-dominated, institutions.[98]

In 1944 epidemics and nutritional diseases followed in the wake of starvation and the priorities of the CPI shifted from the provision of food to medical aid. It was now that communists secured the cooperation of a broader spectrum of Indian political parties by forming the Bengal Medical Coordination Committee (BMCC).[99] The BMCC built on personal and institutional networks that had been established in the context of Indian humanitarian activities during the interwar period.[100] It was initially assembled by communists in February 1944 and included among others the relief committees of both the Mahasabha and the Marwari community, bringing it much closer to the united relief committee the party had originally envisioned.[101] The representation of this broader range of political movements owed much to the initiative and reputation of B.K. Basu, who had just returned to India from China and now supported the CPI's famine relief work in Bengal. Basu had been to China with the Congress Medical Mission.[102]

The INC had sent a Medical Mission to China to demonstrate its solidarity with the Chinese people in the aftermath of the escalation of conflict between the Japanese Empire and China in 1937. The Mission that left India in May 1938 and reached China in early 1939 comprised of an ambulance and five doctors, including Basu. The Chinese political leaders from both the Guomindang and the Chinese Communist Party, who had formed a United Front government in light of the Japanese aggression, welcomed the Congress Medical Mission. For the next year, the Mission worked under the guidance of the Chinese National Red Cross and assisted victims of the war in China.[103] Basu was the last member of the mission to return to India. He arrived in the middle of the Bengal famine, carrying with him letters from Mao Zedong and Zhu De that expressed their solidarity with communists in India.[104] Whereas the Congress, the Muslim League, and the Hindu Mahasabha fostered ties with Chiang Kai Shek, the CPI sided with Mao. Although Basu's party affiliation remains unclear, he seemed to have moved closer to the CPI after his return. He now promoted a second medical mission to China which would also include representatives of the CPI.[105] In support of the mission, P.C. Joshi sent a request for permission to the British government in India, arguing that such a mission would promote cordial relations between the Indian and Chinese peoples: without success.[106] The second Indian medical mission to China did not materialise. Basu instead stepped up the CPI's medical relief efforts in Bengal. He took the lead of the Medical Board of the PRC, which organised mobile units to bring relief into rural Bengal. Shortly after, he approached B.C. Roy, who was the vice-president of the Relief Coordination Committee and a leading member of the Bengal Civil Protection Committee, and had already been able to bring differing relief bodies closer together. Both men led the formation of the Bengal Medical Coordination Committee.

The Art of Communist Campaigning

> The agonised cry of Hungry Bengal, heard all over the country, in many different tongues, was at last rousing the dormant conscience of the whole nation. Food, money, medicines, and above all, brotherly sympathies were beginning to flow in a stream to Bengal, bringing new hopes, new life to forsaken millions. Yet it was too little. And, for tens of thousands, who had already laid themselves on the city pavements and died a homeless death, it was too late.[107]
>
> —From the Script of *Dharti ke lal* (Children of the Earth).

From 1943 to 1945, communist efforts to mitigate the famine in Bengal were accompanied by a turn to culture and art to mobilise assistance. Communist artists like

Sunil Janah (1918–2012), Govind Vidyarthi (1912–2006), D.G. Tendulkar (1909–1972), Chittaprosad Bhattacharya (1915–1978) and Somnath Hore (1921–2006) became chroniclers of the social, economic, and political crisis and produced a broad variety of visual art such as photographs, plays, sketches, drawings, linocuts and sculptures.[108] Their work had parallels with that of Zainul Abedin (1914–1976)[109] and Sudhir Khastgir (1907–1974)—left-wing artists who, although not party members, were frequently featured in communist publications.[110] These Bengali artists created powerful images of human suffering, which were reproduced in the party's mouthpieces and distributed in specially organised exhibitions, and circulated through pamphlets and posters across India and beyond. The party's close ties to the Indian People's Theatre Association (IPTA) also led to the staging of plays on the misery of Bengal's peasantry.

Though the IPTA was founded in 1943, the cultural, theatrical and political movements that birthed the institution were older. The Indian nationalist theatre emerged in the mid-nineteenth century in the socio-political and cultural context of Bengal and soon gained pace elsewhere in India, in particular in Maharashtra. The formation of the Indian Progressive Writers' Association in 1935 preceded the formation of the IPTA by almost a decade. In 1943, the All-India Committee of the IPTA began to support the work of People's Theatre Committees who had already been active in different Indian provinces, especially in Bengal and Bombay.[111] The People's Theatre Movement consisted of a series of novelties that would have a lasting impact on India's theatrical landscape. It established a non-commercial theatre to perform plays on improvised stages, with actors and actresses clothed in modest costumes. The IPTA's slogan "People's Theatre Stars the People" expressed its effort to blur the boundaries between the consumer and the producer of art, and the organisation itself aimed to create equity between them by choosing amateurs for its main cast.[112] It drew inspiration from the use of performing arts in China and the Soviet Union, where culture and art were relied upon to politicise illiterate peasants and labourers in particular. The IPTA's artists, therefore, performed pieces forwarded to them from China and the Soviet Union including "Defend our Homes", "The Doomed Battalion", "The Russian Soul" and "Nervy Deutschland." The IPTA's own plays soon dealt with the economic disintegration of the country, the havoc of famine, and the threat of Japanese aggression.[113] With the formation of IPTA's Central Squad in 1943, the IPTA quickly became involved with the communist efforts to alleviate starvation in Bengal. Its theatrical performances and singing squads successfully solicited funds for the party's famine relief movement.[114] Its biggest success, the widely known play *Nabanna* (New Harvest), was performed by the IPTA for the first time in late October 1944. It was written by Bijon Bhattacharya and directed by Shombhu Mitra, and told the story of a poor peasant family driven by famine to the streets of Calcutta. The main actresses of *Nabanna* were Tripti Mitra and Sova Sen, who had witnessed the famine as it evolved and doled out aid to the starving.[115] Declaring that "the screen [was] the most potent and popular medium for the propagation of the ideals of the People's Theatre," the IPTA gave his film-debut. K.A. Abbas transmuted *Nabanna* into the movie *Dharti ke lal* (Children of the Earth), which premiered in India in 1946.[116] *Dharti ke lal* successfully drew attention to the havoc and suffering created in Bengal and incited further political debate on the eve of India's independence.[117]

Prior to performing *Nabanna* across the country, the IPTA's singing squads moved across Bihar, Bombay, Gujarat, and Maharashtra to solicit donations for Bengal. The IPTA's mobilisation of funds for and solidarity with Bengal often carried international communist symbols, demonstrated the political unity of Indian parties, and tried to incite anger against India's capitalists. In Bombay the organisers of a student rally picked the anniversary of the October Revolution to prepare a van carrying the banner of the Muslim League and the Congress to demonstrate unity. As the van moved through Bombay, loudspeakers amplified the students' slogan "*Bengal Bhooka Nahin Marega*" (Hungry Bengal will not die).[118] Likewise in Bombay, the CPI arranged for a "Bengal Exhibition" that later travelled across the country. It featured sketches, caricatures and photographs to educate visitors about the famine. Visitors learned about the culprits and the making of the disaster in three sections. It began with Sunil Janah's photographs of the hoarder, the victims of famine, and the "heroic efforts of patriotic organisations."[119] The second section contained Chittaprosad's black and white sketches, close-up studies of people's misery in Chittagong and Midnapore. The third and last section exhibited the work of caricaturists from the Bombay Friends of the Soviet Union, R.M. Jambhekar and M. Bharatan, who exposed the hoarder and the political disunity of India's parties as the main drivers of the crisis. To avoid misinterpretation, volunteers guided visitors through the exhibition.[120]

The efforts of communists to evoke solidarity with Bengal in other parts of India also worked the other way around as communists in Bengal pledged support for other Indian provinces. In Bombay, Kerala, and later in Orissa as well, the CPI announced the formation of self-help committees for agricultural and industrial labourers who struggled to access food.[121] The simultaneity of famine conditions in different provinces across India allowed communists to frame their responses as a nation-wide movement. The sending of workers and donations from Bengal to famine-affected areas outside the province and vice-versa substantiated such communist rhetoric. The efforts of the party to summon a nation-wide movement also became visible when the CPI provided medical relief and proved eager to liaise with incoming medical students, whom it perceived as vehicles of inter-provincial and, ultimately, national unity.

> There was the need for emphasising that Bengal's calamity was not an isolated event, that the tragedy of Bengal should be the patriotic concern for Indians wherever they might be,

> that inter-provincial cooperation should emerge out of the immense sufferings of Bengal's millions.[122]

To further cement interprovincial unity, the medical volunteers of the PRC sent a medical mission to Patna to help in treating victims of epidemics.[123] The mission was one of many ways in which Indian communists tried to give credence to a political vision of a united India that aimed to be markedly different, and ultimately more inclusive, than the vision of "Akhand Hindustan."

Solidarity Across and Beyond the British Empire

The efforts of Indian communists to solicit support for Bengal acquired an imperial dimension too. Initially, the Communist Party of Great Britain (CPGB) was reluctant to publicise the famine as it had just begun to enjoy a new popularity at home due to its embrace of the war and its subsequent legalisation. It took some time for the CPGB to find a way to talk about the economic conditions in Bengal without the risk of being seen as engaging in anticolonial, and thus anti-war, rhetoric. British communists' reports about the famine in Bengal concentrated on expressing the solidarity of British workers towards their Indian brethren, publicising the work being carried out by the CPI, and highlighting how the famine weakened India's resistance against Japan. The party later published appeals from P.C. Joshi that insisted that the British assist Bengal; the CPI in turn praised communists in Britain for making the crisis visible to the public.[124]

A common narrative that emerged from these mutual efforts to express British-Indian solidarity was the depiction of alleged good deeds by British soldiers stationed in Bengal, who donated funds, offered their food rations to Bengali children, or empathised in some way with the struggle of Indians. The communist worker Krishna Binode Roy described the arrival of thousands of RAF men in Jessore in West Bengal for the readers of the *People's War*.[125] He described an empathetic encounter with the "RAF boys" that emphasised the soldiers' young age, working-class backgrounds, and longing for home.[126] A picture showing Sergeants Jack Cole, David Price, and Arthur Gilbert sharing their rations with women and children accompanied the article. Similar efforts followed in Britain. In 1944, the CPGB published the letters of Clive Branson, a communist, who had served with the RAF in Bengal during the famine and later died on the Arakan front.[127] British readers could read Branson's sympathy for Indians in his own words, making it possible to identify with the plight of a people that still seemed distant and unfamiliar.[128] The British communist politician Harry Pollitt contributed a foreword to the book, expressing his conviction that the letters, written by Branson to his wife, were meant for a much wider audience. In addition, Arthur Olorenshaw, a *Daily Worker* correspondent, detailed the experiences of British soldiers in Bengal, who had been appalled by the famine conditions in the region and demanded that their countrymen aid the people.[129] With a rise in conscription, the RAF had grown more diverse. It is not unlikely that young British communists found themselves in Bengal in 1943, reading communist literature and siding with the colonised. The newly arrived British Tommies and American GIs had little in common with the *sahibs*, who they viewed as elitist. However, the new arrivals also knew little about Indian politics, were critical of the Congress opposition to the war and saw little sense in defending the Raj.[130] Communist publications deliberately exaggerated British sympathy for Indians. They glossed over the displacement and dispossession that took place as hundreds of thousands of RAF men were moved into India, where agricultural land was cleared, irrigation networks dried, and food requisitioned to accommodate them.[131] They were also silent about the fact that with the arrival of soldiers, prostitution and sexual violence increased to a degree that British and Indian officials likewise voiced concern.[132]

In the second half of 1943, the CPGB intensified its cooperation with the London-based India League. Under the lead of V.K. Krishna Menon the London-based India League strove to unite Indians and British against colonial rule. The influence of the CPGB on the India League had been growing since Stalin propagated the Popular Front Policy in 1934 and communists worldwide started to seek alliances with anti-imperialist forces.[133] British communists now assisted the India League in setting up the India Relief Committee that raised funds for relief work in Bengal.[134] The committee raised funds through the magazine of the India League *NewsIndia*, through pamphlets, meetings, and conferences and it created sub-committees in a plethora of British cities. Menon chaired the London-based committee, while the Welsh politician Clement Davies served as its president and the famous Mulk Raj Anand (1905–2004) was its secretary.[135] During a meeting of the India League in June 1943, the novelist, essayist, and literary critic Mulk Raj Anand announced his theatrical work on the famine conditions in Bengal.[136] Anand's "India Speaks" was performed in London's Unity Theatre and the proceeds collected for Indian famine relief. Apart from its regular show times at the Unity, the piece was also staged for an Indian working-class audience in the east of London. A speech by Menon and an appeal for funds to help famine-stricken Bengal followed these performances.[137] The money gathered by the India League through theatrical performances and other fundraising events was channelled to the founder of the Social Service League, trade union leader and president of the IPTA, N.M. Joshi, who then distributed the money to a selection of relief agencies in India. The Social Service League forwarded money to the FAU, the Ramakrishna Mission, SIS, and the AIWC in addition to the communist PRC.

As time progressed, the display of communist solidarity with comrades in India extended beyond the metropole and across the British Empire to Australia.[138] Like its British counterpart, the Communist Party of Australia (CPA) began to enjoy greater popularity after it declared support for the war in 1941. Following Japan's attack on Pearl Harbour, Australians were increasingly disillusioned about the willingness and ability of Britain, and later the US, to protect its shores.[139] The communist embrace of the war boosted the appeal of the left movement in Australia, and was accompanied by a turn to the Indian Ocean region and the articulation of an alternative vision of Australia's international relations.[140] Efforts to foster international cooperation with India were made, as can be seen in a letter to the CPI from the General Secretary of the CPA, John Bramwell Miles (1888–1969), who expressed his party's commitment to India's quest for self-rule in September 1943:

> We have noted with deep admiration your splendid fight for India's liberty and for the mobilisation of your great nation for the defeat of the fascist marauders. We welcome the opportunity for the development of a closer friendship between our two peoples. Be assured that we will do all in our power to second your efforts for the welfare of the Indian people, the defeat of the Axis and the construction of a world of free democratic nations in the post-war world.[141]

In addition to this display of solidarity, the CPA informed its readers about the efforts of Indian communists to ward off Japanese aggressors and to cater to the needs of the famine-afflicted in Bengal. The official outlet of the CPA, the *Tribune*, published Chittaprosad's and Zainul Abedin's drawings of famine in 1944 to 1946.[142] In March 1944, the CPA invited a graduate named Vaidyanathan from Madras University to tour the northern coalfields of Australia. Speaking about India's war effort and the famine conditions in the country, he solicited money for famine relief in Bengal. The CPA also publicised the work of the Australia-India Association, which was formed by Clarrie Campbell in 1936. Campbell, who had close ties to the CPA but never became a member, was involved in building social-welfare institutions for Indian seamen in Australian port cities.[143] In 1944, the Australia-India Association further expanded and by March it published a monthly bulletin that informed its readers about Indian affairs. It also included reports of famine conditions in Bengal.[144] In June 1944, the bulletin announced the opening of a club for Indian seamen in Sydney along with a famine relief fund for India.[145] The fundraising of the Australia-India Association for Bengal received support from Australian politicians and diplomats.[146] The patron of the India Famine Relief Committee was Bertram Stevens, who had been in New Delhi until 1942 as the Australian representative to the Eastern Group Supply Council. After the war, he went on to become a founding member of the India League of Australia.[147] The Australian Minister of Foreign Affairs, H.V. Evatt, also embraced the work of the India Famine Relief Committee and approved of the "tangible expression to those aims of fostering the closest friendly relations with India."[148] However, the main load of the fundraising was borne by Indian seamen, who in June distributed 50,000 leaflets with pictures, stories and figures of the famine and solicited money from schools and trade unions.[149]

The reporting of the famine conditions in Bengal in Australian leftist forums coincided with the acceleration of Indian seafarer unionism and its politicisation. It would reach its peak in 1945 with the formation of the Indian Seaman's Union in Australia and its subsequent strike in solidarity with the Indonesian independence movement.[150] The deplorable condition of Bengalis served as an emotional subject to politicise and organise Indian seafarers. A similar process was underway in Britain, where different Indian nationalist associations vied for the endorsement of Indian *lascars* (seafarers). With the support of an eminent figure of the working-class movement, Ayub Ali, and its newly formed Women's Section, the India League expanded and opened an office just above Ali's Shah Jolal Restaurant at 76 Commercial Street in June 1943. The restaurant was a meeting point for Indian seamen and workers in the eastern part of London.[151] Over the following months, Indian workers joined public campaigns by the India League and other Indian nationalist associations in bringing demands for food relief to the streets of Britain.[152] This showed, among others, in the activities of the Indian Workers' Association. In October 1943, the Indian Workers' Association organised a protest during the visit of politician and Secretary of State for India, L.S. Amery, to his constituency Sparkbrook in Birmingham. Waiting in front of the venue, the Indian Workers' Association held up banners reading "Churchill and Amery are killing men, women and children in India" and distributed leaflets that placed the Bengal famine within a historical timeline of famines in India, accusing the British of deliberately trying to starve revolutionaries.[153]

6.4 The Indian Women's Movement, Transnational Feminism and Famine Relief

MARS and the Broadening of the Indian Women's Movement in Bengal

The famine in Bengal was a catalyst of the Indian women's movement. As Annie Devenish amongst others has argued, women's various experiences of the economic and social upheaval in Bengal spurred their political mobilisation. Many women in the province organised themselves into new organisations that raised awareness of the special needs of women and articulated the right of women for protection and assistance. The mobilisation of Indian women during the famine was helping the subsequent participation of women in the Tebhaga movement in

1946 that mobilised against landowning elites and money lenders.[154] It was also a step towards the formation of the National Federation of Indian Women in Calcutta in 1954 and Indian women's participation in the globally operating Women's International Democratic Federation (WIDF).[155]

The political activities of women in Bengal in the 1940s led to a significant diversification of the women's movement which had long been dominated by middle- and upper-class women.[156] This diversification owed much to the Mahila Atma Raksha Samiti (MARS) and its ability to enlist the support of women from different social backgrounds.[157] Manikuntala Sen (1911–1987) and Renu Chakravartty (1917–1994), both members of the CPI, and Ela Sen Reid were among the founding members of MARS in 1942.[158] The formation of MARS and other communist women's organisations in India was the consequence of different impulses originating in and beyond South Asia. MARS operated independently of the CPI and yet was closely entangled with the party. The CPI promoted the formation of women's organisations in the 1940s although many male party members continued to oppose the official policy.[159] In her close reading of women's complex experiences of famine, Ania Loomba has shown how women actively carved out a place for themselves in the communist movement in the 1940s. Gargi Chakravartty, in her study of MARS, has also noted that Soviet women's organisations and their call to women worldwide to form anti-fascist institutions inspired the work of women in Bengal.[160] The CPI helped this process by reproducing hagiographic accounts of Soviet women martyrs, of whom the Soviet heroine Zoya Kosmodenskaya was the most prominent example.[161]

MARS attended to the special needs of women who apart from hunger often suffered from gender-based and sexual violence.[162] Alongside conventional famine relief such as food and medical aid, MARS trained women in self-defence, raised awareness of violence committed against women, and prevented sex-work and the trafficking of women and children.[163] To this end, MARS placed volunteers in shops to "prevent the manhandling of the women in queues and to help them collect their legitimate quota" and rescued women and girls who had been sold into prostitution.[164] In July 1943, as migration from rural Bengal to Calcutta peaked, MARS opened canteens in the city to provide meals at subsidised prices or free, and ran milk centres that fed 12,000 women and children daily. In August 1943, MARS initiated the Bengal Women's Food Committee, which combined the efforts of 14 smaller women's organisations in the province.[165] Later, in April 1944, it shifted to the task of rebuilding women's livelihoods in rural Bengal. For this purpose, it assembled more than a dozen social service organisations in the *Nari Seva Sangh*, which operated shelters where women received vocational training.[166] The famine relief work of MARS had intensified its appeal to the women of Bengal and turned it into a mass movement. In early 1944 MARS claimed a membership of 40,000 women from different social strata, classes and castes.[167]

Although MARS defined itself as a women's social service organisation, its proximity to communism embedded it firmly in the political aspirations of the CPI which affected the way its members advocated for women's rights and women's position in Indian society. Since the Soviet entry into the war against Germany, communist organisations mobilised women against the anticipated arrival of Japanese invaders by raising anxieties of sexual violence. MARS and the CPI tapped into well-established symbolism that equated the woman with the nation, and placed the protection of women's "honour" at the heart of India's national defence. Communist writing on women's endangered sexual purity and their need to arm themselves against Japanese invaders was not limited to Bengal, but likewise resonated in Punjab, Madras and Andhra Pradesh.[168] In Andhra Pradesh, where Japanese air raids on Vizagapatnam and Coconada in April 1942 made the diffuse threat of Japan's advance tangible, communists organised special training in schools for young women. Apart from lessons in self-defence, the training offered a set of skills for modern Indian women to master. Lessons included political education, training in public speaking, and—illustrating the role that culture and art had acquired in the communist movement—singing lessons. The training, however, also educated young women in "the principles of maternity, child welfare, personal hygiene and sanitation" to make them "better as mothers and housewives."[169] The emphasis on women's domestic duties alongside their engagement with public political forums occurred more than once. In early 1943, MARS organised women's hunger marches in Bengal. In the *People's War*, Renu Chakravartty depicted women's politicisation as conjuring with their maternal and conjugal obligations. The reason why women left their home was to claim food (for their families) which meant that they kept with their role as mothers and wives or as "pivots of the household."[170]

MARS was clearly at the forefront of the women's involvement in famine relief work in Bengal, but it was not the only women's organisation participating in such work, nor were women visibly active only in Bengal. In 1940, when the CPI was still illegal, more and more communist women had joined the AIWC. MARS and the AIWC worked closely together for several years, although the political divide between the INC and the CPI seemed insurmountable. It was only in 1946 that the AIWC began to lament the growing influence of communist women, leading to rising tensions and the withdrawal of communists from the AIWC.[171] During the famine in Bengal, the AIWC and MARS thus had a small but important overlap in their membership, enabling communist women to flag the needs of Bengal's women in the fora of the AIWC. In addition, the AIWC maintained a network of branches across India and women activists disseminated information about the famine in Bengal through their networks. Naturally, the AIWC branches in Bengal were the first to respond to the famine. The AIWC in Bankura and Calcutta took the lead in April 1943, opening clinics, milk kitchens, and relief centres. They also appealed to families across India to adopt children rendered destitute by the famine, and set up a foster parents' scheme that allowed Indians to pay for the food, shelter and education of orphaned children without actually taking them into their homes. Reports about the famine in the province filtered through to the AIWC Central Assembly in Bombay in early 1943.[172] The AIWC president, Vijaya Lakshmi Pandit (1900–1990), visited Bengal in September 1943, which had a momentous effect on the mobilisation of relief for the region. The social activist, politician, and sister of Jawaharlal Nehru, was repeatedly imprisoned earlier that year because of her affiliation with the Congress and support of the Quit India movement.[173] These incarcerations delayed her personal involvement in the appeals for relief for Bengal. By late September and early October, however, her accounts of the plight of women and children were taken up by the national and international press, adding momentum to the mobilisation of funds already underway for Bengal.[174] In October 1943, the AIWC branch in Srinagar organised an aid drive that raised funds by means of house-to-house collections, charity concerts, and festivals, with the result that five hundred tons of grain were dispatched from Lahore to Calcutta.[175] By November, money was forthcoming from the India League's Famine Relief Committee in London.[176]

In the aftermath of her visit to Bengal, Pandit set up the All-India Save the Children Committee in collaboration with B.C. Roy. The All-India Save the Children Committee continued to assist children after the famine, by aiding during the partition crisis. In 1952, it merged with the Indian Council for Child Welfare.[177] During the famine, the All-India Save the Children Committee opened shelters for children first in Bengal, and later in Malabar and Orissa as well. Pandit's correspondence of the time illustrates the role of famine relief in the political ambitions of the women's organisation. For Pandit, the opening of children's homes tied in with both the nationalist endeavour to produce "useful citizens" and the AIWC's objective of carving out a place for the women's movement in India's nation-building process.[178] This also included the building of permanent institutions that as Pandit noted would serve as a "memorial" of the AIWC's social service activities during the famine, reminding later generations of the importance of the women's organisation.[179]

The Women's Committee of the India League in Britain

At the time women in Bengal and wider India were mobilising relief, Indian women in Britain also were advocating for those afflicted by famine. In June 1943, Asha Bhattacharya opened the inaugural meeting of the India League's East London branch office with a speech that blamed the "short-sighted policy of the British government" for the current famine conditions in India.[180] Bhattacharya's audience included a group of about 80 Indian factory workers and *lascars* who had come to listen to the India League. Before the League's long-time president V.K. Krishna Menon and Mulk Raj Anand rose to speak, Jaikishori Handoo addressed the audience.[181] Speaking in Urdu, she drew further attention to the famine conditions in Bengal. Against the background of the raging naval war between Germany and the allied powers, Handoo questioned the willingness of Indian seamen to risk their lives bringing food to Britain, while in the meantime their families in Bengal were starving. In her efforts to rally Indian seamen behind the India League that continued throughout the following months, Handoo referred frequently to the famine in Bengal. In October 1943, she shocked a group of Indian seafarers who had just returned from North America with reports of Indians starving to death on the streets of Calcutta and called the men to action.[182] At this time, the famine in Bengal had made headlines in British newspapers and Handoo was now translating articles into Bengali and Urdu for Indian seamen in British port cities to read and listen to.[183] Women in Britain had long supported the India League, joined its meetings and took part in its activities. Their voices suddenly acquired greater visibility in March 1943, after Menon's attempts at winning the support of Indian dockworkers and seamen in the east of London resulted in the formation of the India League's Women's Committee. For the task of mobilising Indian labourers, Menon, who was perceived as elitist and lacked the language skills required to address his audience in a more personal tone, depended on women like Handoo and Bhattacharya. The women were fluent in Bengali and Urdu and seemingly found greater acceptance among Indian workers. The current famine conditions in Bengal (as well as in the south of India) gave these women a powerful topic that assisted them in gaining the attention of Indian workers.

Handoo, who was appointed president of the Women's Committee and Bhattacharya, the president of the India League's East End branch, now exchanged with the British liaison committee of the AIWC. The liaison committee of the AIWC in London was formed in 1934 when Grace Lankester and Rajkumari Amrit Kaur were appointed as the first liaison officers between the AIWC and British women's organisations. By 1943 the liaison committee had grown and gained additional prominent support with the joining of the pacifists Agatha Harrison and Vera Brittain.[184]

Brittain's interest in India had been sparked as she listened to a speech by Gandhi in the early 1930s; later, she hosted Jawaharlal Nehru in London. She was invited to the AIWC twice, in 1940 and 1941, but was denied a visa on both occasions because of her anti-war stance.[185] Harrison's involvement with India started around the same time as Brittain's. She was drawn to Gandhi's pacifism upon visiting India as the assistant of a member of the Royal Commission on Labour in 1929–31. On returning to London, Harrison began working with other Quakers in the Indian Conciliation Group (later India Conciliation Group) in their efforts to facilitate India's self-rule by mediating between British and Indian representatives. Harrison attended the meetings of the India League from 1933 onwards, signalling a convergence of white women's feminism and Indian nationalism in the metropole.[186] The importance of such ties showed in April 1943, when members of the Liaison Committee embraced India League's newly founded Women's Committee. Shortly after, its members also joined the India League's Famine Relief Committee, which began to channel a part of its funds to the AIWC.[187] The Bengal Famine thus coincided with the deepening alliance of metropolitan feminists and the AIWC on the one hand, and the greater involvement of British and Indian feminists in the India League and diasporic Indian nationalism on the other. In October 1945, Handoo represented the AIWC at the Board Meeting of the International Alliance of Women in Geneva.[188]

Women's personal and institutional networks that carried information from Bengal to London were part of a larger web of entanglements that subverted the colonial censorship regime and allowed information about the famine to circulate. This ensured that details about the famine in India would, amongst others, filter through to Britain where Indian nationalists, British socialists, communists and pacifists aimed to raise awareness of the plight of people in the province and to mobilise funds and food despite, and against, the inertia of states and governments.

6.5 Conclusion

The breadth of political and social movements in Bengal that were involved in relief activities in the 1940s demonstrates the complexity of the humanitarian response to the famine and its relevance for political mobilisation and nation-building. Without claiming to be exhaustive, the chapter pointed out how relief in Bengal was mobilised through communist, feminist, Hindu nationalist, pan-Asian and anti-imperial networks of various scales—many of which had their origins in the interwar period.

The leader of the Hindu Mahasabha, Syamaprasad Mukherjee, placed himself at the head of four relief committees, using his involvement to strengthen the political influence of the party in the province. The elasticity of Mukherjee's politics allowed him and the Hindu Mahasabha to cultivate very different, and at times, even conflicting political contacts. Although he nourished communal identities through famine relief, the Hindu Mahasabha leader likewise demonstrated the capacity to move beyond communalism and, hence, collaborated with Congress politicians and institutions in another capacity. The famine relief work of the Maha Bodhi Society of India, another organisation Mukherjee was involved in, exemplifies the efforts of Indian, Chinese, Ceylonese and Burmese politicians at fostering pan-Asian solidarity, as well as the Mahasabha's efforts to implement its vision of Greater India. As historians Maria Framke and Esther Möller showed, the

Hindu Mahasabha continued to show a remarkable flexibility in terms of political alliance in the aftermath of the famine, when the relief it gave to Hindus during the partition crisis again catered to multiple audiences.[189] But the Hindu Mahasabha was not the only political force in Bengal to exploit the famine strategically. Indian communists used the famine to nurture and substantiate interprovincial and transnational solidarities. Famine relief enabled communists in India, Britain and Australia to demonstrate commonality with the Congress movement, that was otherwise impeded by the demand of the Comintern to pause anticolonial agitation until the war was won. Famine relief also enabled the CPI to gain back political ground by deflecting criticism that accompanied its opposition against the Quit India movement and to demonstrate its wish to protect Bengal against Japanese aggression.

The famine also galvanised women's political mobilisation in Bengal and, to a much lesser extent, in Britain, where Indian women rose to new prominence within the India League. A growing number of Indian women mitigated famine as members of social service institutions in the first decades of the twentieth century. Following the formation of women's organisations, most notably the AIWC, the participation of women in famine relief markedly grew in the interwar period. It reached an unprecedented scaled in 1943–44 when the AIWC and MARS successfully tapped into national and transnational feminist networks to mobilise aid. Famine relief, as part of the broader social service activities of Indian women's organisations, helped the women's movement to claim space in the political arena by highlighting the importance of women's participation in Indian nation-building and drawing attention to the special needs of women and children.[190]

CHAPTER 7

American Food Aid for Independent India

Abstract

The political alliances between American and Indian civil society actors had intensified in the interwar period but came to full fruition in the 1940s and 1950s. In response to famines in Bengal, Bihar and Madras, organisations formed in the US to push the American government to assist India. The chapter provides a new perspective on the history of US food aid for India by examining the involvement of non-state actors and the role of Indian activists who began to cater to Indian national development in the 1940s.

Keywords: Food aid, development, postcolonial India, South Asian Americans

7.1 American Anti-imperialism and the Bengal Famine, 1943–1945

> I found myself eating a candy bar, disinterestedly watching a woman die. [...] There is something that happens to all who live through it that leaves a mark on the body and mind. Surely this is one of the dangers of imperialism, this subtle poison that eats away whatever it touches. Surely it has left us all a little less human.[1]
>
> —John Frederick Muehl, *American Sahib* (1946).

Frederick Muehl travelled from Detroit to India with the American Field Service in 1943. Upon his return to the United States after the war, he embarked on a university career in the Department of English Language and Literature at the University of Michigan and put his experiences in India on paper.[2] *American Sahib* (1946) paints a grim and appalling picture of the conduct of GIs in India and gives his personal account of the Bengal Famine. Muehl recalled that one evening, on his return from the Grand Hotel in Calcutta—one of the many places that still served lavish meals while people died on the streets— he was confronted with the agony of a famished man who tried to approach Muehl in his despair. Muehl instinctively ran away and threw up his as yet undigested five-course meal on the doorstep of his hotel. The shock of seeing a man dying on the street made him nauseated for two more days but he soon became accustomed to the sights.[3] He explained his fading empathy with the impact of British imperialism, the "subtle poison" that had corrupted his soul as well as those of other American soldiers stationed in

India. Muehl thus found an explanation of why "his America", which had liberated Europe and promised peace and prosperity to the world, stood by as famine killed millions in Bengal. He likewise reasoned why it was important for Americans to stop defending the Raj now that the war had been won. Muehl was a late addition to the American critics of British imperialism and colonial rule in India.

When British intelligence services sought to enumerate the sources of support for Indian nationalism in North America in September 1944, the list proved extensive. It began with "republicans – isolationists – imperialists – big business – and elements who are anti-New Deal, anti-Roosevelt, and anti-British." It went on to cite "liberals, religious and social pacifists, idealist antagonists of anything savouring British imperialism and champion of colonial and coloured people," and it ended with "communists" and "fellow travellers."[4] Though the documents produced by the British intelligence services have to be read with a pinch of salt, they were not mistaken in their estimate of the diverse assortment of American support for Indian nationalism.

Indian nationalism gained visible momentum in North America in the early 1940s, but the face of this nationalism had changed tremendously. It did not have much in common anymore with the revolutionary activities of Ghadrites. Instead it resembled the moderate long-distance nationalism of the India Home Rule League, but with differences in its methods. Apart from a group of Indian nationalists around Syed Hossain, who felt uncomfortable placing the quest for Indian self-rule in the hands of Americans, Indians in the US began to mingle extensively with American sympathisers of Indian nationalism and to draw on their support. One institution that was particularly successful in making Indian nationalism more appealing in the US was the India League of America (India League).[5] Under the lead of Sirdar Jagjit Singh, better known to his contemporaries as JJ Singh, the India League turned into an institution of mass appeal.[6] Singh had left India in 1922, driven by his disappointment with Gandhi's decision to pause the civil disobedience movement. Seeking distance from the Indian nationalist movement, he began to immerse himself in the world of trade and business, first in Great Britain and, after 1926, the United States as well. In the following decades JJ Singh and his brother built a business venture in New York called Singh, Singh & Co. At the same time, he strengthened economic ties between the United States and India in his function as the first president of the Indian Chamber of Commerce. Singh eventually returned to politics in the late 1930s, when, encouraged by Jawaharlal Nehru's political ascendance, he drew on his business contacts to boost American support for India.[7] In 1941, Singh joined the India League and began to transform the organisation from a fringe group of mainly Indian supporters of Indian self-rule to an institution of much wider appeal. This transformation started in 1942 with a new campaign that solicited the support of well-known religious, labour and media representatives. Pearl Harbour provided the canvas against which the India League formulated a compelling demand for US support of India's quest of self-rule. Its campaign boiled down to the slogan "winning the war requires a free India."[8]

It is somewhat ironic that the campaign was drafted by the editor and journalist Sidney Hertzberg (1910–1983). As a former member of the America First Committee, Hertzberg had opposed American involvement in the war; now he appealed to the American desire to win the war to generate support for India's quest for self-rule. The America First Committee united a group of fierce isolationists and pacifists, but Sidney Hertzberg was motivated by his disapproval of British imperialism and his deep-seated belief that British colonial rule was not much better than fascism.[9] Hertzberg began to help the League land advertisements in leading American newspapers. One result of this cooperation was a full-page advertisement in the *New York Times*, published in September 1942 (figure 4).[10] The ad depicted the League's demand that Roosevelt use his influence over Britain to press for India's right to self-rule. It listed over fifty prominent Americans who supported the League by lending their names.[11] The signatories illustrate the different sections of American society that supported the India League in 1942. The support was proof of the connections between the African-American civil rights movement and Indian anticolonialism, evidenced in this case by the name Walter White, secretary of the National Association for the Advancement of Colored People (NAACP).[12] The extent to which anti-racism and anti-imperialism provided a common ground for supporters of Indian nationalism also showed in Clare Boothe, who signed the India League appeal in 1942 and elaborated at length on her support of India's demand for self-rule a year later. During the India League's "mass meeting" in August 1943, she delivered a speech in support of the Indian anticolonial movement wherein she declared that her hope was that US involvement in the war was to safeguard the freedom of all peoples.[13] Despite her close friendship with Winston Churchill, Boothe sharply criticised the British Prime Minister and asked him if he was truly convinced that "freedom was a white man's monopoly" in her exhortation to recognise the right of South Sea Islanders, Malays and Indians to attain freedom from colonial rule.[14] Boothe had first attained fame as the author of a number of plays including *The Women*, which became a Broadway hit and later a Hollywood movie. She also worked as a correspondent for *Life* during World War II and embarked on a political career as a Republican representative for Connecticut in 1942. Boothe used her popularity in these various public roles to raise her voice in support of anticolonial movements in Asia. In addition, she co-drafted the Luce-Celler Act in 1943, which would ease US immigration restrictions on Indians three years later.[15]

After 1942, the India League also drew substantial support from a section of anti-imperial protestants, many of whom had served as missionaries in India. Amongst the first to step up was Sherwood Eddy, who signed the appeal in the

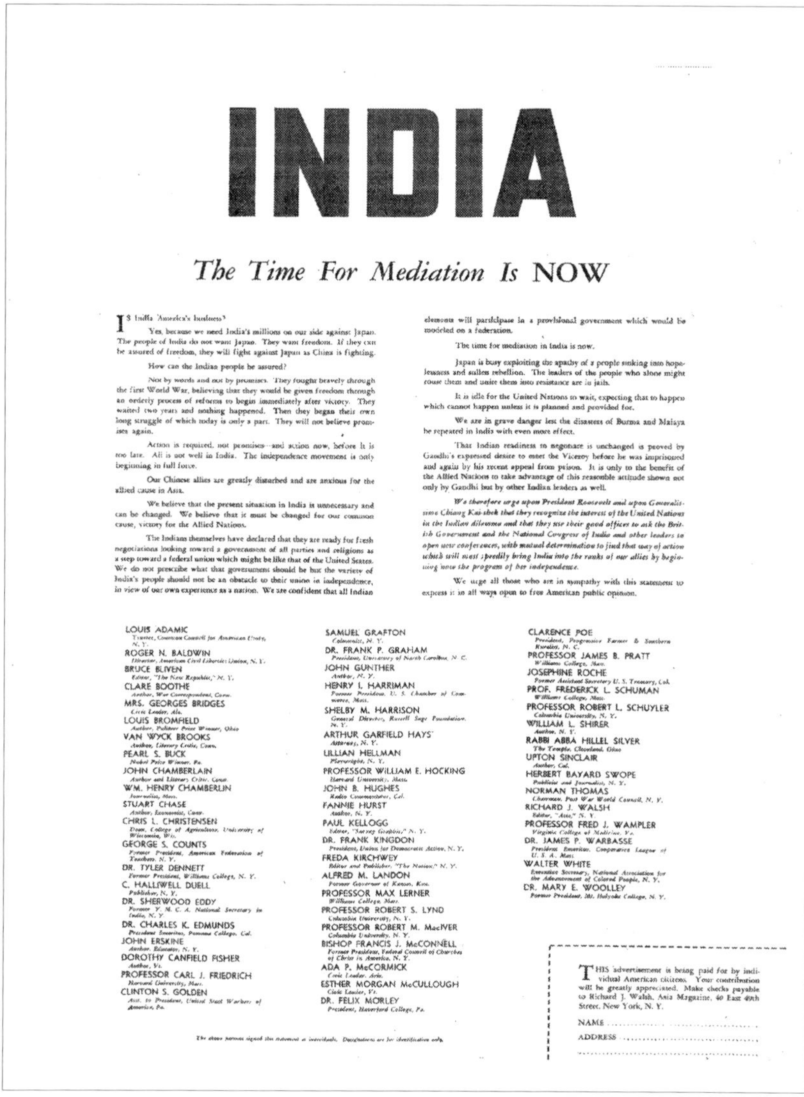

INDIA

The Time For Mediation Is NOW

IS India America's business?

Yes, because we need India's millions on our side against Japan. The people of India do not want Japan. They want freedom. If they can be assured of freedom, they will fight against Japan as China is fighting.

How can the Indian people be assured?

Not by words and not by promises. They fought bravely through the first World War, believing that they would be given freedom through an orderly process of reforms to begin immediately after victory. They waited two years and nothing happened. Then they began their own long struggle of which today is only a part. They will not believe promises again.

Action is required, not promises—and action now, before it is too late. All is not well in India. The independence movement is only beginning in full force.

Our Chinese allies are greatly disturbed and are anxious for the allied cause in Asia.

We believe that the present situation in India is unnecessary and can be changed. We believe that it must be changed for our common cause, victory for the Allied Nations.

The Indians themselves have declared that they are ready for fresh negotiations looking toward a government of all parties and religions as a step toward a federal union which might be like that of the United States. We do not prescribe what that government should be but the variety of India's people should not be an obstacle to their union in independence, in view of our own experience as a nation. We are confident that all Indian elements will participate in a provisional government which would be modeled on a federation.

The time for mediation in India is now.

Japan is busy exploiting the apathy of a people sinking into hopelessness and sullen rebellion. The leaders of the people who alone might rouse them and unite them into resistance are in jails.

It is idle for the United Nations to wait, expecting that to happen which cannot happen unless it is planned and provided for.

We are in grave danger lest the disasters of Burma and Malaya be repeated in India with even more effect.

That Indian readiness to negotiate is unchanged is proved by Gandhi's expressed desire to meet the Viceroy before he was imprisoned and again by his recent appeal from prison. It is only to the benefit of the Allied Nations to take advantage of this reasonable attitude shown not only by Gandhi but by other Indian leaders as well.

We therefore urge upon President Roosevelt and upon Generalissimo Chiang Kai-shek that they recognize the interest of the United Nations in the Indian dilemma and that they use their good offices to ask the British Government and the National Congress of India and other leaders to open new conferences, with mutual determination to find that way of action which will most speedily bring India into the ranks of our allies by beginning now the program of her independence.

We urge all those who are in sympathy with this statement to express it in all ways open to free American public opinion.

LOUIS ADAMIC
Trustee, Common Council for American Unity, N. Y.
ROGER N. BALDWIN
Director, American Civil Liberties Union, N. Y.
BRUCE BLIVEN
Editor, "The New Republic," N. Y.
CLARE BOOTHE
Author, War Correspondent, Conn.
MRS. GEORGES BRIDGES
Civic Leader, Ala.
LOUIS BROMFIELD
Author, Pulitzer Prize Winner, Ohio
VAN WYCK BROOKS
Author, Literary Critic, Conn.
PEARL S. BUCK
Nobel Prize Winner, Pa.
JOHN CHAMBERLAIN
Author and Literary Critic, Conn.
WM. HENRY CHAMBERLIN
Journalist, Mass.
STUART CHASE
Author, Economist, Conn.
CHRIS L. CHRISTENSEN
Dean, College of Agriculture, University of Wisconsin, Wis.
GEORGE S. COUNTS
Former President, American Federation of Teachers, N. Y.
DR. TYLER DENNETT
Former President, Williams College, N. Y.
C. HALLIWELL DUELL
Publisher, N. Y.
DR. SHERWOOD EDDY
Former Y. M. C. A. National Secretary in India, N. Y.
DR. CHARLES K. EDMUNDS
President Emeritus, Pomona College, Cal.
JOHN ERSKINE
Author, Educator, N. Y.
DOROTHY CANFIELD FISHER
Author, Vt.
PROFESSOR CARL J. FRIEDRICH
Harvard University, Mass.
CLINTON S. GOLDEN
Asst. to President, United Steel Workers of America, Pa.
SAMUEL GRAFTON
Columnist, N. Y.
DR. FRANK P. GRAHAM
President, University of North Carolina, N. C.
JOHN GUNTHER
Author, N. Y.
HENRY I. HARRIMAN
Former President, U. S. Chamber of Commerce, Mass.
SHELBY M. HARRISON
General Director, Russell Sage Foundation, N. Y.
ARTHUR GARFIELD HAYS
Attorney, N. Y.
LILLIAN HELLMAN
Playwright, N. Y.
PROFESSOR WILLIAM E. HOCKING
Harvard University, Mass.
JOHN B. HUGHES
Radio Commentator, Cal.
FANNIE HURST
Author, N. Y.
PAUL KELLOGG
Editor, "Survey Graphic," N. Y.
DR. FRANK KINGDON
President, Union for Democratic Action, N. Y.
FREDA KIRCHWEY
Editor and Publisher, "The Nation," N. Y.
ALFRED M. LANDON
Former Governor of Kansas, Kan.
PROFESSOR MAX LERNER
Williams College, Mass.
PROFESSOR ROBERT S. LYND
Columbia University, N. Y.
PROFESSOR ROBERT M. MacIVER
Columbia University, N. Y.
BISHOP FRANCIS J. McCONNELL
Former President, Federal Council of Churches of Christ in America, N. Y.
ADA P. McCORMICK
Civic Leader, Ariz.
ESTHER MORGAN McCULLOUGH
Civic Leader, Vt.
DR. FELIX MORLEY
President, Haverford College, Pa.
CLARENCE POE
President, Progressive Farmer & Southern Ruralist, N. C.
PROFESSOR JAMES B. PRATT
Williams College, Mass.
JOSEPHINE ROCHE
Former Assistant Secretary U. S. Treasury, Col.
PROF. FREDERICK L. SCHUMAN
Williams College, Mass.
PROFESSOR ROBERT L. SCHUYLER
Columbia University, N. Y.
WILLIAM L. SHIRER
Author, N. Y.
RABBI ABBA HILLEL SILVER
The Temple, Cleveland, Ohio
UPTON SINCLAIR
Author, Cal.
HERBERT BAYARD SWOPE
Publicist and Journalist, N. Y.
NORMAN THOMAS
Chairman, Post War World Council, N. Y.
RICHARD J. WALSH
Editor, "Asia," N. Y.
PROFESSOR FRED J. WAMPLER
Virginia College of Medicine, Va.
DR. JAMES P. WARBASSE
President Emeritus, Cooperative League of U. S. A., Mass.
WALTER WHITE
Executive Secretary, National Association for the Advancement of Colored People, N. Y.
DR. MARY E. WOOLLEY
Former President, Mt. Holyoke College, N. Y.

The above persons signed this statement as individuals. Designations are for identification only.

THIS advertisement is being paid for by individual American citizens. Your contribution will be greatly appreciated. Make checks payable to Richard J. Walsh, Asia Magazine, 40 East 49th Street, New York, N. Y.

NAME ....

ADDRESS ....

....

Fig. 4. India. The Time for Mediation is now. An ad placed by the India League of America in the *New York Times*. Source: *New York Times* (28 Sep 1942), 9.

New York Times as the former national secretary of the YMCA India chapter. Eddy, who was now in his mid-seventies, had been a long-time proponent of Christian internationalism.[16] Further personal contact underpinned the India League's connections to the American missionary movement. Pearl S. Buck's missionary past put her in a unique position to mediate between "the east" and "the west". Her reputation as a cultural interpreter gave her support of Indian independence a certain weight. Buck grew up in China as the daughter of Presbyterian missionaries and lived there until her return to the US in 1932. Taking up a career as a writer and leading expert on China, she underwent a considerable personal transformation. Buck grew increasingly sceptical of the tenets of her Christian fundamentalist upbringing and eventually upset her church by divorcing her first husband and marrying her publisher Richard Walsh (1886–1960).[17] As the owner of the John Day Company, Richard Walsh published Buck's work and the writings of Jawaharlal Nehru and M.K. Gandhi. Joining forces, the couple founded the East and West Association in 1942 that strove to enhance the cultural exchange and understanding between Asia and the US. It published a magazine called *Asia and the Americas*. The East and West Association brought Walsh and Buck in ideological and geographical proximity to the India League. Even the offices of both organisations were housed in the same building. This foreshadowed the formal ties that would emerge in 1944, when Buck and Walsh joined the India League as honorary director and chairman respectively.[18] The support the India League received from the East and West Association illustrated a larger trend. In the 1940s, organisations that promoted "cultural exchange" between India and the US aimed to knit the two countries closer together, and fed American opposition to British colonialism.

Time-wise, American support for India's quest for self-rule began to grow noticeably with the India League's efforts to make the American public aware of Bengal's suffering. Since the outbreak of World War II, the interception of letters and telegrams that contained information of potential harm to the British war effort had thinned the stream of information that connected India and the United States.[19] In addition, the attention of the American press for events in South Asia was diluted by the war in Europe.[20] The economic disruption caused by the war in India, and the suffering it began to produce in Bengal towards the end of 1942, remained largely invisible to the general American public throughout 1943. This changed to some extent in late November 1943, when the US American press eventually provided snippets of the famine in Bengal. *Life* magazine printed the drawings of its war-artist correspondent Millard Sheets that pictured death and agony in Bengal.[21]

A month earlier, in October 1943, JJ Singh had appealed directly to President Roosevelt for food aid to India. The Roosevelt administration had known about the severity of the crisis in Bengal long before the American press picked up on the famine. US diplomats in India had cabled a series of straightforward descriptions about the unfolding tragedy and forwarded the famous photographs taken by journalists of the Calcutta-based newspaper the *Statesman*. Roosevelt however demonstrated no concern for the famine-afflicted population of Bengal and decided to prioritise the Anglo-American alliance, aiming to ward off the further rise of fascism in

Europe. Instead of the President, Singh's letter was answered by Paul Alling, Chief of the State Department's Division of Near East Affairs, who declared that the US was making efforts to alleviate the crisis. Although there is no evidence of such official involvement, Singh's success in eliciting an official response is nevertheless remarkable. The American and Indian press covered Singh's correspondence with Alling and helped draw attention to the India League's demands.[22] This was when the thanksgiving season gave the India League further opportunities to harness the guilt of the well- and the over-fed. Members of the League went out on the streets and distributed leaflets that asked American readers, "Can America stuff while India starves? [...] Will you reach a hand of sympathy to the hungry multitudes?" British intelligence noted with dismay that the India League appropriated the famine for a political purpose.[23]

UNRRA

"Humanitarianism was entering a new phase of global governance" in November 1943, as the United Nations Relief and Rehabilitation Administration, better known as UNRRA, became the sole institution to administer state-led food assistance.[24] It was the outcome of an initiative by British and Canadian diplomats who argued for the necessity of cooperation in the administration of aid to liberated countries in the aftermath of the devastation caused by the war. UNRRA was rooted in earlier Anglo-American cooperation in coordinating relief and drew from the experience of World War I; however, it was in many ways an unprecedented attempt to internationalise aid. Its underlying motivations were multiple and included strong economic incentives such as the calculation that relief would allow for the distribution of surplus commodities and would stabilise Europe's economy in favour of Anglo-American trade interests. In addition, it was the product of a refutation of American isolationism, and the belief that cooperation in the area of relief would enable engagement in other internationalist projects. Though the UNRRA comprised of forty-four member states, including India, the US was its largest monetary donor and exerted a tight control over the institution.[25]

The efforts of the India League to mobilise US support for Bengal in its fight against famine coincided with the formation of UNRRA. Seeking to take advantage of this moment, Singh was among the first to argue that India was a legitimate recipient of UNRRA's assistance. He opined that India had not only profoundly contributed to the British war effort, but the economic and political disruptions of the war were central ingredients in the Bengal famine, bringing the famine within the remit of UNRRA.[26]

India had become a member of UNRRA in November 1943, when Indian representatives under the lead of the Agent General for India Girja Shankar Bajpai signed the organisation's founding agreement. India thereby also agreed to contribute eighty million rupees to UNRRA's relief budget. Although UNRRA was conceived as a new international forum where India's national aspirations could be put into practice, the Indian delegation and Bajpai in particular failed to demand assistance for Bengal. Bajpai, an Indian civil servant and representative of the British Indian government, merely echoed the British representative John J. Llewellin, who reiterated that Britain was already offering aid to the province.[27]

In light of this opposition, Singh's efforts to bring UNRRA to Bengal seemed futile. But when he, in a final effort, turned to the Congressman Karl J. Mundt, he struck a victory. Singh convinced Mundt of the legitimacy of his demand and the latter pushed an amendment that allowed for famine relief in India through UNRRA. At this time, the Congress was currently preparing the allocation of US food grains to UNRRA and as the largest contributor, the influence of the US Congress on the body was significant. In spite of the opposition of the Assistant Secretary of Agriculture, Dean Acheson, the Congress followed Mundt, and UNRRA was forced to change its statutes in March 1944. Although the amendment made no explicit reference to India, the wording adopted allowed for the provision of relief to countries affected by famine. This could have paved the way for the shipment of American grain to Bengal, but UNRRA's help never reached Indian shores. UNRRA depended on the official British request of assistance, which Churchill was unwilling to give.[28] On the one hand, the Churchill administration had successfully sat out the crisis for over a year, and, with the conditions in Bengal easing, there was no incentive to alter course. On the other hand, the British were alarmed by US attempts to access the Indian market. Even as American goods provided through the Lend-Lease Agreement sustained the British war effort, US politicians had used the negotiations to push for the imperial preference system to be dismantled.[29] This added to the reluctance on the part of the British administration to seek assistance via international relief mechanisms, as the demand for food aid provided an opportunity for the US to access British colonial markets.[30]

The India Famine Relief Committee

Buck and Walsh, who were not yet official members of the India League in late 1943 but had spoken frequently about the legitimacy of India's demand for self-rule, watched the inertia of the US government with increasing impatience. In November 1943 they formed a voluntary agency, the India Famine Relief Committee (IFRC), to raise money to assist famine-stricken Bengal.[31] The couple was experienced in running aid campaigns. Buck previously set up the China Emergency Relief Committee in 1938 to cater to the needs of China's civilian population suffering under the Sino-Japanese war. In 1941, her Committee joined United China Relief, an umbrella organisation that combined private American aid for China, and Buck became its public face.[32] Her dual engagement for China and India might account for the overlap of members of aid committees. One case in point was Henry Luce (the husband of Clare Boothe), who was one of the first to be elected to the Board of Directors of IFRC.[33] Like Buck, Luce had grown up as the child of missionaries in China, and had established himself as a China expert in the interwar period.[34] This coincided with his growing media empire: Luce was owner and editor of *Life, Fortune,* and *Time* magazines, which enabled him to exert tremendous influence over American perceptions of China and now India. *Life* ran its first article on the famine at the same time as Luce joined the IFRC. Luce then assigned the Magnum photographer William Vandivert to Bengal to take pictures of human misery.[35] Vandivert's images of the dead corpses and the skeletonised bodies of Indian men and women appeared in *Life* in December 1943, and were the first photographs of the famine to be printed in the American press. Although the images were meant to shock, Vandivert refrained from criticising the British famine management.[36] The omission of criticism was congruent with Luce's own understanding of the famine. Luce was among those who saw no contradiction between promoting Anglo-American solidarity on the one hand, and recognising the need for American assistance in Bengal on the other. As he had famously heralded the ascendance of the "American century" in 1941, Luce expressed his conviction that America would turn into the world's "Good Samaritan."[37] Thus according to Luce, helping starving Indians clearly was an American task and he was not alone in this estimation. The IFRC now had over a hundred members, a notable fraction of whom were prominent American protestants like the YMCA leader John R. Mott, the aforementioned Sherwood Eddy, and the Christian philanthropist and pastor of the famous New York Riverside Church, Harry Emerson Fosdick. Other names included Gandhi biographer Louis Fischer as well as Albert Einstein, who supported the Committee as part of his wide-ranging social and political engagement.

By this time, the IFRC was prepared to jump into action, but as a charitable institution it was bound to comply with the official regulatory mechanisms that controlled and directed American philanthropy since mid-1942.[38] As a first step in this process, the committee appealed to the President's War Relief Control Board, an agency connected to the State Department but operating independently of the government. Since its establishment in July 1942, any relief agency was required to get a license from the Board, which was therefore in firm control of the private charitable sector.[39] The IFRC managed to clear this first obstacle and was officially recognised "as a channel for American contributions for relief in India."[40] Seeking to turn over its funds to the AFSC, it worked out a relief budget jointly with the Quaker organisation. The budget estimated Bengal's needs at 900,000 US dollars, allocated at a monthly rate of 100,000 US dollars over the next nine months.[41] Complying with official war regulations of a private charity, which permitted independent fundraising campaigns only in exceptional cases, the IFRC turned to the National War Fund for money. The National War Fund took up its work in October 1943, aiming to make American private philanthropy more efficient and to tie it closer to the US war needs. To this end, it conducted annual aid drives, soliciting donations mainly through community chests, women's organisations, and business groups. It allocated money to registered relief agencies following established quotas, with the goal of providing "welfare and recreational services to the US armed forces, their auxiliaries, and the merchant marine; to provide relief to occupied countries" and "to give assistance to unoccupied areas and aid to refugees."[42] Among the largest contributors to the National War Fund was American labour.[43] Since the US entered the war, the labour movement used its well-oiled fundraising mechanisms to raise substantial funds for both national and overseas relief work. The IFRC enjoyed labour support from the beginning. The secretaries of the still-separate American Federation of Labor (AFL) and the Congress of Industrial Organisations (CIO), William Green and Philip Murray, were both members of the IFRC. As a result, the IFRC had the backing of a vital section of the organisation even before its appeal to the National War Fund. Nevertheless, it soon faced considerable obstacles.

"British Censorship" in the United States

The famine in Bengal gave the India League a powerful and emotive topic to generate public interest in India. However, institutional and individual attempts to mobilise relief for Bengal on humanitarian grounds were incapacitated by the IFRC's association with Indian nationalism. Critics disavowed the IFRC as overtly political and anti-British, and hence, anti-war.

The British War Relief Society (BWRS) was a section of the National War Fund, and administered American private donations for relieving the distress of people under the jurisdiction of the British Empire. Problems arose when it requested the IFRC to submit its publicity material for prior approval. In its own estimation, the BWRS aimed to prevent that "publicity regarding India […] cause[d] offense to other parts of the Empire."[44] Though the BWRS was an American institution, a British faction on its board exerted a vital influence on the organisation. The BWRS therefore showed a conspicuous preference for relieving the distress of the war primarily in Britain. It was also sceptical of the IFRC's motives, and assumed that it would use the National War Fund as a platform to promote Indian political demands.[45] Walsh, who led the negotiations with the BWRS, sharply rebuffed the request to control the IFRC's publicity material, calling it an attempt to enact "British censorship."[46] At this point, the IFRC merely secured enough money to finance its intended relief in Bengal for an initial four months. During that time the conflict with the BWRS further escalated. After an initial advancement through the National War Fund, the BWRS handicapped the IFRC by delaying budget decisions. When the hearing before the budget committee eventually took place in July 1944, eight months after the IFRC had commenced its work, the BWRS presented contradictory arguments to discourage funding. It declared that the famine in Bengal was now over and thus relief was no longer required. At the same time, it emphasised that Bengal's needs exceeded the capacities of both the IFRC and the National War Fund. Conspicuously, the BWRS referred to endemic hunger and poverty in India, conveniently placing Bengal's needs outside the mandate of the National War Fund, which was exclusively for emergencies. As a result of lobbying of the BWRS, the IFRC was denied further funds. Representatives of American labour insisted that the money they had raised for India (200,000 US dollars) be forwarded to the IFRC, a demand the National War Fund fulfilled.[47]

The IFRC remained in close contact with the AFSC throughout the process as the Quaker organisation awaited funding for its relief work in Bengal.[48] The fact that little money was forthcoming threatened to jeopardise the aid efforts. Although the decision of the National War Fund to withdraw its support came as a shock, it had a beneficial side-effect. After the official refusal of the National War Fund, Buck and Walsh were allowed to seek funds independently from it. In the process, the IFRC became American Relief for India which began to work outside of the structure of the National War Fund. Articulating their contempt for the politics of the National War Fund that had seemingly given in to pressure of the BWRS, Buck and Walsh officially joined the India League of America.[49] Ironically, the National War Fund, which had tried to contain Indian nationalism in the United States, drove prominent public figures closer to the India League. The decision of the National War Fund against financing famine relief in Bengal also caused a larger outcry. The War Relief Control Board considered forbidding foreign nationals from serving on the National War Fund to prevent imperial politics from influencing the administration of American relief.[50] In addition, those forces in the National War Fund who supported financing relief for India (mainly labour representatives, leading Quakers and businessmen) joined the newly formed American Relief for India to secure funds for the AFSC.[51] As American supporters for India reorganised into American Relief for India, the representation of American business interests grew significantly. The shipping agent and government counsellor Henry F. Grady became the chairman of American Relief for India. Grady had visited India in 1942 as delegate of an American Technical Mission, which advised on the expansion of India's war production and at the same time sought to push into the Indian market. As British power in India was diminishing, American business was hopeful of gaining greater access. Grady would become the first US ambassador to India after independence, when he would use his presence in New Delhi to foster American investments in the subcontinent.[52] The *New York Times* drew attention to the appeal for funds of American Relief for India, aptly summarising the US business interest in attending to the Indian famine:

> For American business men India is one of the mightiest potential markets on the globe, about to enter upon an industrial era that will release the latent energies of one fifth of the human race. She will want machinery for farm and factory; she will want tens of thousands of products that America can provide. We of this country have a stake in India. For our well-being, if for no higher reason, we can no longer think of India as outside our world. We cannot deny her our interest or leave her ills and misfortunes to others to cure.[53]

Although not an entirely new phenomenon, the entanglement of capitalism and aid acquired an extra dimension in World War II, given India's nearing independence and the dominant role of the US in the international administration of food relief. There was no effort to conceal this link.

7.2 Indo-US Alliances on the Eve of Independence and After (1946)

Indian Food Needs and US Negligence

The end of World War II contributed to a temporary easing of the food shortage in India as the import of rice from India's eastern neighbour Burma resumed in late November 1945 following Britain's regaining control of the region. However, a series of meteorological events soon tipped the balance again. The failure of the monsoons in Madras and Mysore in late 1945, coupled with poor rains in the Deccan region and cyclones in the Godavari-Krishna Delta, resulted in poor wheat and paddy harvests at the beginning of the next year. Food scarcities first appeared in the South and the East and threatened to spill over into a nation-wide famine in the subsequent months. The Government of India, still under British control, enacted a food-rationing scheme to keep famine at bay.[54] Rationing being a makeshift measure at best, the British colonial government of India planned to offset the food scarcity at least partly by importing grain from the US. The British colonial administrations in Whitehall and Delhi officially requested US assistance via the Combined Food Board in January 1946: they received a negative response. The visits of both the British food minister and the food secretary in Washington likewise failed to generate support from the Truman administration. A British-Indian delegation of politicians and scientists was then put together in February 1946. The head of the delegation, A. Ramaswami Mudaliar, who would later be the first president of the UN Economic and Social Council, reminded the US public that India was not asking for charity but rather merely permission to purchase American grain. The Agent General for India Girja Shankar Bajpai reiterated Mudaliar's urgent request in March 1946, aligning with official British statements by stating "the outlook for India is famine."[55] The mission eventually returned to India believing an agreement had been reached with the Truman administration; however, the latter demonstrated no efforts to send grain.[56]

The formation of the Combined Food Board, which began to administer the flow of food commodities between North America and its European allies in 1942, had cemented US dominance in the global food trade. The latter further extended its control over food in the allied nations' colonial territories, including India, through the London Food Committee, a subordinate body of the Combined Food Board. This made the Combined Food Board "responsible for food for more than half of the world's population", as Lizzie Collingham points out.[57] Although, theoretically, the Board was run from 1943 onwards by the US, Britain, and Canada, the economic strength of the US, which offered most of the commodities administered by the Combined Food Board, meant its overwhelming influence on decisions taken on matters of international food relief.[58] Though the International Emergency Food Council eventually replaced the Combined Food Board in July 1946, this did little to weaken the food monopoly of the US. The new Council included a much larger membership, but given that the US still produced the majority of the world's food surplus, it remained in control of the international food trade.[59] New hopes for a more egalitarian international body to administer food surpluses were stirred in September 1946, when John Boyd Orr, the first Director-General of the United Nations Food and Agriculture Organisation (FAO), suggested forming a World Food Board to create a supra-national institution to balance worldwide scarcities. The newly sworn-in Indian interim government, represented in the FAO by an Indian delegation led by K.N. Katju, strongly supported Orr's plans.[60] Although Orr was also supported by other countries including France, Austria, Poland, and Greece, his plans met with strong opposition from the US, which rejected the potential interference in its trade.[61]

The dominance of the US in the world's food surplus administration is a crucial factor explaining why little international assistance was provided both to remedy the famine in Bengal and in response to the British colonial government's request to meet India's food needs in 1946. The US had no incentive for allowing its food surpluses to reach India until India became important in the US government's efforts to ward off the spread of communism in Asia.[62] During the famine in Bengal, the US government echoed the Viceroy of India, who declared that there was no need for US involvement in famine relief.[63] In 1946, rather than widen the scope of its commitments, the US administration focussed on the colossal task of rehabilitating war-devastated Europe and therefore channelled its food commodities to UNRRA and its voluntary relief agencies.[64] With the US joining the allied fight against the axis powers in 1942, North American agriculture was geared to meet the needs of a nation at war. As the war drew to a close, US agricultural production recorded a new high. However, when peace became imminent in 1945, the overflowing granaries that had previously sustained the war threatened to cause an economic slump. The US Department of Agriculture pre-emptively slashed harvests targets and began to feed farm animals large quantities of the wheat and maize that had originally been produced to nourish human bodies. As the US worked towards reducing its surpluses, allied troops moved across war-ravaged Europe to find that the need for food aid was greater than expected. In addition, droughts affected vast land areas in Africa and Asia, creating a further deficit in the world's food reserves. A year after the war ended, famine threatened one-third of the world's population. The world looked to the Truman administration for assistance but the Food Department's hasty efforts to reduce the nation's agricultural output had already yielded results. The US government needed to re-enact austerity policies in order to meet the world's food requirements but feared upsetting its voters. As the US government found itself ill prepared to meet the task of attending to the world's food crisis, it offered little grain to India.[65] Instead, the US administration adhered to the idea that hunger in India fell within the purview of the (former) British colonial administration in Whitehall and Delhi.

The India Famine Emergency Committee and its Mission to India

In 1946, news of the impending famine spread like bushfire across India and gave way to panic, hoarding, and protests. In February 1946, the *Hindustan Times* reported on "a spectacular hunger march" of 50,000 people in Lucknow demanding the end of rationing.[66] A month later, in a tragic re-enactment of scenes documented during the Bengal Famine, officials in Calcutta ordered the removal of people from the cities' streets, followed in April by the collection of the unidentified dead bodies of victims of starvation. Indian Congress leaders, recently released from prison, made the food crisis a part of their negotiations for self-rule. They opposed the Viceroy's efforts to jointly form a Food Council, because of the fact that Indian representation at the centre of the government had still not been granted.[67] The situation grew dire over the subsequent months because of the failure to secure imports. Officials responded by further cutting the ration amounts, leading to widespread under- and malnutrition.

In April 1946, the cofounder and key ideologue of the CPGB, Rajani Palme Dutt, arrived in India to report on the Cabinet Mission. The latter was to oversee the formation of an interim government for India, which would assist in the smooth withdrawal of the British from the subcontinent. The Cabinet Mission attracted a plethora of correspondents like Dutt, and increased the visibility of India to international audiences. As he conferred with Indian politicians and assisted the CPI in carving out a space for the party in post-colonial India, Dutt used his four months in India to collect first-hand information about famine in the country, which he cabled to the *Daily Worker* in London. Dutt predicted "the death of many millions from starvation" if immediate action was not undertaken. On behalf of the SIS, he travelled across Mysore, which was gravely affected by the famine, to inspect the government's rationing system and to make recommendations for its improvement.[68] In Bombay, Dutt also attended the premier of IPTA's *Dharti ke lal,* K.A. Abbas's famous appraisal of the famine in Bengal, which was released in time to exploit the increased presence of an international audience. IPTA invited delegates and accompanying journalists for a second screening of the movie in Shimla in May. While members of the mission remained absent, Dutt reported that North American and British journalists "were genuinely impressed."[69] Other artists who had previously chronicled the famine in Bengal again lent their pens to the communist *People's Age* to portray the current food crisis. In June 1946, in one of the many biting commentaries he produced during the year, Chittaprosad visualised an alleged statement of president Truman. Answering the question of whether India should be assisted with US grain, the president presumably answered: "the world is a bitch with too much a litter. We have to decide which puppies to drown first."[70] The Australian communist *Tribune* published the Chittaprosad's caricature a few months later, followed by photographs taken by Sunil Janah, who had travelled across the South of India to capture the famine conditions for the CPI.[71] Janah travelled with the famous American photographer Margaret Bourke-White, a correspondent for *Life* magazine. The two assisted each other: Bourke-White financed the trip (which meant, as Janah noted, that *Life* financed communist activity); conversely, Janah played intermediary and facilitated Bourke-White's work in India.[72]

In light of the deteriorating food situation, former supporters of food aid for Bengal in the US concentrated their efforts on the new task of securing grain for India. The India League contacted Pearl Buck who agreed to spearhead the efforts and set up the India Famine Emergency Committee (IFEC). It seemed evident at the time that diplomatic efforts had failed to convince the Truman administration to overhaul its India policy. Continuing JJ Singh's earlier critique of UNRRA, IFEC demanded that the Combined Food Board end its preferential treatment of Europe and base its grain allocations on a principle of equality. This would oblige the food regime to allow India to purchase grain. IFEC carried out a neatly designed campaign modelled after the government's own propaganda efforts over the following months, into early 1947.

The US government had started appealing to American humanitarian sentiments to persuade customers to voluntarily scale back their consumption. US president Truman created the International Famine Relief Committee in early 1946 to this end, which subsequently embarked on a propaganda campaign to convince Americans of their duty to assist the world. The former champion of American food relief (and later US president) Herbert Hoover was appointed as the government's special food envoy. Accompanied by an entourage of agricultural economists, Hoover was sent to inspect the world's food needs and to produce simple messages to mould public opinion favourably towards American food aid.[73]

One of the first acts of IFEC was to address letters of protest to the US administration after Herbert Hoover's global travel itinerary revealed that India was omitted from his world tour. Hoover had followed orders from President Truman, who strove to exclude India from American efforts to ward off the global food crisis. IFEC protested publicly against this neglect of India and claimed its first victory: Hoover was to visit India briefly. In April 1946, he clarified the US position towards India in a speech in Bombay, broadcast over the radio. He publicly acknowledged that India was in dire need of additional grain imports to avoid a nation-wide calamity, but asserted that wheat should be forthcoming from Australia instead of North America.[74] Indian press responses in the aftermath of Hoover's visit to India bore witness to the widespread disappointment at the lack of assistance from the US. Due to the growing nervousness of US officials about the spread of communism in Asia, the Commissioner of India, George Merrell, kept the Secretary of State particularly well informed about the publications of the communist sections in India, which vociferously criticised American responses to the Indian food crisis and received support from abroad. Merrell reported to the Secretary of State:

> During the past few months virtually all dailies read by the Mission—pro-Hindu, pro-Muslim and pro-British—have exhibited a remarkably Anti-American bias in their editorials and in their handling of news stores on the subject of food [...] the reader gains the impression that the average American is not only a glutton, but a racketeer, or at best a candidate for an institution for the feeble-minded.[75]

The brevity of Hoover's visit to India caused disappointment not only in India but also among proponents of American food aid in New York. This disappointment was all the greater because Hoover had visited India without his entourage of experts. Many believed that without this expertise, his report on India's food needs lacked authority and would therefore exert little influence on the Combined Food Board. This dissatisfaction sparked IFEC's decision to send its own delegation of experts to inspect India's food situation.

The American Famine Mission to India consisted of nine prominent American men and women, most of whom were longstanding members of the India League. Among them were Gandhi biographer Louis Fischer and the journalist and editor Sidney Hertzberg. Other members included Lilian Smith, who was a well-known writer, feminist and civil rights activist, Joseph Willen, vice-president of the Federation of Jewish Philanthropies in New York, and Henry Smith Leiper, Ecumenical Secretary of the Federal Council of the Churches of Christ. Likewise travelling to India were Mary Jean Kempner, who wrote for *Vogue* magazine, and Dorothy Dunbar Bromley, journalist and feminist who joined the mission as a correspondent of the *Herald Tribune*. Lastly, the scientific authority of the mission was the economist and author of *Food for the World* (1945) Theodore W. Schultz.[76] Schultz was also chairman of the Department of Economics at the University of Chicago and advisor to both the US government and the newly founded Food and Agricultural Organisation (FAO) at this time. The Committee counted on Schultz's authority as a renowned American economist and agricultural expert, but more importantly, on his ability to influence the Secretary of Agriculture Clinton P. Anderson to act on behalf of India. The publicity campaign focussed on the mission's core group of nine, but they were accompanied by additional journalists and the group grew while travelling in India. Bradley Smith, a professional photographer who would later attain fame for his pictures of jazz icons, was paid to take pictures and make films of the mission.[77]

The American Famine Mission to India arrived in Karachi at the end of June 1946, on the day the Cabinet Mission announced their failure to negotiate the formation of an interim government. Whereas the British Cabinet Mission returned to London, the Americans began to travel across India in a chartered British army plane and visited, among other places, New Delhi, Bhopal, Poona, Calcutta, and Madras. They arranged meetings with government officials as well as inspected regions most affected by the recent scarcities. The mission succeeded in attracting some attention, not least because of its far-reaching networks. The Quaker Horace Alexander and the journalist Phillips Talbot joined the mission temporarily and assisted in spreading the word; the mission also crossed paths with Margaret Bourke-White.[78] Furthermore, K.A. Abbas contacted the India Famine Emergency Committee to arrange for the screening of *Dharti ke lal* prior to the leaving of the Americans.[79] In addition, the editor of the *Statesman*, who had previously played an eminent role in bringing the Bengal Famine to the attention of the world, sent copies of his famine reports from 1943 and 1944 to the mission members and met them in India.[80] Most importantly, however, India's Congress politicians, who were soon to form India's independent government, were interested in using this opportunity to reach American audiences. Representatives of the INC including Jawaharlal Nehru, Mohandas K. Gandhi, and Sarojini Naidu welcomed the mission in Delhi.[81]

The mission members held press conferences in every town they visited, explaining its purpose by emphasising the dire need for first-hand information gathered by "prominent Americans […] to report to the American people" so that the scales may be tipped in the US and wide support for India unleashed.[82] Schultz summarised the problem: "there is [...] some suspicion that these big figures do not mean anything unless they can be confirmed by eye-witnesses. We have come here to study the statistics and see for ourselves what is happening."[83] After months of press reports and diplomatic efforts in India, the US, and Britain, it is puzzling that the need to send a group of Americans for confirmation was felt. Indeed, expert opinion on India's food needs was readily available. Rather than proving a lack of information, the American Famine Mission to India illustrates that the battle for food was primarily fought in the American press. And this is what IFEC did: before, during, and after the mission's visit to India, those members of IFEC who remained in New York made full use of any means of advertising to broadcast the mission's objectives. A total of three hundred items were brought out in US newspapers and in the organisational publications of the AFSC, Church World Service, the labour movement, and NAACP. In addition, IFEC released a full report of the mission's findings, 15,000 copies of which were circulated widely. IFEC also made extensive use of radio broadcasts and secured fifteen minutes on air for each member of the mission in different local and national radio shows.[84] Upon the mission's return, IFEC urged President Truman to meet the former's members. Truman delegated Hoover to meet the returned mission, knowing that the latter would be in California at the time. The mission delegates finally conveyed their report to the Secretary of Agriculture. The report reiterated India's food deficit and the need to raise the allotment of food for India considerably so as to avoid a deterioration of the food crisis.[85] Although the US government seemed unimpressed by IFEC, food assistance for India slowly materialised.

In June 1946, the Department of Agriculture announced the sending of 500,000 tons of grain, which fell short of IFEC's recommendation by 250,000 tons. Less than half the promised grain was shipped to India in late September. After that, a disruption of transportation succeeded the initial problem of allocation. The shortage of boxcars together with maritime strikes brought American food exports to India to a standstill. In the following months, until the end of 1946, the Department of Agriculture failed to keep its promise. Arguing that more urgent demands arose in the meantime, it used the grain allocated for India to assist other countries. As a consequence, India had received only 100,000 tons by December 1946.[86]

From One Famine to the Next

In May 1947 Congress politician Rajendra Prasad, Food Minister of India's interim government, and later Indian president estimated India's grain deficit to constitute three to four million tons. Knowing that the much-awaited wheat harvest could only bring partial relief, Prasad concluded that "the position [was] very difficult if not desperate from another point of view."[87] The partition crisis that accompanied

India's political independence in 1947 had exacerbated India's food needs as it disrupted trade, destroyed harvests, and uprooted and displaced millions who now needed to be attended to. While the government of independent India, under the lead of Jawaharlal Nehru now faced the question of sustenance, the planning of the nation's economic and political future had begun earlier.[88] The National Planning Committee (NPC) of the INC was established in 1938. Headed by Jawaharlal Nehru and Subhas Chandra Bose, the NPC's twenty-nine expert subcommittees began to sketch out post-independence policies.[89] The achievement of food security soon assumed a prominent role in the discussions of the NPC and the latter proved unequivocal in the acceptance of a strong role of the state in India's "national food planning."[90] While World War II and the imprisonment of the Congress leadership temporarily disrupted the work of the NPC, it did not prevent elites from devising plans for post-colonial India. The experience of the Bengal famine provided a strong impetus to national planning and left an imprint on the schemes for India's economic development that were published in 1944.[91] India's business elite, which had secured a prominent position in the NPC, released the Bombay Plan in 1944. It acknowledged the state's need for economic planning, however, set limits to its encroachment on the private sector.[92] Indian communists who were themselves occupied with detailing India's path to economic prosperity resented the plan.[93] "The People's Plan", drafted under the auspices of M.N. Roy and the Federation of Labour, prioritised investments in India's agricultural sector.[94] Meanwhile, a disciple of Gandhi, S.N. Agarwal, lamented the lack of the plan's "cultural and sociological foundations, "renounced them for "merely copying Western plans" and called for a truly Indian scheme for economic development.[95] Aiming to reconcile economic development with the pursuit of cultural and spiritual progress, "the Gandhian Plan" reiterated the pillars of Gandhian thought that advocated the renunciation of luxuries and placed its focus on India's villages as the nucleus of the Indian nation. While these three plans do not capture the polyphony of ideas on India's postcolonial development, they nevertheless can be used to illustrate that the urgency of solving the food question had been widely shared. Hence, India's industrialists set out to reach a national diet that was balanced in its nutritional composition and rich in its caloric value.[96] The authors of the People's Plan stated that "the provision of an adequate nutritive diet to the entire people, undernourished for centuries together, must therefore be the first concern of the planning authority."[97] In accord with Gandhi's longstanding interest in nutritional science, the Gandhian Plan advocated the improvement of Indian diets through national planning and took the recommendations of India's leading nutritional scientist, Wallace Aykroyd, as a benchmark.[98]

When in 1947 the Indian government addressed the national food needs, the measures implemented continued some of the schemes that had been enacted under colonial rule during World War II. In 1943, the colonial administration had instigated a Grow More Food programme that had envisioned the expansion of food cultivation. Although it had achieved little, the government expanded the programme in September 1946 and a year later, it became part of India's post-independence food planning.[99] In addition, the last years of the war had seen the introduction of the world's largest food procurement and rationing systems, which however, remained focussed on urban populations and failed to secure a sufficient diet for the bulk of the Indian population. Except for a brief period of decontrol from December 1947 to September 1948, the government of India continued the food control measures enacted during the war.[100] Despite the focus on rising agricultural productivity and managing food procurement, India's food economy remained dependent on imports that threatened to exhaust monetary reserves. Realising that India's hard-won political independence was likely to become the victim of new economic dependencies, Nehru pushed the country to speed up the nation's food production and announced the national target of attaining self-sufficiency in food in the season of 1951–52. His hopes were ultimately shattered when in late 1950 signs of famine in Bihar and Madras rendered the national target unachievable.

In mid-1950, while the Indian government drafted the first Five Year Plan, unfavourable weather conditions, leading to floods in the East and droughts in the South of India, threatened the goal of attaining self-sufficiency in food by 1951. Widespread crop failure affected people in Hyderabad, Kashmir and Uttar Pradesh. However, the poor harvest yields hit Bihar and Madras the hardest.[101] What would become the worst food scarcity since India had attained independence caused political turmoil in the country as Nehru and his cabinet strove to belittle the crisis. The INC was well aware that famine could cause a debacle in the soon approaching first general elections.[102]

"What kind of dictionary does Mr. Nehru have in his library?" asked David Cohen, editor of *Unity*, the magazine of the IPTA, which during and in the wake of the Bengal Famine had brought the suffering of the people to the stage.[103] Cohen went on:

> Mr. Nehru has a very fine dictionary, but he has not got a dictionary with the word 'famine' in it, but instead, words like 'distress' and 'acute scarcity', because obviously if Mr. Nehru's dictionary did contain the word 'famine' he would certainly—since he uses a great many words every day—have that word in one of his great many speeches.[104]

Unity's polemic against Nehru reveals anecdotal evidence for the continuity in the responses of political activists and groups who wished to hold the government accountable as well as to exploit famine to gain political ground. Although the tables had turned and India gained independence from Britain, the government was careful of publicly admitting the existence of famine. Congress politicians were well aware that the famine could cause it to lose in the next general elections and further diminish its political influence in Madras, where it faced a strong communist opposition.[105] On the eve of freedom, India's political elite had reiterated the claim that self-government would free India from famine, which left it shouldering a heavy political burden. Many of those who had criticised the colonial state's failure to provide sustenance to India's famine-affected were now part of the government, in charge of implementing the changes they had demanded for decades. It weighed heavy on the Congress government that people in Bihar and Madras were as hungry as they had been under colonial rule.

While famine conditions assumed an alarming scope in Madras and in Bihar, Nehru was opposed to the idea of requesting foreign assistance. Sceptical of the world's super powers, Nehru embarked on the course of non-alignment, trying to keep India out of the Cold War. At the same time, he aimed to foster ties with India's Asian neighbours, hoping to take a lead in the post-war pan-Asian movement that could establish Asia as a counterweight to the dominance of the superpowers.[106] The Minister of Food and Agriculture K.M. Munshi, and the Minister of Finance C.D. Deshmukh differed. Not only did they think that the time was ripe for negotiating with the US about assistance in food matters, the ministers also aimed to foster closer ties to the US administration. Nehru gave in grudgingly to the heightened demand to turn to the US for food assistance. In late 1950, the Indian request of a loan to purchase food grains from the US was officially submitted.[107] Unlike earlier responses of the US government to Indian requests for food aid, the ascendance of the Cold War had led the Truman administration to envision the use of US food aid to draw Independent India closer to the US.[108] In addition, food loans had become a means of the US government to disperse its agricultural surpluses in the after-war period. Hence, in contrast to the prior reluctance of the US president to support India in meeting its food needs, a group of US politicians, including Truman himself, now embraced the request. In February, however, as the draft of the bill that sought to authorise the food loan for India was submitted to Congress, a viable section of its members delayed the passing of the act. Nehru's public endorsement of socialism and his politics of non-alignment had antagonised a fraction of the Congress, which wanted assurance that India would not side with either China or the Soviet Union. It was only in June that the US Congress would eventually pass the India Emergency Food Aid Act of 1951 (also known as Public Law 48), which granted India a loan to purchase two million tons of US grain.

While the US Congress still debated whether it should send relief, the pressure on the Indian government grew. In early 1951, the Reserve Bank of India surveyed the Grow More Food campaign in Bombay and announced its failure to enlist the support of the peasantry who were hardly touched by the glossy campaign posters hanging in Bombay's urban centres. In response to the criticism, Munshi, who had been ridiculed for his demand that starving peasants in Bihar should turn to roots and leaves to ward off the threat of food imports a year earlier, now felt induced to defend the Grow More Food campaign.[109] Yet, while Munshi answered the questions of Members of Parliament in April 1951, he admitted that peasants were unlikely to grasp the goals of the campaign: "Ordinary people do not understand it. They simply say 'There you are. 12 oz. have been reduced to 9 oz. Here is scarcity, here is famine. What do you mean by saying that you are growing more food?'"[110] In light of mounting criticism, the urgency of raising food imports to meet the nation's food needs was widely acknowledged. For many politicians in India, however, turning to the United States for food was not the first choice. For months, communists had been advertising Chinese and Soviet offers of food and now accused the Food Secretary, Vishnu Sabhai and the Minister of Food and Agriculture, K.M. Munshi of safeguarding US interests in India by halting negotiations with China and the Soviet Union over the import of rice and wheat.[111]

Since the attainment of independence, the CPI had struggled to position itself in the political landscape of post-colonial India. The INC had expelled communists from its ranks in 1945, despite efforts of P.C. Joshi to move the party closer to the Congress. In 1948, when the Soviet leadership began to renounce the political legitimacy of the INC because of its bourgeois character, the CPI went along with it. The CPI replaced Joshi as its long-time General Secretary with the hardliner B.T. Ranadive. Ranadive took a central role in the party's involvement in peasant revolt, resulting in the ban of the party in Hyderabad, West Bengal and Madras.[112] In light of the partial ban of the CPI, the party again aimed to mobilise political support through its cultural front.[113] However, it had lost touch with most of its artists, who had disassociated themselves from the CPI. Independence removed the unifying moment of anticolonialism. Many of the artists who had previously supported communist efforts now embraced the Congress government.[114] Artists who had earlier documented hunger in British India, however, did not stop making demands for food security and advocating for social change after independence. In 1951, the IPTA was disenchanted with the Congress government and advocated a strong role for Indian artists in securing the well-being of Indians at large: "to-day the writers and artists of India will no longer be silent, but will enter the battle for life with all the talent and genius and resources at their command, so that the people may be fed, so that the people may be clothed, so that the people may have prosperity."[115] Chittaprosad who had chronicled the suffering of Bengal's rural population for the CPI in the 1940s, continued to produce artistic criticism of the political failure to attend to the hunger of the people after he had left the party in 1948. His art, however, had changed, from focussing on agony and despair to picturing the possibility of attaining agricultural plenitude.[116] For India's left, and particularly for Indian communists, this vision of plenitude entailed a move towards China and the Soviet

Union.[117] When in 1951, India needed to raise its food imports to meet the scarcity in parts of the country, communists emphasised the ability of China and the Soviet Union to render assistance.

In mid-April, the Soviet Union announced to provide China with shipping space, and allegedly passed its own grain reserves through Manchuria to allow China to send one million tons of grains to India as a Chinese-Soviet gesture of goodwill.[118] Communists in and beyond India celebrated the Chinese-Soviet offer and claimed moral victory over the US Congress, which was still torn over the intricacies of a wheat loan for India. Those in the US who were in favour of sending food swiftly to India intended the grain to be a gift. However, in the process of passing the respective bill in the Congress, the gift was converted into a loan, with India being asked to repay part of it through strategic materials.[119] It was against the background of the debate that the Indian communist weekly *Atom* emphasised: "While the American Congress continues to debate and sit in judgment over India's foreign policy, China has already shipped large quantities of food-grains and she is determined to implement her offer of one million tons."[120]

Although China struggled to feed its own population, making it difficult to meet the offer of one million tons of grains, food from China and the Soviet Union soon arrived in India. China sent 10,000 tons of milo on its on vessel, followed by 6,700 tons of Ukrainian wheat on board the Soviet *Krasnodor* in late May.[121] The Chinese and Soviet ships were part of 130 vessels that arrived in May and June and to bring food to India, either as an advanced loan, regular import, or gift. The little over 600,000 tons that arrived at Indian ports in this period was nominal compared to the country's food gap. It was, however, a powerful symbolic gesture. Apart from an expression of Soviet and Chinese communist commitment, the sending of food to India exhibited a moment of pan-Asian solidarity that was also upheld by the Indian diaspora.

In early May, Munshi visited Rangoon and during his stay, posed with the Gujarati community in Burma who had collected grain for India. Its representatives poured grain into the hand of the Indian Minister, who was seen standing behind a bag of grain, which carried the imprint "Burma" in bold letters.[122] Similar traces of pan-Asian solidarity can be found with regard to China. In May 1951, the China Friendship Association in Bombay was inaugurated and at its first meeting, Chinese diplomats assured that peasants at home wanted to give rice "to the land that gave them Buddha."[123]

Rallying Support for India's National Development

Many Indians who had made North America their home, returned to India after 1947, seeking to take part in the momentous task of nation-building and decolonisation.[124] Not all of them, however, chose this path. Those remaining in the United

States often used their position and networks to contribute to India's quest for development from afar. Indians who had formerly demanded political independence now facilitated cooperation between India and the United States in areas they conceived as important to India's national development. The India League created a Service Bureau in the fall of 1951 that aimed to become a nodal point between American institutions seeking to work, or already working in India, and the Indian government.[125] To facilitate the work of American volunteer organisations in India, the Service Bureau answered queries of American voluntary organisations and supplied them with relevant information. For that purpose, *India Today* began a section called Service Bureau Notes that informed its readers about current events of interest in the United States and India as well as about the visits of Indian officials and non-governmental visitors to the US. After independence even the chairman of the National Committee for India's Freedom, Syed Hossain, previously hostile to American participation in Indian nationalist organisations in the US, now enthused about the potential of Indo-US cooperation:

> There is much that India can learn from America in modern techniques of industry, commerce, technology and agriculture and we can reasonably hope that America will respond in genuine goodwill and co-operate, to the extent that may be called for, in the modernizing and industrialisation of Indian economy.[126]

Another example of the growing American-based Indian support for national development was the Watumull Foundation. The music teacher Ellen Jensen and her husband Gobindram "Goma" Watumull, owner of two flourishing department stores in Honolulu, had founded the Watumull Foundation in 1942. The couple had moved from Hawai'i to the US West Coast in the 1940s, offering its support to a group of Indian activists who were currently rallying Americans behind the demand for Indian self-rule. The commercial success of Gobindram Watumull enabled the spouses to invest money in political, philanthropic and educational work that gained them wide recognition. They financed the creation of the National Committee for India's Freedom in Washington, which intended to influence policy-makers on decisions of importance to India.[127] The Watumull Foundation remained a meeting point of Indian intellectuals on the US West Coast after independence. Through a pioneering scholarship programme that enabled Indian researchers to visit the US for their postgraduate and postdoctoral studies, the Watumull Foundation now sponsored research on subjects of alleged importance to India's national development. A number of prominent Indian and US citizens sat on its board of directors that assessed applications for the programme, among them the historian Merle Curti, the senior Indian nationalist firebrand Taraknath Das, former president of the SIS, H.N. Kunzru, and the globally acclaimed statistician and Jawaharlal Nehru's chief planner P.C. Mahalanobis. Many of the men and women sponsored by the Watumull Foundation later assumed high posts in the Indian government, the WHO and FAO. A well-known beneficiary of the Watumull sponsorship programme was the Indian demographer and the first Indian minister of Health and Family Planning (1967–70) Sripati Chandrasekhar.[128] Chandrasekhar, who had earned his PhD in sociology from Columbia University with a thesis on India's population problem in 1944, had also travelled the US for a lecture tour on Indian nationalism in the name of Pearl S. Buck's East and West Association in the 1940s.[129] Chandrasekhar would deem the use of inducements and pressure on Indian women and men a legitimate means to increase sterilisation quotas during his tenure as Indian minister. He was, however, also amongst a group of scholars that claimed that the best way to limit fertility in the long-run was to improve the nutritional health (and hence living standards) of the poor.[130] Advocates of population control frequently supported nutritional reform and feeding programmes alongside other, often coercive, population policies. This convergence also showed in the activities of the Watumull Foundation in the 1950s that began to promote famine relief for India, at the same time that it became involved in family planning in the country.[131]

In January 1951, the Watumull Foundation teamed up with MFM that promoted the food supplement MPF to counter global hunger, to form the United Emergency Committee on Famine Relief for India (United Emergency Committee). The committee combined a series of small-scale institutions, mostly from California and with a longstanding history of mediating contacts between India and the United States.[132] Among them were American Wives for India and the India Students Association as well as a series of Christian organisations, such as the National Conference of Methodist Youth. Ellen Jensen Watumull became the spokesperson of the United Emergency Committee, after she had met Dhanvanthi Rama Rau in India. Rama Rau had started to devote herself to the cause of family planning in the 1940s. At the time she concluded that limiting the poor's fertility was central to the achievement of "social and economic development", and in doing so joined a chorus of proponents of birth control who lamented the differential fertility of India's poor.[133] Eager to give birth control a prominent place in Indian national planning, Rama Rau set up the All India Family Planning Association in Bombay in 1949 and convened the first All India Conference on Family Planning in 1951. A year later, international birth control activists assembled in Bombay for the Third International Conference of Planned Parenthood—a milestone in the consolidation of family planning in India. Still under the impression of the conference, the Indian Planning Commission acknowledged the need for a population policy towards the end of 1952 and the Indian government committed itself to family planning in its First Five Year Plan (1951–56). In 1950, however, Rama Rau struggled to rally supporters behind her crusade, facing fierce opposition from the Minister of Health Rajkumari Amrit Kaur, who rejected the idea of family planning through modern contraceptives. Kaur, who embraced Gandhian teachings of sexual abstinence and rejected the common assumption that population growth constituted a major impediment to development in India, was finding no common ground with Rama Rau.[134] The latter looked elsewhere for support. At the annual conference of the Institute of Pacific Relations in Lucknow in 1950, Rama Rau inquired with Jensen Watumull about the availability of funds in the US to distribute cheap contraceptives in India. Jensen Watumull who had recently joined the board of directors of MFM (a position she would hold until the late 1970s) used the opportunity that her new networks provided to discuss Rama Rau's request with Margaret Sanger.[135] The meeting that took place in January 1951 was the starting point for the Watumulls' long-time investment in plans to advance family planning in Asia.[136] They generously sponsored the Third International Conference on Planned Parenthood in Bombay in 1952, as well as the visit of Indian physicians to the Margaret Sanger Research Institute in New York around the same time.[137]

The timing of the Watumulls' growing interest in family planning on the one hand and in food relief on the other was not coincidental, and the activities of the foundation were indeed interrelated. The Watumulls began to invest in population control in India as the debate on India's food situation gained momentum. The famine in Bihar and in Madras and the ensuing debates on US food aid for India created publicity for the failure of the Indian state to provide substance to its population. In addition, the 1951 Indian census revealed that the Indian population was growing despite war, partition and famine. This alarmed those who had identified India's growing population as a principal cause of mal-and undernutrition in the country and who now felt that the implementation of population policies was urgently needed. That such debates influenced the activities of the Watumull Foundation was demonstrated, amongst other things, in an essay contest in 1951 that invited Indian scholars to submit their studies on "Population Control in Relation to Food in India" and awarded prize money to the best papers.[138] The winner of the contest was Amolak Ram Mehta (1895–1986), a London-trained physician and former director of the Public Health Department in Punjab. He seemingly impressed the jurors with his demand to introduce family planning to India's poor and his warning that in case of failure "future generations may suffer from a predominantly inferior stock."[139] Apart from scholars' rising concern with the quantity of the Indian population, a concern with "the quality" of the latter surfaced in debates on population and food in the post-war era.[140] This was in itself not a new development. Interest in eugenics had been growing both in India and outside the country over the last decades, inciting Indian nutritionists to conclude that maternal malnutrition was likely to produce physically and mentally weakened offspring.[141] Studies on malnutrition in India fuelled fears of racial decay that allegedly undermined India's ability to stand as a nation, and, hence, jeopardised political independence.

7.3 Conclusion

Indian nationalism in the United States did not wane in the 1940s but transformed under the growing influence of a wide network of American supporters. Mitigating famine and food insecurity in India remained central topics of organisations and individuals in the United States who had long engaged in efforts to deepen the contacts between India and America, embracing Indian demands for political self-determination. Among them were (former) North American missionaries turned Christian internationalists, the American civil rights movement and American labour, which rallied behind Indian political demands for different reasons. Tapping into this vital support, South Asians in the United States sought to shape official policies towards India and to resolve the problem of food scarcity.

In early 1946, India's food situation looked bleak. Only a few years after famine had claimed a million lives in Bengal, scarcities threatened to throw India into an even greater crisis. At this time IFEC assembled in the US in protest against the decision of the US-dominated Combined Food Board to deny India a higher quota of food imports. Aiming to "secure a fair share" for India in the post-war international scramble for food assistance, IFEC sent a delegation to India that was accompanied by a systematic media campaign, thereby creating awareness for India's quest for food security in the United States. The American Famine Mission to India, which toured the subcontinent in 1946 to mobilise American opinion in favour of US food aid, followed in the wake of decades of efforts by civil society organisations to secure greater US involvement in India.

The end of colonial rule in India in 1947 did not solve the country's food needs overnight. The Indian government undertook multiple efforts to attain food security; the burden it had to shoulder was immense. Freedom from famine was not attained so soon: in late 1950, famines occurred in Madras and Bihar. Despite the fact that India was now governed by its own political elite, the occurrence of starvation continued to be a sensitive political subject after independence and threatened to weaken the legitimacy of the state. The request of the Nehru government for US food assistance in early 1951 raised fears that India's hard-won political independence from British colonial rule was lost to a new economic dependency on the United States. The attempts of Indian politicians at securing food from Russia, China and other regional powers illustrate the continued efforts to tap into pan-Asian and communist solidarities for famine relief.

Seemingly unaffected by the political reconstitution of South Asia, the India Lobby in the United States reassembled again in late 1950 to mobilise food aid for the famine-afflicted people of Bihar and Madras. The United Emergency Committee on Famine Relief for India was a striking amalgam of (former) Indian nationalists, birth control activists and North American do-gooders. It was dominated by two institutions with long-standing ties to the Indian subcontinent: The Watumull Foundation and MFM. The former ran a scholarship exchange programme between India and the United States, hoping to stimulate research on topics of relevance for India's national development. The latter had been formed in 1946 to promote the food supplement MPF among US relief agencies. Together, they aligned to send food aid, including MPF and CARE packages, to India. Prominent members of the birth control movement rendered indispensable support to the United Emergency Committee. US-sponsored famine relief in post-war India was carried out against the backdrop of the perceived threat that India's poor and malnourished posed to post-war utopias of global prosperity. Such fears drove, and reinforced, the heightened involvement of US and Indian birth control activists in mobilising famine relief and nutritional supplements for India in the 1950s.

Conclusion

Instead of repeating the findings of this book which are set out in the conclusions of each individual chapter, I end with some general observations. It is not self-evident that this study foregrounds cooperation and connections. Historians and anthropologists have demonstrated that disconnections dominate the experience of mass starvation. Famines destroy lives and throw communities and societies into disarray as they disrupt and terminate social, political and economic relations.[1] But famines in South Asia also prompted the responses of a vast group of individuals and organisations, who intervened in the lives of the famine-afflicted to their benefit and detriment. In so doing, they enforced connections, networks and institution building. As shown in this book, the quest of ending famine in India attracted a growing cast of domestic and international actors in the late nineteenth and early twentieth centuries when relieving famines and curbing malnutrition increasingly aligned with the aims and agendas of historical actors and movements. Mitigating famine allowed Indian nationalists to undermine colonial authority and to discipline the poor, enabled American missionaries to expand their presence in South Asia, and fuelled innovation in food and nutritional science. Despite the fact that, in the decades between the "late Victorian Holocausts" and the Bengal Famine of 1942–44, famine decreased in scope and scale, activities undertaken to mitigate starvation and promote visions of an India free from hunger grew in multiple ways. In this interim period, famines continued to influence the social and political history of the subcontinent through the various debates and activities they inspired. To encompass the activities of actors who promised to set an end to famine in India, the spatial framework that commonly undergirds historical research on famines in South Asia needs to expand. Famine occasioned deepening contact between India and the United States which I investigate against the background of South Asian migration to North America, American missionary activity in South Asia, and the dominance of the United States in the global food aid regime. South Asia became a site of knowledge production on famine relief and rural reform in the twentieth century and a contact zone of British, Indian and American "famine experts"—a finding that anchors South Asia in the global history of food aid, humanitarianism and development.

The topicality of some of the debates covered in this book is glaring. It remains a challenge to bridge the gap between this historical study, which ends in the 1950s, and an analysis of the present without oversimplifying. And yet some continuities seem too significant not to be mentioned, even if only in the closing paragraphs of the book. Contemporary disagreement about the extent and nature of hunger in India is reminiscent of the political manoeuvrings that accompanied the fight against famine in the nineteenth and twentieth centuries, and that tapped into science to reveal as well as to minimize the prevalence of hunger. Such continued disagreement showed vividly in a controversy sparked by the publication of the annual Global Hunger Index (GHI) in October 2021. In the report prepared by Welthungerhilfe and Concern Worldwide, India dropped seven points, landing at 101 out of 116 countries.[2] By confirming that India's food situation was "serious", the GHI questioned the success of the Indian government's recent national campaign to improve the nutritional health of mothers and children.[3] The BJP government launched POSHAN Abhiyaan in 2018, an ambitious campaign to combat malnutrition as part of prime minister Narendra Modi's election promise of a "New India" by 2022.[4] The Indian Ministry of Women and Child Development that oversees the implementation of POSHAN Abhiyaan renounced the GHI, arguing that the report lacked accuracy.[5] An unnamed government official also recommended the GHI to change its name, as the term hunger was supposedly too drastic to describe the condition of the Indian population.[6] Critics of the government's social welfare schemes admitted that the statistical basis of some indicators of the GHI could be improved, but found the ministry's blanket rejection of the report unfortunate and unjustified.[7] Oxfam India, reaffirming the report, also pointed out the delay between the collection of the data and the publication of the GHI: the report did not reflect the impact of the Covid-19 pandemic.[8] At the time the GHI 2021 was published, the pandemic had intensified India's hunger drastically. The scale of hunger and death in India that followed the lockdown even invited critics of the BJP government to compare it to the Bengal Famine of the 1940s.[9]

As hunger retains its political significance, debates on the appropriateness of the scales used to measure hunger and the terms for describing the nutritional health of Indians continue unabated. Meanwhile, the prevalence of malnutrition and undernourishment in India undermines the political success to offset famine.[10] Despite the disagreement between activists and politicians on India's ranking in the GHI, there is sufficient evidence to support the claim that mal- and undernutrition had already been a major killer in India before Covid.[11] As the lockdown confined labourers to the cities, leaving millions of Indians without income, starvation reached a new tragic scope in the country.[12] The political response was insufficient: Although public food stocks hit a new high at the beginning of the pandemic, the government did not release stocks in sufficient quantities to meet the needs of the population at risk. The Public Distribution System (PDS), the main instrument to offset food scarcities in India, remains insufficient and exclusionary, and the recent rollback of social security is likely to cause further hardship.[13] The many forms of hunger that continue to afflict the most vulnerable parts of the population in the aftermath of *Ending Famines in India* may be less sensational. They are no less tragic, nor are they inevitable.

Notes

Introduction

1 Recent studies of the entangled histories of India and the United States include N. Slate, *Lord Cornwallis is Dead: The Struggle for Democracy in the United States and India* (Cambridge: Harvard University Press, 2019); R. Kaur and A. Arora (eds), *India in the American Imaginary, 1780s-1880s* (New York: Palgrave Macmillan, 2017); H. Fischer-Tiné and N. Slate (eds), *The United States and South Asia from the Age of Empire to Decolonization: A History of Entanglements* (Leiden: Leiden University Press, 2022).

2 On the contrary, histories of food have evolved beyond such frames. See K. Ray and T. Srinivas (eds), *Curried Cultures: Globalization, Food, and South Asia* (Berkeley: University of California Press, 2012); R. Berger, "Alimentary affairs: historicizing food in Modern India," *History Compass* 16:2 (Feb 2018), e12438.

3 A non-exhaustive list of literature on nineteenth-century famines in India includes M. Davis, *Late Victorian Holocausts: El Niño Famines and the Making of the Third World* (London: Verso, 2007); S. Sharma, *Famine, Philanthropy, and the Colonial State: North India in the Early Nineteenth Century* (New Delhi: Oxford University Press, 2001); D. Hall-Matthews, *Peasants, Famine and the State in Colonial Western India* (Basingstoke: Palgrave Macmillan, 2005); D. Hall-Matthews, "Inaccurate conceptions: disputed measures of nutritional needs and famine deaths in Colonial India," *Modern Asian Studies* 42:6 (Nov 2008), 1189–1212; D. Hall-Matthews, "Historical roots of famine relief paradigms: ideas on dependency and free trade in India in the 1870s," *Disasters* 20:3 (Sep 1996), 216–30; D. Arnold, "Social crisis and epidemic disease in the famines of nineteenth-century India," *Social History of Medicine* 6:3 (1993), 385–404; A. Major, "British humanitarian political economy and famine in India, 1838–1842," *The Journal of British Studies* 59:2 (April 2020), 221–44; A similarly incomplete list of literature on the Bengal famine includes: J. Mukherjee, *Hungry Bengal: War, Famine and the End of Empire* (Oxford: Oxford University Press, 2015); M. Mukerjee, *Churchill's Secret War: The British Empire and the Ravaging of India during World War II* (New York: Basic Books, 2010); Srimanjari, *Through War and Famine. Bengal, 1939-45* (New Delhi: Orient BlackSwan, 2009); P.R. Greenough, *Prosperity and Misery in Modern Bengal: The Famine of 1943-1944* (New York: Oxford University Press, 1982); A. Sen, "Starvation and exchange entitlements: a general approach and its application to the Great Bengal Famine," *The Cambridge Journal of Economics* 1:1 (March 1977), 33–59; C. Ó Gráda, *Eating People Is Wrong, and Other Essays on Famine, Its Past, and Its Future* (Princeton: Princeton University Press, 2015), 38–91.

4 The described divide has only seldom been breached by studies on famine in twentieth-century colonial India, see for instance D. Arnold, "Looting, grain riots and government policy in South India 1918," *Past and Present* 84:1 (1979), 111–45; For a rare long-term perspective of famine and food politics see S.S Amrith, "Food and welfare in India, c. 1900–1950," *Comparative Studies in Society and History* 50:4 (Oct 2008), 1010–35.

5 In addition to the works already cited, other literature dealing with the causes of famine in India includes: D. Arnold, *Famine: Social Crisis and Historical Change* (Oxford: Basil Blackwell, 1988); B.M. Bhatia, *Famines in India. A Study of the Aspects of the Economic History of India* (1860–1965) (New Delhi: Asia Publishing House, 1963); V. Damodaran, "Famine in a forest tract: ecological change and the causes of the 1897 famine in Chotanagpur, Northern India," *Environment and History* 1:2 (1995),

129–58; V. Damodaran, "The East India Company, Famine and Ecological Conditions in Eighteenth-Century Bengal" in V. Damodaran (ed.) *The East India Company and the Natural World* (Houndsmill: Palgrave Macmillan, 2015), 80–101; T. Roy, *Natural Disasters and Indian History* (Oxford: Oxford University Press, 2012); M.B. McAlpin, *Subject to Famine: Food Crisis and Economic Change in Western India, 1860–1920* (Princeton: Princeton University Press, 1983).

6 J. Edkins, *Whose Hunger?: Concepts of Famine, Practices of Aid* (Minneapolis: University of Minnesota Press, 2000), 67–102; For a brief, and necessarily selective, overview of the historical genesis of definitions of famine, see C. Ó Gráda, *Famine: A Short History* (Princeton: Princeton University Press, 2009), 1–39.

7 P. Howe and S. Devereux, "Famine intensity and magnitude scales: a proposal for an instrumental definition of famine," *Disasters* 28:4 (2004), 356–57.

8 Cf. Famine Relief Code. Bombay Presidency (Bombay: Government of India, 1912), 8; Bihar and Orissa Famine Code (1930) (Calcutta: Jharna Supply Syndicate, 1950), 11–12.

9 Cf. Report of the Indian Famine Commission (Calcutta: Office of the Superintendent of Government Printing, 1901), 19.

10 Howe and Devereux, "Famine Intensity and Magnitude Scales," 357.

11 For some controversies related to famine deaths: "The famine in India. Through the stricken districts. Concerning death from starvation," *West Australian* (6 June 1900), 3; "Famine in Puri and elsewhere," *Modern Review* 28 (July 1920), 110; see also Davis, *Late Victorian Holocaust*, 21–22.

12 Cf. V.K. Krishna Menon et al., *The Condition of India. Being the Report of the Delegation sent to India by the India League, in 1932* (London: Essential News, 1933), 343.

13 Amrith, "Food and welfare in India, c. 1900–1950," 1012.

14 Historians have explored nutritional debates in the context of the South Indian Famine of 1876–78: D. Hall-Matthews, "Inaccurate Conceptions," 1189–1212; L. Sami, "Famine, Disease, Medicine and the State in Madras Presidency (1876–78)" (Unpublished dissertation, London, UCL, 2006); On the history of food and nutrition see D. Arnold, "The 'discovery' of malnutrition and diet in Colonial India," *The Indian Economic & Social History Review* 31:1 (March 1994), 1–26; D. Arnold, "The Good, the Bad, and the Toxic: Moral Foods in British India" in A. Ki Che Leung et al. (eds), *Moral Foods* (University of Hawaii Press, 2020), 111–29; C. Ellis, "'If you cannot feed the body of a child you cannot feed the brain': education and nutrition in late colonial Madras," *South Asia: Journal of South Asian Studies* 44:1 (2021), 135–51; A. Malhotra, "Cutting edge research in the contact zone? The establishment of the nutritional research laboratories in Coonoor (1925–27)," *South Asia: Journal of South Asian Studies* 44:1 (Jan 2021), 117–34; U. Ray, "The body and its purity: dietary politics in Colonial Bengal," *The Indian Economic & Social History Review* 50:4 (Oct 2013), 395–421; N. Slate, *Gandhi's Search for the Perfect Diet: Eating with the World in Mind* (Seattle: University of Washington Press, 2019); The interplay of nutritional science and famine relief has been studied in other regional contexts: Y. Slobodkin, "Famine and the science of food in the French Empire, 1900–1939," *French Politics, Culture & Society* 36:1 (Jan 2018), 52–75; I. Miller, *Reforming Food in Post-Famine Ireland: Medicine, Science and Improvement, 1845–1922* (Manchester: Manchester University Press, 2014), 21–40; A. Grossmann, "Grams, calories, and food: languages of victimization, entitlement, and human rights in Occupied Germany, 1945–1949," *Central European History* 44:1 (2011), 118–48; C. Aldous, "Contesting famine: hunger and nutrition in Occupied Japan, 1945–1952," *The Journal of American-East Asian Relations* 17:3 (2010), 230–56.

15 N. Durbach, *Many Mouths: The Politics of Food in Britain from the Workhouse to the Welfare State* (Cambridge: Cambridge University Press, 2020).

16 The famine codes, despite their centrality in the history of famine relief in India, are rarely studied in detail. Notable exceptions include M. Jha, "'Men diggers and women carriers': gendered

work on famine public works in colonial North India," *International Review of Social History* 65:1 (April 2020), 71–98; Hall-Matthews, "Historical roots of famine relief paradigms," 216–30; The famine codes' historical evolution is dealt with in L. Brennan, "The Development of the Indian Famine Codes Personalities, Politics, and Policies" in B. Currey and H. Graeme (eds), *Famine as a Geographical Phenomenon* (Dordrecht: Springer, 2012), 91–111.

17 On the development and influence of the British imperial system of famine relief shaped by colonialism in India: T. Sasson, "From empire to humanity: the Russian famine and the imperial origins of international humanitarianism," *The Journal of British Studies* 55:3 (July 2016), 519–37; T. Sasson and J. Vernon, "Practising the British way of famine: technologies of relief, 1770–1985," *European Review of History: Revue Européenne d'histoire* 22:6 (2015), 860–72; Recent publications on the global history of food aid include N. Götz, G. Brewis and S. Werther, *Humanitarianism in the Modern World: The Moral Economy of Famine Relief* (New York: Cambridge University Press, 2020); T. Scott-Smith, *On an Empty Stomach: Two Hundred Years of Hunger Relief* (Ithaca: Cornell University Press, 2020).

18 H.D. Curtis, *Holy Humanitarians: American Evangelicals and Global Aid* (Cambridge: Harvard University Press, 2018); A.M. Wilson, "In the name of God, civilization, and humanity: the United States and the Armenian massacres of the 1890s," *Le Mouvement Social* 227:2 (2009), 27; I. Tyrrell, "Connections, Networks, and the Beginnings of a Global America in the Gilded Age and Progressive Era" in C. McKnight Nichols and N.C. Unger (eds), *A Companion to the Gilded Age and Progressive Era* (West Sussex: Wiley, 2017), 379–98 in particular page 392.

19 B. Little, "An explosion of new endeavours: global humanitarian responses to industrialized warfare in the First World War era," *First World War Studies* 5:1 (2 Jan 2014), 1–16; J. Irwin, *Making the World Safe: The American Red Cross and a Nation's Humanitarian Awakening* (Oxford: Oxford University Press, 2013); B. Cabanes, *The Great War and the Origins of Humanitarianism, 1918–1924* (Cambridge: Cambridge University Press, 2014).

20 On American relief in the Near and Middle East: D. Rodogno, *Night on Earth: A History of International Humanitarianism in the Near East, 1918–1930* (New York: Cambridge University Press, 2022); K.D. Watenpaugh, *Bread from Stones: The Middle East and the Making of Modern Humanitarianism* (Oakland: University of California Press, 2015).

21 I. Tyrrell, *Reforming the World: The Creation of America's Moral Empire* (Princeton: Princeton University Press, 2010), 98–119.

22 S.R. Porter, *Benevolent Empire: U.S. Power, Humanitarianism, and the World's Dispossessed* (Philadelphia: University of Pennsylvania Press, 2017).

23 D.A. Hollinger, *Protestants Abroad: How Missionaries Tried to Change the World but Changed America* (Princeton: Princeton University Press, 2017); I. Tyrrell, "Vectors of Practicality: The Social Gospel, the North American YMCA in Asia and the Global Context" in H. Fischer-Tiné, S. Huebner, and I. Tyrrell (eds), *Spreading Protestant Modernity*, 39–60.

24 S. Sinha, "Lineages of the developmentalist state: transnationality and Village India, 1900–1965," *Comparative Studies in Society and History* 50:1 (Jan 2008), 57–90; N. Sackley, "Village models: Etawah, India, and the making and remaking of development in the Early Cold War," *Diplomatic History* 37:4 (2013), 749–78; N. Sackley, "The village as Cold War site: experts, development, and the history of rural reconstruction," *The Journal of Global History* 6:3 (Nov 2011), 481–504.

25 H. Fischer-Tiné, "The YMCA and low-modernist rural development in South Asia, c.1922–1957," *Past & Present* 240:1 (Aug 2018), 193–234; H. Fischer-Tiné, *The YMCA in Colonial India: Modernization, Philanthropy and American Soft Power in South Asia* (London: Bloomsbury Academic, 2022).

26 P. Kumar, "'Modernization' and agrarian development in India, 1912–52," *The Journal of Asian Studies* 79:3 (2020), 633–58.

27 R. Jachertz and A. Nützenadel, "Coping with hunger? Visions of a global food system, 1930–1960," *The Journal of Global History* 6:1 (March 2011), 99–119; A. Ribi Forclaz, "A new target for international social reform: the international labour organization and working and living conditions in agriculture in the inter-war years," *Contemporary European History* 20:3 (2011), 307–29.

28 S.S. Amrith and P. Clavin, "Feeding the world: connecting Europe and Asia, 1930–1945," *Past & Present* 218:8 (Jan 2013), 29–50.

29 An exception is the work of Maria Framke: M. Framke, "'We must send a gift worthy of India and the Congress!'," War and political humanitarianism in Late Colonial South Asia', *Modern Asian Studies* 51:6 (Nov 2017), 1969–98; M. Framke, "Political humanitarianism in the 1930s: Indian aid for Republican Spain," *European Review of History: Revue Européenne d'histoire* 23:1–2 (Jan 2016), 63–81.

30 C.A. Watt, *Serving the Nation: Cultures of Service, Association, and Citizenship* (New Delhi: Oxford University Press, 2005).

31 H.A. Gould, *Sikhs, Swamis, Students, and Spies: The India Lobby in the United States, 1900–1946* (New Delhi: Sage 2006); M. Ramnath, *Haj to Utopia: How the Ghadar Movement Charted Global Radicalism and Attempted to Overthrow the British Empire* (Berkeley: University of California Press, 2011); S. Sohi, *Echoes of Mutiny: Race, Surveillance, and Indian Anticolonialism in North America* (New York: Oxford University Press, 2014).

32 On the link between anti-Asian immigration legislation and Indian anticolonialism in the US and Canada, see also S. Sohi, "Race, surveillance, and Indian anticolonialism in the transnational Western U.S.-Canadian borderlands," *The Journal of American History* 98:2 (Sep 2011), 420–36.

33 Although 1943 is usually given as the year of the famine in Bengal, hunger and displacement began earlier and lasted longer, which is also reflected in the mobilisation of relief efforts by civil society actors before and after 1943. For this reason, the years 1942–44 are given as the timeframe of the Bengal Famine throughout the book, although one could certainly draw the boundaries of this periodisation even further. See J. Mukherjee, *Hungry Bengal*, 10.

34 Siegel, *Hungry Nation*, 23.

35 On the conflation of communalism and famine relief, see A. Roy, *Making Peace, Making Riots: Communalism and Communal Violence, Bengal 1940–1947* (New Delhi: Cambridge University Press, 2018), 69–103; A. Sarkar, "Fed by famine: the Hindu Mahasabha's politics of religion, caste, and relief in response to the Great Bengal Famine, 1943–1944," *Modern Asian Studies* 54:6 (2020), 2022–86; On communalism in Bengal, see J. Chatterji, *Bengal Divided: Hindu Communalism and Partition, 1932–1947* (Cambridge Cambridge University Press, 1994); R. Batabyal, *Communalism in Bengal: From Famine to Noakhali, 1943–47* (New Delhi: Sage Publications, 2005).

36 R. Ahuja, "Authoritarian shadows," *Social Scientist* 46:5–6 (2018), 12.

37 Siegel, *Hungry Nation;* B. Siegel, "'Fantastic Quantities of Food Grains': Cold War Visions and Agrarian Fantasies in Independent India" in L. James and E. Leake (eds), *Decolonization and the Cold War: Negotiating Independence* (London: Bloomsbury, 2015), 21–42; J. Loveridge, "Between hunger and growth: pursuing rural development in partition's aftermath, 1947–1957," *Contemporary South Asia* 25:1 (2 Jan 2017), 56–69; T.C. Sherman, "From 'grow more food' to 'miss a meal': hunger, development and the limits of post-colonial nationalism in India, 1947–1957," *South Asia: Journal of South Asian Studies* 36:4 (Dec 2013), 571–88.

38 For a recent debate of the ranking of India in the Global Hunger Index 2021, see the conclusion of the book.

Chapter 1. The Limits of Famine Relief: Colonialism, Nutritional Science, and the Indian Social Service Movement, 1890s–1930s

1 On British responses to famines in this period and in subsequent decades: V. Damodaran, "Famine in Bengal: a comparison of the 1770 famine in Bengal and the 1897 famine in Chotanagpur," *The Medieval History Journal* 10:1–2 (Oct 2006), 143–81; Sharma, *Famine, Philanthropy, and the Colonial State*; Hall-Matthews, *Peasants, Famine and the State in Colonial Western India*; Major, "British Humanitarian Political Economy," 221–44.

2 R. Møller Stahl, "The Economics of Starvation: Laissez-Faire Ideology and Famine in Colonial India," in M. Thorup (ed.), *Intellectual History of Economic Normativities* (New York: Palgrave Macmillan, 2016), 169–84; S. Rashid, "The policy of laissez-faire during scarcities," *The Economic Journal* 90:359 (Sep 1980), 493–501; Hall-Matthews, "Historical Roots of Famine Relief Paradigms," 216–30.

3 S. Ambirajan, "Malthusian population theory and Indian famine policy in the nineteenth century," *Population Studies* 30:1 (March 1976), 5–14; J.C. Caldwell, "Malthus and the less developed world: the pivotal role of India," *Population and Development Review Population and Development Review* 24:4 (Dec 1998), 675–96.

4 Major, "British Humanitarian Political Economy," 226.

5 R. Ahuja, "State formation and 'famine policy' in early colonial South India," *The Indian Economic & Social History Review* 39:4 (Dec 2002), 351–80.

6 D. Arnold, "Social crisis and epidemic disease in the famines of nineteenth-century India," *Social History of Medicine* 6:3 (1993), 385–404; D. Arnold, "Looting, grain riots and government policy in South India 1918," *Past and Present* 84:1 (1979), 111–45; Sharma, *Famine, Philanthropy, and the Colonial State*, 79–134.

7 See the second chapter of James Vernon, *Hunger: A Modern History* (London: The Belknap Press of Harvard University Press, 2007), 17–40.

8 Major, "British humanitarian political economy," 221–44; H. Curtis, "Depicting distant suffering: evangelicals and the politics of pictorial humanitarianism in the age of American Empire," *Material Religion* 8:2 (June 2012), 153–82; C. Twomey, "Framing atrocity: photography and humanitarianism," *History of Photography* 36:3 (Aug 2012), 255–64; C. Twomey and A.J. May, "Australian responses to the Indian famine, 1876–78: Sympathy, photography and the British Empire,", *Australian Historical Studies* 43:2 (June 2012), 233–52; Götz, Brewis, and Werther, *Humanitarianism in the Modern World*, 68–143.

9 S.S. Amrith, *Unruly Waters: How Rains, Rivers, Coasts, and Seas Have Shaped Asia's History* (New York: Basic Books, 2018), 65–89.

10 For a more detailed discussion of the emergence of this three-pronged approach, see Bhatia, *Famines in India*, 102–23.

11 Hall-Matthews, "Historical Roots of Famine Relief Paradigms," 216–30.

12 R. Ahuja, "'The Bridge-Builders': Some Notes on Railways, Pilgrimage and the British 'Civilizing Mission' in Colonial India" in H. Fischer-Tiné and M. Mann (eds), *Colonialism as Civilizing Mission: Cultural Ideology in British India* (London: Anthem Press, 2004), 95–116.

13 S. Sweeney, "Indian railways and famines, 1875–1914: magic wheels and empty stomachs," *Essays in Economic & Business History* 26 (2008), 147–58.

14 U.P. Mukherjee, *Natural Disasters and Victorian Empire: Famines, Fevers and the Literary Cultures of South Asia* (Basingstoke: Palgrave Macmillan, 2013), 35–37.

15 D. Arnold, *Colonizing the Body: State Medicine and Epidemic Disease in Nineteenth-Century India* (Berkeley: University of California Press, 1993a); G. Prakash, *Another Reason: Science and the Imagination of Modern India* (Princeton: Princeton University Press, 1999), 123–58.

[16] Sasson and Vernon, "Practising the British way of famine: technologies of relief," 860–72; U. Kalpagam, *Rule by Numbers: Governmentality in Colonial India* (Lanham: Lexington Books, 2014), 162–66; A. Nag, "Managing Hunger: Famine, Science and the Colonial State in India, 1860–1910" (Unpublished dissertation, Los Angeles, UCLA, 2010); S. Hodges, "Governmentality, population and reproductive family in Modern India," *Economic and Political Weekly* 39:11 (2004), 1157–63.

[17] On the tenets of British poor relief, see Durbach, *Many Mouths*, 18–48.

[18] Arnold, "The 'discovery' of malnutrition," 1–26.

[19] Durbach, *Many Mouths*, 52; S. Sharma, "Poorhouses and Gratuitous Famine Relief in Colonial North India" in A. Mukherjee (ed.), *A Cultural History of Famine: Food Security and the Environment in India and Britain* (London: Routledge, 2019), 129–48.

[20] D. Arnold, "The Colonial Prison: Power, Knowledge and Penology in Nineteenth Century India" in D. Hardiman and D. Arnold (eds), *Subaltern Studies VIII* (New Delhi: Oxford University Press, 2008), 148–87, in particular page 179.

[21] Arnold, *Colonizing the Body*, 112.

[22] W.R. Cornish, *Observations, on the Nature of the Food of the Inhabitants of Southern India: And, on Prison Dietaries in the Madras Presidency* (Madras: Adelphi Press, 1864); W.R. Cornish, *Reports on the Nature of the Food of the Inhabitants of the Madras Presidency and on the Dietaries of Prisoners in Zillah Jails* (Madras: United Scottish Press, 1863).

[23] E. Neswald, "Nutritional Knowledge between the Lab and the Field. The Search for Dietary Norms in the Late Nineteenth and Early Twentieth Centuries" in E. Neswald, D.F. Smith, and U. Thoms (eds), *Setting Nutritional Standards: Theory, Policies, Practices* (Rochester: University of Rochester Press, 2017), 35–36.

[24] Cornish, *Observations*, 30.

[25] J. Ewart, *The Sanitary Condition and Discipline of Indian Jails* (London: Smith, Elder and Co., 1860), 75.

[26] Arnold, "The 'discovery' of malnutrition," 5.

[27] Note that Sanjay Sharma traced debates on the size of famine relief rations and wages during the famine of 1837–38 which indicates that the search for adequate diets started earlier. S. Sharma, "Measuring Hunger. Debates on an 'Adequate' Diet in Colonial North India" in K. Bhushi (ed.) *Farm to Fingers. The Culture and Politics of Food in Contemporary India* (New York: Cambridge University Press, 2018), 184–214.

[28] Davis, *Late Victorian Holocausts*, 25–33; Amrith, *Unruly Waters*, 65–89.

[29] Temple's response to the famine and its consequences are explicated in Hall-Matthews, "Inaccurate Conceptions," 1189–1212; Davis, *Late Victorian Holocausts*, 33–60.

[30] W. Digby, *The Famine Campaign in Southern India, 1876–1878*, vol. II (London: Longmans, Green, and Co., 1878), 190.

[31] On the public dispute between Temple and Cornish: Sami, "Famine, Disease, Medicine and the State in Madras Presidency (1876–78)".

[32] Davis, *Late Victorian Holocausts*, 7.

[33] Brennan, "The Development of the Indian Famine Codes Personalities, Politics, and Policies", 91–111.

[34] Report of the Indian Famine Commission (London: Commission of Inquiry on Indian Famines, 1880), 59.

[35] This was noted in the Report of the Indian Famine Commission (London: Printed for H.M. Stationery Office by Darling & Son, 1898), 256.

[36] N. Cullather, "When Is Famine Not a Famine? Gauging Indian Hunger in Imperial and Cold War Contexts" in Neswald, Smith and Thoms (eds), *Setting Nutritional Standards*, 199.

[37] Neswald, "Nutritional Knowledge between the Lab and the Field. The Search for Dietary Norms in the Late Nineteenth and Early Twentieth Centuries," 32–37; Durbach, *Many Mouths*, 54–95.

38 On the administration of women's diets in British prisons, see Durbach, *Many Mouths*, 59–68.

39 Report of the Indian Famine Commission (1880), 59.

40 Jha, "'Men diggers and women carriers'," 71–98.

41 "The Government of India's Circular No. 44F, 9 June 1883" reprinted in parts in Report of the Indian Famine Commission (1901), 36. Emphasis added.

42 Report of the Indian Famine Commission (1898), 255–56; J.E. Scott, *In Famine Land. Observations and Experiences in India during the Great Drought of 1899–1900* (New York and London: Harper and Brothers Publishers, 1904), 42–53.

43 Report of the Indian Famine Commission (1898), 67.

44 On the punitive nature of labour on public works, see also A. Shahid, "Famine Labour and Coercion in Relief-based Public Works Construction in Colonial India in the Late Nineteenth Century" in K. Ekama, L. Hellman and M. van Rossum (eds), *Slavery and Bondage in Asia, 1550–1850* (Berlin: De Gruyter, 2022), 231–50.

45 Report of the Madras Famine Code Revision Committee, 1938 (Madras, 1940), 23; V. Nash, *The Great Famine and Its Causes* (London: Longmans, Green, and Co., 1900), 42.

46 Report of the Indian Famine Commission (1901), 30.

47 Scott, *In Famine Land. Observations and Experiences in India during the Great Drought of 1899–1900*, 53.

48 An observation made by Nash as well. Nash, *The Great Famine and Its Causes*, 34.

49 Report of the Indian Famine Commission (1898), 260.

50 Ibid., 45.

51 Report of the Indian Famine Commission (1901), 43.

52 Nash, *The Great Famine and Its Causes*, 30.

53 Report of the Indian Famine Commission (1898), 283.

54 Ibid., 283–84.

55 Sharma, "Poorhouses and Gratuitous Famine Relief in Colonial North India," 129–48.

56 A. Forth, *Barbed-Wire Imperialism: Britain's Empire of Camps, 1876–1903* (Oakland: University of California Press, 2017), 100–112.

57 Famine Relief Code. Bombay Presidency (Bombay: Government Central Press, 1904), 46; 111–12.

58 Sharma, "Poorhouses and Gratuitous Famine Relief in Colonial North India," 134.

59 On the reform of prison diets, see Arnold, "The Colonial Prison," 148–87; A.A. Yang, "Disciplining 'natives': prisons and prisoners in early nineteenth century India," *South Asia: Journal of South Asian Studies* 10:2 (Dec 1987), 29–45.

60 Report of the Indian Famine Commission (1898), 286.

61 Report of the Indian Famine Commission (1901), 41.

62 Report of the Indian Famine Commission (1898), 288.

63 Famine Relief Code. Bombay Presidency (1912), 210.

64 The Bihar and Orissa Famine Code (1930) (Calcutta: Jharna Supply Syndicate, 1950), 123.

65 G.G. Raheja, "'Hear the tale of the famine year': famine policy, oral traditions, and the recalcitrant voice of the colonized in nineteenth-century India," *Oral Tradition* 31:1 (2017), 155–98.

66 Report of the Indian Famine Commission (1898), 287.

67 Sharma, "Poorhouses and Gratuitous Famine Relief in Colonial North India," 129–48.

68 Durbach, *Many Mouths*, 105–13. Cf. page 105.

69 H. Woodd Nevinson, *The New Spirit of India* (London: Harper & Brothers, 1908), 272.

70 G.K. Devadhar, "The famine of 1908 in India and the work done by non-official agencies," *Modern Review* 6:3 (Sep 1909), 250.

71 Resolution on the Administration of Famine Relief in the United Provinces of Agra and Oudh during the Year 1907 and 1908 (Allahabad: Government Press, 1908), 26.

72 Resolution on the Administration of Famine Relief in the United Provinces, 85.

73 Ibid., 87.

74 Ibid., 49.

75 I discuss the work of the SIS and other Indian relief providers in Chapter 5.

76 G. Brewis, "Education for service: social service and higher education in India and Britain, 1905–1919," *History of Education Review* 42:2 (Oct 2013), 119–36; C.A. Watt, *Serving the Nation: Cultures of Service, Association, and Citizenship* (New Delhi: Oxford University Press, 2005), 1–5; 211–15; E. Valdameri, *Indian Liberalism Between Nation and Empire. The Political Life of Gopal Krishna Gokhale* (Abingdon: Routledge, 2022).

77 Watt, *Serving the Nation*, 214.

78 Resolution on the Administration of Famine Relief in the United Provinces, 97.

79 G.K. Devadhar, "The famine of 1908 in India and the work done by non-official agencies," *Modern Review* 6:3 (Sep 1909), 260.

80 G.K. Devadhar, "The present famine in the United Provinces (continued)," *Indian Social Reformer* XIX:5 (4 Oct 1908), 52. Emphasis in the original.

81 Devadhar, "The famine of 1908 in India and the work done by non-official agencies," 260.

82 T.N Jagadisan and Shyamlal (eds), *Thakkar Bapa. Eightieth Birthday Commemoration Volume* (Madras: Diocesan Press, 1949).

83 "Topics of the week," *Servant of India* III:12 (22 April 1920), 1; M.K. Gandhi, CWMG, vol. 20 (Delhi: Ministry of Information and Broadcasting, 1958), 302–3.

84 "The muddle of the Puri famine," *Servant of India* III:23 (8 July 1920), 272–73.

85 "Famine in Puri and elsewhere," *Modern Review* 28 (July 1920), 109.

86 Report of the Madras Famine Code Revision Committee of 1938, 3.

87 Cf. "Famine in Bellary," *Servant of India* IV:37 (13 Oct 1921), 434.

88 C.J. Baker, *The Politics of South India. 1920–1937* (Cambridge: Cambridge University Press, 1976), 169–211.

89 Report of the Madras Famine Code Revision Committee of 1938, 2–3.

90 A Brief Account of the Work of the Servants of India Society, Poona. From 1st of Jan 1917 to 30th June 1923 (Pune: Servants of India Society, April 1924), 40.

91 "Revision of the Famine Code," *Servant of India* IV:45 (15 Dec 1921), 545–48.

92 Ibid., 547.

93 W.J. Buchanan, "A note on the nutritive values of the dietaries of Bengal jails," *Indian Medical Gazette* 31:5 (May 1896), 153–58.

94 Arnold, "The 'discovery' of malnutrition," 8–9; On prison diets, see also D. Arnold, "India: The Contested Prison" in F. Dikötter and I. Brown (eds), *Cultures of Confinement: A History of the Prison in Africa, Asia, and Latin America* (Ithaca: Cornell University Press, 2018), 147–84.

95 On the elections, see Baker, *The Politics of South India*, 293–316.

96 Report of the Madras Famine Code Revision Committee of 1938, 3.

97 Ibid., 4.

98 Ibid., 21.

99 Ibid., 22.

100 Ibid., 15.

101 Famine Inquiry Commission. Final Report (Madras, 1945), 508.

102 Servants of India Society Report for 1939–40 (Poona, 1940), 25.

103 W.R. Aykroyd and B.G. Krishnan, "The carotene and Vitamin A requirements of children," *The Indian Journal of Medical Research* 23 (1936), 741–45.

104 On Reginald Passmore, see M.L.G. Bizzo, "Postponing Equality: From Colonial to International Nutritional Standards, 1932–1950" in A. Widmer and V. Lipphardt (eds), *Health and Difference: Rendering Human Variation in Colonial Engagements* (New York: Berghahn Books, 2016), 143 fn. 5.

105 Report of the Madras Famine Code Revision Committee of 1938, 130.

106 Ibid., 127–28.

107 Durbach, *Many Mouths*, 109–13.

108 Report of the Madras Famine Code Revision Committee of 1938, 128.

109 Ibid., 128.

110 This differentiation was in place from 1883 and is reiterated in the report of the Third Indian Famine Commission. Report of the Indian Famine Commission (1901), 36.

111 On this wider phenomenon, see V. Fildes, L. Marks, and H. Marland (eds), *Women and Children First: International Maternal and Infant Welfare, 1870–1945* (London: Routledge, 1992).

112 Amrith, "Food and welfare in India," 1016–17; R. Nair, "The construction of a 'population problem' in colonial India 1919–1947," *The Journal of Imperial and Commonwealth History* 39:2 (2011), 232–47.

113 Nair, "The construction of a 'population problem' in colonial India 1919–1947," 232–33.

114 N. Gangulee, *Health and Nutrition in India* (London: Faber and Faber, 1939), 130–31.

115 B. Kim, "Lakshmi and the scientific housewife: a transnational account of Indian women's development and production of an Indian modernity," *Economic and Political Weekly* 38:11 (2003), 1055–68.

116 G.K. Devadhar (ed.), *Poona Seva Sadan Society. Review & Report with an Album* (Poona: Seva Sadan Society, 1934), 26–27.

117 P. Anagol, *The Emergence of Feminism in India, 1850–1920* (Burlington: Ashgate, 2005), 531–32.

118 A. Basu and B. Ray, *Women's Struggle. A History of the All India Women's Conference, 1927–2016* (New Delhi: Manohar, 2018), 95–97.

119 See Chapter 3 of S. Zurbrigg, *Malaria in Colonial South Asia: Uncoupling Disease and Destitution* (London: Routledge, 2019).

120 Vernon, *Hunger*, 81–117; Arnold, "The 'discovery' of malnutrition," 1–26.

121 A. Malhotra, "Race, diet, and class: Robert McCarrison's laboratory rat experiments in Coonoor, 1925–27," *Food Studies: An Interdisciplinary Journal* 9:1 (2019), 17–27; Arnold, "The Good, the Bad, and the Toxic," 111–29.

122 W.R. Aykroyd and B.G. Krishnan, "The effect of skimmed milk, soya bean, and other foods in supplementing typical Indian diets," *The Indian Journal of Medical Research* 24 (1937), 1093–1206.

123 "Soya bean or skim-milk for India?" *Lancet* 229:5938 (June 1937), 1474–75; R.K. Pal and N. Singh, "The effect on rats of supplementing a North Indian diet with vegetable proteins (sprouting pulses) and calcium," *The Indian Medical Gazette* 73:8 (Aug 1938), 469–71.

124 C. Ellis, "If you cannot feed the body of a child," 135–51.

125 "Soya bean or skim-milk for India?," 1474–75; Aykroyd and Krishnan, "The effect of skimmed milk, soya bean, and other foods in supplementing typical Indian diets," 1093–1206.

126 W. Aykroyd et al., "The rice problem in India," *Indian Medical Research Memoirs* 32 (1940), 78.

127 N.N. Godbole, *Zur Kenntnis der Höchstmolekularen Gesättigten Säuren des Erdnussöles* (Berlin: Emil Dreyer's Buchdruckerei, 1925).

128 N.N. Godbole, *Milk. The Most Perfect Food* (Benares: Benares Hindu University, 1936); On Malaviya, see J.P. Misra, *Madan Mohan Malaviya and the Indian Freedom Movement* (New Delhi: Oxford University Press, 2016); H.K. Patel, "Aspects of socio-political thoughts of Madan Mohan Malaviya, a liberal Hindu nationalist'," *Proceedings of the Indian History Congress* 74 (2013), 581–86; K. Maclean, "Hybrid Nationalist or Hindu Nationalist? The Life of Madan Mohan Malaviya (1861–1946)" in

K. Brittlebank (ed.), *Tall Tales and True: India, Historiography, and British Imperial Imaginings* (Clayton: Monash University Press, 2008), 107–24.

129 Godbole, *Milk. The Most Perfect Food*, v–xiii.

130 On the role of vegetarianism in colonial India, see L. Gandhi, *Affective Communities: Anticolonial Thought, Fin-De-Siècle Radicalism, and the Politics of Friendship* (Durham: Duke University Press, 2006), 67–114; J. Hauser, "Körper, Moral, Gesellschaft. Debatten über Vegetarismus zwischen Europa und Indien, ca. 1850–1914" in N. Aselmeyer and V. Settele (eds), *Geschichte des Nicht-Essens: Verzicht, Vermeidung und Verweigerung in der Moderne* (Berlin: De Gruyter Oldenbourg, 2018), 259–87.

131 Godbole, *Milk. The Most Perfect Food*, 72–74.

132 Slate, *Gandhi's Search for the Perfect Diet*, 67; P. Roy, "Meat-eating, masculinity, and renunciation in India: a Gandhian grammar of diet," *Gender & History* 14:1 (April 2002), 62–91; On Gandhi's efforts to create a national diet, see Amrith, "Food and welfare in India," 1010–35.

133 C.D. Williams, "Kwashiorkor: a nutritional disease of children associated with a maize diet," *Lancet* (1935), 1151–52.

134 J. Stanton, "Listening to the Ga: Cicely Williams' discovery of Kwashiorkor on the Gold Coast," *Clio Medica* 61 (2001), 151.

135 The Problem of Nutrition. Report on the Physiological Bases of Nutrition (London: The League of Nations. Technical Commission of the Health Committee, 1936), 19.

136 W. Aykroyd and E. Burnet, *Nutrition and Public Health*, Bulletin of the LNHO IV:2 (June 1935); On the LNHO, see I. Borowy, *Coming to Terms with World Health: The League of Nations Health Organisation 1921–1946* (Frankfurt am Main: Peter Lang, 2009); Amrith and Clavin, "Feeding the world," 29–50.

137 W.R. Aykroyd, *The Nutritive Value of Indian Foods and the Planning of Satisfactory Diets*, Health Bulletin 23 (Delhi, 1959); Also see BL, APAC, IOR/V/26/830/8, "Nutrition Advisory Committee, Indian Research Fund Association. Report of the Sub-committee on Nutritional Requirements (23 Nov 1944)" in *Report of the Eleventh to Fifteenth Meetings of the NAC of the IRFA, 1944–1946*, Vol. II (Cawnpore, 1947), 109–17.

138 M.V. Radhakrishna Rao, *Diet Manual. A Compilation of Diets Recommended in Health and Disease* (Bombay, 1959).

139 N. Cullather, "The foreign policy of the calorie," *The American Historical Review* 112:2 (1 April 2007), 355.

140 Bizzo, "Postponing Equality: From Colonial to International Nutritional Standards," 136.

141 Ibid., 136.

142 R. Mukerjee, *Food Planning for Four Hundred Millions* (London: Macmillan, 1938), 55–68.

143 E. Messer, "The 'small but healthy' hypothesis: historical, political, and ecological influences on nutritional standards," *Human Ecology* 14:1 (1986), 62.

144 Mukerjee, *Food Planning for Four Hundred Millions*, 82–84.

145 R. Ahuja, "A Beveridge Plan for India? Social insurance and the making of the 'formal sector'," *International Review of Social History* 64:2 (2019), 215.

146 D.C. Lall, *Coolie: The Story of Labour and Capital in India* (Lahore: Oriental Publishing House, 1932), 6.

147 The reports on the administrative responses to famine in India that were published in 1945 were not concerned with the Indian famine codes. See Famine Inquiry Commission. Report on Bengal (Madras, 1945); The Famine Inquiry Commission. Final Report (Madras, 1945). In 1950–51, the Madras government relied on the Madras Famine Code to mitigate hunger. See the Famine Code Madras Province 1950 (Madras, 1950).

148 S.E. Sukthankar, *Report of the Fact-finding Committee for Survey of Scarcity Areas Maharashtra State 1973* (Bombay: Govt. of Maharahstra, 1973), 86–87.

149 M.D. Morris, "Needed: a new famine policy," *Economic and Political Weekly* 10:5/7 (1975), 283–94.

150 J. Drèze, "Famine Prevention in India" in J. Drèze and A. Sen (eds), *The Political Economy of Hunger: vol. 2* (Oxford: Oxford University Press, 1991), 65–99.

151 Morris, "Needed: a new famine policy," 288.

152 Ibid., 289.

153 Ibid., 293.

154 Ibid., 288.

155 D. Seckler, "The 'small but healthy?' hypothesis: a reply to critics," *Economic and Political Weekly* 19:44 (1984), 1886–88; E. Messer, "The 'small but healthy' hypothesis: historical, political, and ecological influences on nutritional standards," *Human Ecology* 14:1 (1986), 57–75, in particular page 67; C. Sathyamala, "Sukhatme's Legacy and the Indian Exceptionalism" in M. Arsel, A. Dasgupta, and S. Storm (eds), *Reclaiming Development Studies: Essays for Ashwani Saith* New York: Anthem Press, 2021), 179–96.

156 See also J. Glasman, "Measuring malnutrition: the history of the MUAC tape and the commensurability of human needs," *Humanity: An International Journal of Human Rights, Humanitarianism, and Development* 9:1 (2018), 19–44.

Chapter 2. Food Technology, Nutritional Science, and Indo-US Entanglements in the 1940s and 1950s

1 Nehru Memorial Museum and Library [NMML], All-India Women's Conference Papers, Correspondence Relating to Relief Work done by AIWC in Bengal Famine of 1943, "The AIWC to members of the Standing Committee," 9 Oct 1943, unpag.

2 Reprint from the *Hindustan Times* (7 Nov 1943) (unpag.) in K. Santhanam, *The Cry of Distress. A First-Hand Description and an Objective Study of the Indian Famine of 1943* (New Delhi: The Hindustan Times, 1943), unpag; Also see Vernon, *Hunger. A Modern History*, 148–50.

3 "Halibut oil flown to aid Bengal's starving," *Evening Telegraph and Post* (4 Nov 1943), 8; "Action to end famine," *Press and Journal* (5 Nov 1943), 3.

4 J. Simonow, "The Great Bengal Famine in Britain: metropolitan campaigning for food relief and the end of empire, 1943–44," *The Journal of Imperial and Commonwealth History* 48:1 (2020), 168–97.

5 For mortality estimates, see Mukherjee, *Hungry Bengal*, 184–85; Mukerjee, *Churchill's Secret War*, 267–72; A. Sen, "Starvation and exchange entitlements," 35.

6 K. Smith-Howard, *Pure and Modern Milk: An Environmental History since 1900* (Oxford: Oxford University Press, 2014), 76.

7 B.M. Patenaude, *The Big Show in Bololand: The American Relief Expedition to Soviet Russia in the Famine of 1921* (Stanford: Stanford University Press, 2002), 511.

8 Smith-Howard, *Pure and Modern Milk*, 77–79; T. Scott-Smith, "Beyond the 'raw' and the 'cooked': a history of fortified blended foods," *Disasters* 39:s2 (Oct 2015), 244–60.

9 Report on Friends' Famine Relief in India 1945. Together with a Report of American Relief for India, Inc. (1945), 11; American Relief for India (New York: American Relief for India, Inc., 1945), 19. Copies are kept in Haverford College Quaker & Special Collection, John F. Rich Papers, box 1, folder 14.

10 Famine Inquiry Commission. Report on Bengal, 130.

11 H. Wieters, *The NGO CARE and Food Aid from America 1945–80: "Showered with Kindness"?* (Manchester: Manchester University Press, 2017).

12 B. Siegel, "'The claims of Asia and the Far East': India and the FAO in the age of ambivalent internationalism," *The International History Review* 41:2 (March 2019), 427–50.

13 New York Public Library [NYPL], Manuscripts and Archives Division, Sidney Hertzberg [SH] Papers, box 23, "Aykroyd to Whitman" (2 April 1946).

14 NYPL, Manuscripts and Archives Division, SH Papers, box 23, "Notes on conversation between H. Hastburn Thompson, AFSC, and Hazel Whitman, Regarding Food Packages for India," 26 March 1946.

15 UCLA Library, Special Collections, Meals for Millions Foundation [MFM] Records, box 17, "Information Bulletin, Committee on Relief and Gift Supplies. National Christian Council of India, May 1955," 2.

16 B.A. Scholten, *India's White Revolution: Operation Flood, Food Aid and Development.* (London: I.B. Tauris & Co., 2010); R. Heredia, *The Amul India Story* (New Delhi: Tata McGraw-Hill, 1999); M. Black, *Children First: The Story of UNICEF. Past and Present* (Oxford: Oxford University Press, 1996).

17 "Famished People of Bengal. Report on Scientific Methods of Treating," *Amrita Bazar Patrika* [ABP] (7 Oct 1943), 2.

18 Mukherjee, *Hungry Bengal*, 174.

19 V.N. Patwardhan, "The Problem of Nutrition in India & Asia" (Asian Relations Conference, New Delhi, 1944), 6.

20 "Famished people of Bengal. Report on scientific methods of treating," *ABP* (7 Oct 1943), 2.

21 Ibid.

22 Also published as "Treatment and management of starving sick destitutes," *The Indian Medical Gazette* 79:2 (Feb 1944), 74–81.

23 Ibid., 74–81.

24 This method of treatment of patients in Bengal and later in Europe is also discussed in Vernon, *Hunger. A modern History*, 148–54.

25 K.V. Krishnan, E.K. Narayanan, and G. Sankaran, "Protein hydrolysates in the treatment of inanition," *The Indian Medical Gazette* 79:4 (April 1944), 160.

26 R. Elman, "The intravenous use of protein and protein hydrolysates," *Annals of the New York Academy of Sciences* 47:3 (Oct 1946), 345–57; "Conference on therapy: the use of protein hydrolysates," *The American Journal of Medicine* 3:4 (Oct 1947), 472–85.

27 "Treatment and management of starving sick cestitutes," 74–81.

28 The figure is given in Krishnan, Narayanan, and Sankaran, "Protein hydrolysates in the treatment of inanition," 160.

29 B.N. Bhandari, "Œdema and scabies in a famine hospital," *The Indian Medical Gazette* 79:12 (Dec 1944), 574.

30 M.V. Chari, "Inanition cases: a report" *The Indian Medical Gazette* 79:12 (Dec 1944), 578–82.

31 Chari, "Inanition Cases," 581.

32 Examples include Bhandari, "Œdema and scabies in a famine hospital," 574–78; Chari, "Inanition cases: a report," 578–582; M.L. Chakrabarty, "Blood-sugar levels in slow starvation," *Lancet* 251:6503 (April 1948), 596–97.

33 NMML, Syamaprasad Mookerjee Private Papers [SPM Private Papers], Subject File 108, "Bhattacharjee to Syama Prasad Mukherjee" (28.09.1943).

34 "Bhattacharjee to Syama Prasad Mukherjee" (28.09.1943).

35 B. Prasad and B.L. Raina (eds), *Official History of the Indian Armed Forces in the Second World War, 1939-45. Medical Services. Administration* (Delhi: Combined Inter-Services Historical Section, 1953), 405–14.

36 On the importance of control, see D. Arnold, "The medicalization of poverty in Colonial India," *Historical Research* 85:229 (Aug 2012), 494; Forth, *Barbed-Wire Imperialism*, 43–73.

37 D. Simmons, "Starvation Science: From Colonies to Metropole" in A. Nützenadel and F. Trentmann (eds), *Food and Globalization: Consumption, Markets and Politics in the Modern World* (Oxford: Oxford, 2008), 173–92.

38 T. Tucker, *The Great Starvation Experiment: Ancel Keys and the Men Who Starved for Science* (Minneapolis: University of Minnesota Press, 2007); L.M. Kalm and R.D. Semba, "They starved so that others be better fed: remembering Ancel Keys and the Minnesota experiment," *The Journal of Nutrition* 135:6 (June 2005), 1347–52.

39 L. Collingham, *The Taste of War: World War II and the Battle for Food* (Penguin Group USA, 2012), 476–81.

40 "Discussion: the physiology and treatment of starvation," *Proceedings of the Royal Society of Medicine* 38:7 (1945), 388–98; "Treatment of severe starvation," *The British Medical Journal* 1:4400 (May 1945), 634–35.

41 Scott-Smith, *On an Empty Stomach*, 103–5.

42 "Discussion: the physiology and treatment of starvation," 395–97.

43 This was recalled by a British medical student who worked in Bergen-Belsen. See Imperial War Museum, London, Bradford, David Cordley (Oral History).

44 J. Brooks, "'Uninterested in anything except food': the work of nurses feeding the liberated inmates of Bergen-Belsen: nursing work and the liberation of Bergen-Belsen," *The Journal of Clinical Nursing* 21:19–20 (Oct 2012), 2951; P. Weindling, "'Belsenitis': liberating Belsen, its hospitals, UNRRA, and selection for re-emigration, 1945–1948," *Science in Context* 19:3 (Sep 2006), 407–8.

45 "New diets for old," *APB* (7 Dec 1943), 2.

46 Ibid.

47 On Indian food planning, see Siegel, *Hungry Nation*; Sherman, "From 'grow more food' to 'miss a meal'," 571–88; Loveridge, "Between hunger and growth: pursuing rural development in partition's aftermath," 56–69.

48 D.C. Engerman, *The Price of Aid: The Economic Cold War in India* (Cambridge: Harvard University Press, 2018), 53.

49 *The CFTRI Mysore* (New Delhi: Council of Scientific & Industrial Research, 1950), 8. The institute was set up in 1948 and was officially inaugurated two years later.

50 *The CFTRI Mysore*, 4–5; Siegel, *Hungry Nation*, 104–7.

51 C.P. Natarajan and J.V. Shankar, "Food science and technology development in India: concepts and contributions of Dr. V. Subrahmanyan," *The Journal of Food Science and Technology* 17:1&2 (April 1980), 17–23.

52 M.V.L. Rao, "Dr Vaidyanatha Subrahmanyan – A pioneering food scientist," *The Journal of Food Science and Technology* 17:1&2 (April 1980), 1–8; N. Bhima Rao, "Prof. V. Subrahmanyan – some personal reminiscences," *The Journal of Food Science and Technology* 17:1&2 (April 1980), 121–23.

53 Siegel, *Hungry Nation*, 96–97.

54 NYPL, Manuscripts and Archives Division, SH Papers, box 23, "Smith to Singh" (July 15, 1946).

55 Ibid.

56 Collingham, *The Taste of War*, 490–95.

57 T. Robertson, "New Frontiers: World War II Technologies and the Opening of Tropical Environments to Development" in S.J. Macekura and E. Manela (eds), *The Development Century: A Global History* (Cambridge: Cambridge University Press, 2018), 107–29.

58 H. Borsook, "We could feed the world," *Engineering and Science Monthly* (Dec 1948), 7.

59 The Caltech Institute Archives, "Interview with Henry Borsook, Transcript, 5 April 1978".

60 H.R. Roberts, "The Meals for Millions Foundation: A Study of a Non-Governmental Organization" (Unpublished dissertation, Washington, University of Washington, 1967), 8–24; "Life visits Clifton's Cafeteria," *Life* (27 Nov 1944), 102–5.

61 UCLA Library Special Collections, MFM Records, box 34, "Fight Hunger! Win the Peace!" (leaflet of Meals for Millions).

62 "Report on the Navajo," *Desert Magazine* (March 1948), 5–8.

63 NYPL, Manuscripts and Archives Division, HS Papers, box 24, "Table II. Powdered milk as compared to Multi-Purpose Food, Annex of Comparison of Cereal Mix and Multi-Purpose Food."

64 Roberts, "The Meals for Millions Foundation," 50–114, 161–96.

65 K.J. Leong, *The China Mystique: Pearl S. Buck, Anna May Wong, Mayling Soong, and the Transformation of American Orientalism* (Berkeley: University of California Press, 2005), 12–56.

66 Interview with Henry Borsook, 53.

67 Interview with Henry Borsook, 52; UCLA Library Special Collections, MFM Records, box 46, "Holmes to Rose" (24 April 1951).

68 UCLA Library Special Collections, MFM Records, box 46, "Rose to Roy" (2 April 1951); "Rose to Roy" (24 April 1951).

69 Engerman, *The Price of Aid*, 52–88; R.J. McMahon, "Food as a diplomatic weapon: the India wheat loan of 1951," *Pacific Historical Review* 56:3 (Aug 1987), 349–77.

70 USNA, Record Group 220, Subject Files Regarding India, 1951–53, "Hubert H. Humphrey. Press Release 'Friendship Grain'" (2 May 1951).

71 USNA, Record Group 220, Subject Files Regarding India, 1951–53, "Special Press Feature. American Voluntary Agencies distribute Relief Supplies in India"; "Summary of Relief Shipments made to India by Voluntary Relief Agencies".

72 M.D. Lincoln and D. Karp, *Vice President in Charge of Revolution* (New York: McGraw-Hill Book Company, 1960), 253–54; On the failure of CARE to spark enthusiasm of Indian farmers for its plough, see D.F. Ebright, *Free India. The First Five Years: An Account of the 1947 Riots, Refugees, Relief and Rehabilitation* (Nashville: Pantheon Press, 1954), 164–65.

73 UCLA Library Special Collections, MFM Records, box 17, "Around the world with MPF: Sept. through June 1953" and "How Meals for Millions cooperates with churches and religious agencies"; Ebright, *Free India. The First Five Years: An Account of the 1947 Riots, Refugees, Relief and Rehabilitation*, 79–81.

74 UCLA Library Special Collections, MFM Records, box 46, "Rose to Munshi" (24 April 1951).

75 On the National Council for Child Welfare, see A. Devenish, *Debating Women's Citizenship in India, 1930–1960* (Delhi: Bloomsbury, 2019), 122–28.

76 UCLA Library Special Collections, MFM Records, box 17, "Meals for Millions News Letter, Fall 1953," 2; On the AIWC's rural health programmes, see Basu and Ray, *Women's Struggle*, 127–29.

77 UCLA Library Special Collections, MFM Records, box 17, "Digest of Report submitted by R. Rajagopalan of the Indian Institute of Science Bangalore," 1 June 1953.

78 V. Subrahmanyan et al., "The effect of a supplementary multipurpose food on the growth and nutritional status of schoolchildren," *The British Journal of Nutrition* 11:4 (Dec 1957), 382; V. Subrahmanyan, T.R. Doraiswamy, and R.K. Bhagavan, "The effect of supplementary processed protein foods fortified with essential vitamins and minerals on the growth and nutritional status of undernourished weaned infants," *Indian Journal of Paediatrics* 26:4 (1959), 406–13.

79 V. Subrahmanyan et al., "Processed Protein Foods of Vegetable Origin Fortified with Vitamins and Minerals" in R.C. Bhutiani, K.L. Raghakrishnan, and B.V. Subbarayappa (eds), *Some Aspects of Food Technology in India* (Mysore: CFTRI, 1959), 119–23.

80 UCLA Library Special Collections, MFM Records, box 43, Meals for Millions. Association of India, *A Decade of Progress* (Delhi: Meals for Millions India, 1963), 61–73.

81 UCLA Library Special Collections, MFM Records, box 43, "Indian Multi Purpose Food Recipes," 6; V. Subrahmanyan and H.A.B. Parpia, 'Some Aspects of the Utilization of Food Research in India', in R.C. Bhutiani, K.L. Raghakrishnan, and B.V. Subbarayappa (eds), *Some Aspects of Food Technology in India*, 9 (Mysore: CFTRI, 1959).

82 B. Siegel, "'Self-help which ennobles a nation': development, citizenship, and the obligations of eating in India's austerity years," *Modern Asian Studies* 50:3 (May 2016), 988; *A Decade of Progress*, 77.

83 *A Decade of Progress*, 41.

84 B.M. Popkin and M.C. Latham, "The limitations and dangers of commerciogenic nutritious foods," *The American Journal of Clinical Nutrition* 26 (1973), 1016; A.E. Bender, "Problems of human protein nutrition," *The Royal Society of Health Journal* 89:5 (Sep 1969), 224–26.

85 *A Decade of Progress*, 45–48.

86 "Making up for deficiency in nutrition," *Indian Express* (4 June 1955), 22.

87 UCLA Library Special Collections, MFM Records, box 42, "Meals for Millions" (leaflet published by Meals for Millions India).

88 UCLA Library Special Collections, MFM Records, box 43, "India 'Pattern' for Meals for Millions Program."

89 UCLA Library Special Collections, MFM Records, box 42, "Introductory Remarks at Opening of Mysore Branch, Meals for Millions Association of India at Bangalore by V. Subrahmanyan" (22 Dec 1963).

90 W. Shurtleff and A. Aoyagi, *History of Meals for Millions, Soy, and Freedom from Hunger (1946–2011): Extensively Annotated Bibliography and Sourcebook* (Lafayette: Soyinfo Centre, 2011), 76 (citation); 70; 87; UCLA Library Special Collections, MFM Records, box 42, "Nehru Quote in Clinton's Letter" (8 April 1957).

91 V. Subrahmanyan et al., "Plan for the Manufacture of Indian Multipurpose Food" in R.C. Bhutiani, K.L. Raghakrishnan, and B.V. Subbarayappa (eds), *Some Aspects of Food Technology in India*, 96 (Mysore: CFTRI, 1959).

92 *A Decade of Progress*, 41; "Introductory Remarks at Opening of Mysore Branch," (22 Dec 1963).

93 On US investments in the development of Balahar, see Appendix A of Alan Berg, *The Nutrition Factor: Its Role in National Development* (Washington: Brookings Institution, 1973); J. Levinson, "Vital to the creation: interview with Alan Berg," *Development* 56:1 (March 2013), 24–36; "Balahar... and the kitchen's children," *War on Hunger* 11:19 (1968), 9–13.

94 Siegel, *Hungry Nation*, 206–7; W. Scarfe and A. Scarfe, *Tiger on a Rein: Report on the Bihar Famine* (London: G. Chapman, 1969), 148.

95 UCLA Library Special Collections, MFM Records, box 43, "India 'Pattern' for Meals for Millions Program".

96 UCLA Library Special Collections, MFM Records, box 17, Progress Report of Feb. 1953.

97 UCLA Library Special Collections, MFM Records, box 42, "Message from Dr. V. Subrahmanyan on the occasion of the First Meeting of Meals for Millions Association of India" (3 June 1955).

98 "Introductory Remarks at Opening of Mysore Branch by V. Subrahmanyan" (22 Dec 1963).

99 UCLA Library Special Collections, MFM Records, box 17, "Chamberlain to Subrahmanyan, 5 July 1961".

100 Ibid.

101 Rao, "Dr Vaidyanatha Subrahmanyan – a pioneering food scientist," 1.

102 For a list of such products see W.R. Aykroyd, *The Conquest of Famine* (New York: Reader's Digest Press, 1975), 126; Bender, "Problems of human protein nutrition," 225.

103 The beginning of the demystification of protein food supplements is commonly attributed to the nutritionist Donald McLaren and his *Lancet* article of 1974. D.S. McLaren, "The great protein fiasco," *Lancet* 304:7872 (1974), 93–96. Also see P.V. Sukhatme, "Protein strategy and agricultural development," *The Indian Journal of Agricultural Economics* XXVII:1 (1972), 1–24.

Chapter 3. Worldly Needs and Religious Opportunities: The Famine Relief of American Missionaries in Bombay, 1870s–1920s

1 D. Hall-Matthews, "Colonial ideologies of the market and famine policy in Ahmednagar District, Bombay Presidency, c. 1870–1884," *The Indian Economic & Social History Review* 36:3 (1999), 303–33 here in particular page 304; N. Charlesworth, *Peasants and Imperial Rule. Agriculture and Agrarian Society in the Bombay Presidency, 1850–1935* (Cambridge: Cambridge University Press, 1985), 10–16 here in particular page 12; M.N. Kamat, "'A relief camp in the Deccan countryside': the Sholapur cotton textile industry, Western India," *Proceedings of the Indian History Congress Proceedings of the Indian History Congress* 66 (2006/2005), 818–26.

2 S. Sarkar, *Modern India, 1885–1947* (Basingstoke: Macmillan, 1992), 31.

3 Hall-Matthews, "Colonial ideologies," 308–18; Charlesworth, *Peasants and Imperial Rule*, 10–16.

4 S. Chhabria, *Making the Modern Slum: The Power of Capital in Colonial Bombay* (Seattle: University of Washington Press, 2019), 62; Hall-Matthews, "Colonial ideologies," 306.

5 Chhabria, *Making the Modern Slum*, 59–66; Charlesworth, *Peasants and Imperial Rule*, 221–22.

6 W. Digby, *The Famine Campaign in Southern India, 1876–1878*, vol. II (London: Longmans, Green, and Co., 1878); Sami, "Famine, Disease, Medicine and the State in Madras Presidency (1876–78)"; D. Hall-Matthews, "Inaccurate Conceptions," 1189–1212; Davis, *Late Victorian Holocausts*, 25–59.

7 Davis, *Late Victorian Holocausts*, 143–143.

8 Davis, *Late Victorian Holocausts*, 158–60; "Report on the Famine in the Bombay Presidency, 1899–1902" (Bombay: Government of Bombay, 1903).

9 Report of the American Mission among the Marathas for 1877 (Bombay, 1878), 7–9.

10 D.E. Haynes, *Small Town Capitalism in Western India: Artisans, Merchants and the Making of the Informal Economy, 1870–1960* (Cambridge: Cambridge University Press, 2012), 69.

11 Kamat, "'A relief camp in the Deccan countryside'" here in particular page 821.

12 E. Conroy-Krutz, *Christian Imperialism: Converting the World in the Early American Republic* (Ithaca: Cornell University Press, 2015), 82.

13 Conroy-Krutz, *Christian Imperialism*, 57–101; On the expansion of missionary schools in Bombay, see J. Tschurenev, *Empire, Civil Society, and the Beginnings of Colonial Education in India* (Cambridge: Cambridge University Press, 2019), 245–74.

14 Tschurenev, *Empire, Civil Society, and the Beginnings of Colonial Education in India*, 254–55; Andover Harvard Theological Library Boston, ABCFM, India, Records, 1813–1966, bMS 1264, box 3, folder 11, "100 Years 1838–1938. American Mission Girls' High School Ahmednagar India" (Ahmednagar, 1938), 4; On the wider context of such protest, see J.S. Palsetia, "Parsi and Hindu traditional and nontraditional responses to Christian conversion in Bombay, 1839–45", *The Journal of the American Academy of Religion* 47:3 (2006), 615–45.

15 Conroy-Krutz, *Christian Imperialism*, 79.

16 W. Hazen, *A Century in India: A Historical Sketch of the Marathi Mission of the ABCFM from 1813–1913* (Bombay: AMM, 1913), 70.

17 E.F. Kent, "'Mass Movements' in South India, 1877–1939" in D.C. Washburn and A.K. Reinhart (eds), *Converting Cultures: Religion, Ideology, and Transformations of Modernity* (Leiden: Brill, 2007), 371–81, the number is given on page 372.

18 Hazen, *A Century in India*, 68–69.

19 On the spatial concentration of the famine-affected population in camps, see Forth, *Barbed-Wire Imperialism*, 100–158.

20 Report of the AMM for the Year 1896 (Bombay, 1897), 22; "Our missionaries and famine relief," *Missionary Herald* XCVI:IX (Sep 1900), 356–7; G. Lambert, *India, the Horror-Stricken Empire* (Elkhart: Mennonite Publishing, 1898), 94.

21 Centennial Report of the AMM (Ahmednagar: AMM, 1913), 49–52; For biographical information on Devadhar see P. Kidambi, *The Making of an Indian Metropolis: Colonial Governance and Public Culture in Bombay, 1890–1920* (Aldershot: Ashgate, 2007), 208–9.

22 Yale Divinity Library Special Collections, New Haven, "General Report of the Americo-India Famine Relief Committee, Bombay", 1901; H. Fischer-Tiné, "Prarthana Samaj" in D. Crush, C.A. Robinson, and M.A. York (eds), *Encyclopedia of Hinduism* (New York: Routledge, 2008) 627–28; Kidambi, *The Making of an Indian Metropolis*, 162.

23 Such donations are documented: A Year's Work in India. Report of the American Marathi Mission of Western India of 1900 (Bombay: ABCFM, 1901), 55; Report of the AMM for the Year 1898 (Bombay, 1899), 15, 32.

24 Palsetia, "Parsi and Hindu Traditional and Nontraditional Responses", 615–45.

25 L. Vevaina, "Good deeds: Parsi trusts from 'the womb to the tomb'," *Modern Asian Studies* 52:1 (2018), 238–65.

26 Soni, "Learning to labour: 'native' orphans in colonial India, 1840s–1920s," *International Review of Social History* 65:1 (April 2020), 35–39; K.W. Jones, *Socio-Religious Reform Movements in British India* (Cambridge: Cambridge University Press, 1989), 94–103; D. Hardiman, "Purifying the nation: the Arya Samaj in Gujarat 1895–1930," *The Indian Economic & Social History Review* 44:1 (March 2007), 41–65; H. Fischer-Tiné, *Der Gurukul-Kangri oder die Erziehung der Arya-Nation: Kolonialismus, Hindureform und "nationale Bildung" in Britisch-Indien (1897–1922)* (Würzburg: Ergon, 2003).

27 "Lajpat Raj to the editor of the *Tribune* of 24 Oct 1899", published in B.R. Nanda, *The Collected Works of Lala Lajpat Rai*, vol. 1 (New Delhi: Manohar, 2008), 123.

28 "From Rev. R.A. Hume", *Life and Light of Women* XXVIII:6 (June 1898), 266.

29 "Feeding India's starving people", *Christian Herald* 20:37 (15 Sept 1897), 694.

30 In 1897, missionaries of the AMM received 5 to 10 percent less than in the previous years, while the allowances for Indian mission members were even cut by 30 to 45 percent. "Memorial from the Prudential Committee," *Missionary Herald* XCIV:I (Jan 1898), 2; Report of the American Marathi Mission for the Year 1896 (Bombay, 1897), 2; The annual budget application for the year 1897 was due before the famine drove up food prices. The approved budget as a result fell short of the actual needs of the AMM. For details see "Appropriations for India," *Missionary Herald* XCIII:I (Jan 1897), 1; "Indian famine relief," *Missionary Herald* XCIII:III (March 1897), 86.

31 "Indian Famine Relief," *Missionary Herald* XCIII:III (March 1897), 86; "Needs of the native agency," *Missionary Herald* XCIII:II (Feb 1897), 44; "High prices in India," *Missionary Herald* XCIV:IV (April 1898), 128.

32 "Appropriations for India," *Missionary Herald* XCIII:I (Jan 1897), 1; "Financial," *Missionary Herald* XCIV:IV (April 1898), 127.

33 D.G. Dawson, "Mission philanthropy: selected giving and Presbyterians," *American Presbyterians* 68:2 (1990), 121–32; Annual Report of the ABCFM 1897 (Boston, 1897), 18.

34 "A most successful year," *Missionary Herald* XCIV:V (May 1898), 176.

35 The Middle District included Connecticut, Ohio, and the Middle and South Atlantic States. Annual Report of the ABCFM 1897 (Boston, 1897), 7.

36 This argument has also been made in Curtis, *Holy Humanitarians*, see pages 12–14 for a summary of the argument that is fleshed out throughout the book.

37 Curtis, *Holy Humanitarians*, 124–25.

38 Annual Report of the ABCFM 1900 (Boston, 1900), 20, 69.

39 Tyrrell, *Reforming the World*, 43.

40 For a study of the *Christian Herald's* humanitarian campaigns in book-length see Curtis, *Holy Humanitarians*. See also H. Curtis, "Depicting distant suffering: evangelicals and the politics of pictorial humanitarianism in the age of American empire," *Material Religion* 8:2 (June 2012), 153–82.

41 Tyrrell, *Reforming the World*, 110.

42 Ibid., 110–11.

43 Ibid., 98–119 here in particular 114.

44 Lambert, *India, the Horror-Stricken Empire*, 194.

45 Curtis, *Holy Humanitarians*, 125–34; Moreover, Ian Tyrrell pointed out the role of Mary and Margaret Leitch, missionaries of the ABCFM who had worked in Ceylon, in convincing Klopsch to raise funds for famine relief in India in 1899. Tyrrell, *Reforming the World*, 111–12.

46 On US Anti-Imperialism, see the introduction of I. Tyrrell and J. Sexton (eds), *Empire's Twin: U.S. Anti-Imperialism from the Founding Era to the Age of Terrorism* (Ithaca: Cornell University Press, 2015), 1–18.

47 Curtis, *Holy Humanitarians*, 126, 138.

48 It became the second largest donor of famine relief in 1900 forwarding $252,988 to Bombay. NYPL, Manuscripts and Archives Division, Richard Roger Bowker Papers, Letters (Committee of the One Hundred on India Famine Relief 1900), "Final Report of the Committee of one Hundred on India Famine Relief."

49 Report of the Ecumenical Conference on Foreign Missions (New York: Ecumenical Missionary Conference, 1900), 48; "To give aid to famine sufferers," *New York Times* (10 May 1900), 6.

50 Curtis, *Holy Humanitarians*, 131; 236–38.

51 "Final Report of the Committee of one Hundred on India Famine Relief."

52 "The India Famine Fund," *New York Times* (12 June 1900), 9.

53 The Burke Library at Union Theological Seminary, Columbia University in the City of New York, MRL3: American Marathi Mission Records, box 1, folder 1, "Hume of Ahmednagar—His story, III. Philanthropic and Causal Work," unpaginated; "General Report of the Americo-India Famine Relief Committee, Bombay," 1901.

54 Houghton Library, Boston, ABCFM Papers film A467 (Robert A. Hume), "R.A. Hume to J. Barton" (9 Oct 1900).

55 "Many missionaries return from India," *Boston Daily Globe* (4 June 1919), 10; A.C. Hunsberger, "From Brimstone to the World's Fair: A Century of 'Modern Missions' as seen through the American Hume Missionary Family in Bombay" in C.W. Putney and P.T. Burlin (eds), *The Role of the American Board in the World: Bicentennial Reflections on the Organization's Missionary Work, 1810–2010* (Eugene: Wipf & Stock, 2012), 114–23.

56 B. Ruswick, *Almost Worthy. The Poor, Paupers, and the Science of Charity in America, 1877–1917* (Bloomington: Indiana University Press, 2013).

57 G. Brewis, "'Fill full the mouth of famine': voluntary action in famine relief in India 1896–1901," *Modern Asian Studies* 44:4 (July 2010), 887–918; G. Brewis, "The Making of an Imperial Ideal of Service: Britain and India before 1914" (Unpublished dissertation, University of East London, 2009), 149–65.

58 The colonial efforts to carve out a space for private charity during famines were part of broader debates on charity in India, and of a set of colonial interventions that aimed to discredit "Indian" philanthropy and to encourage ostensibly modern forms of charity. M. Kasturi, "'All gifting is sacred': the Sanatana Dharma Sabha Movement, the reform of *dana* and civil society in late colonial India," *The Indian Economic & Social History Review* 47:1 (Jan 2010), 107–39; M. Kasturi, "*Gurus* and gifting: *dana*, the *math* reform campaign, and competing visions of Hindu *Sangathan*

in twentieth-century India," *Modern Asian Studies* 52:1 (Jan 2018), 99–131; N. Chatterjee, *The Making of Indian Secularism: Empire, Law and Christianity, 1830–1960* (Basingstoke: Palgrave Macmillan, 2011), 51–74; F. Osella, "Charity and philanthropy in South Asia: an introduction," *Modern Asian Studies* 52:1 (Jan 2018), 4–34.

59 On the perpetuation of caste in colonial famine policy, see G.G. Raheja, "'Hear the tale of the famine year'," 155–98. On orphan care, see Soni, "Learning to labour," 15–42.

60 Report of the Central Executive Indian Famine Charitable Relief Fund, 1897 (Madras: Superintendent of Government Printing, 1898), 15; Report of the Central Executive Committee, Indian Famine Charitable Relief Fund, 1900 (Calcutta: Superintendent of Government Printing, 1901), 6.

61 A Year's Work in India. Report of the American Marathi Mission of Western India of 1900 (Bombay: ABCFM, 1901), 36, 41.

62 A Year's Work in India, 41–42.

63 R. Ward, *Some Fruits of the Great Famine* (Madras, 1902), 5.

64 The Burke Library at Union Theological Seminary, Columbia University in the City of New York, MRL3: AMM Records, box 1, folder 1, "Hume of Ahmednagar – His story," unpag.

65 Tyrrell, *Reforming the World*, 114.

66 P.A. Kramer, "Empires, exceptions, and Anglo-Saxons: race and rule between the British and United States empires, 1880–1910," *The Journal of American History* 88:4 (March 2002), 1341–42.

67 Reprinted in "England's war and famine in India," *San Francisco Call* (14 May 1900), 4; Since Anglophobia was much more widespread among Democrats, the case of the *Cleveland Leader*—a daily associated with the Republican Party—stands out. See Kramer, 1341–42.

68 "Famine v. war," *Illustrated Missionary News* (18 June 1900), 104.

69 Cf. "To give aid to famine sufferers," *New York Times* (10 May 1900), 6.

70 Cf. Ibid.

71 M. Kupiec Cayton, "Canonizing Harriet Newell. Women, the Evangelical Press, and the Foreign Mission Movement in New England, 1800–1840" in B. Reeves-Ellington, K. Kish Sklar and C.A. Shemo (eds), *Competing Kingdoms: Women, Mission, Nation, and the American Protestant Empire, 1812–1960* (Durham: Duke University Press, 2010), 69–93.

72 D.L. Robert, "The 'Christian Home' as a Cornerstone of Anglo-American Missionary Thought and Practice" in D.L. Robert (ed.), *Converting Colonialism: Visions and Realities in Mission History, 1706–1914* (Grand Rapids: William B. Eerdmans Pub. Co, 2008), 134–65; M. Chawla Singh, *Gender, Religion, and "Heathen Lands": American Missionary Women in South Asia, 1860s–1940s* (New York: Garland Pub, 2000) in particular pages 88–92; E.F. Kent, *Converting Women: Gender and Protestant Christianity in Colonial South India* (Oxford: Oxford University Press, 2004).

73 R. Wollons, "Travelling for God and adventure: women missionaries in the late 19th century," *The Asian Journal of Social Science* 31:1 (2003), 57.

74 Hunsberger, "From Brimstone to the World's Fair," 112; Tschurenev, *Empire, Civil Society, and the Beginnings of Colonial Education in India*, 287; The number is provided in Report of the American Marathi Mission for the Year 1896 (Bombay, 1897), 5; On the context of Indian mission members and their important role as intermediaries, see H. Liebau,"Country Priests, Catechists, and Schoolmasters as Cultural, Religious, and Social Middlemen in the Context of the Tranquebar Mission" in R.E. Frykenberg and A.M. Low (eds), *Christians and Missionaries in India: Cross-Cultural Communication since 1500, with Special Reference to Caste, Conversion, and Colonialism* (Grand Rapids: Routledge, 2003), 70–92.

75 For a comprehensive discussion of their careers see S. Ross Pripas-Kapit, "Educating Women Physicians of the World: International Students of the Woman's Medical College of Pennsylvania, 1883–1911" (Unpublished dissertation, UCLA, 2015), 107–53.

76 Ross Pripas-Kapit, "Educating Women Physicians of the World," 120–22.

77 M. Kosambi, "A prismatic presence: the multiple iconisation of Dr Anandibai Joshee and the politics of life-writing," *Australian Feminist Studies* 16:35 (2001), 157–73; N. Patwardhan, *Radical Spirits. India's First Woman Doctor and Her American Champions* (Raleigh: Story Artisan Press, 2020).

78 A. Abbott, "Gurubai Karmarkar, M.D." in *Life Stories of Native Helpers in India, Ceylon, China, Japan, Africa, Turkey, Mexico, the Balkans* (Boston: Women's Board of Missions, 1917), 5–10.

79 Ross Pripas-Kapit, "Educating Women Physicians," 127; Report of the AMM of the ABCFM of 1912 (Ahmednagar, 1913), 12; Report of the AMM of 1909 (Ahmednagar, 1910), 159.

80 A Year's Work in India, 19–20.

81 M. Sebastian, "Reading archives from a postcolonial feminist perspective: 'native' Bible women and the missionary ideal," *The Journal of Feminist Studies in Religion* 19:1 (2003), 5–25; E.F. Kent, "Tamil bible women and the zenana missions of colonial South India," *History of Religions* 39:2 (1999), 117–49.

82 A Year's Work in India, 59.

83 "From the report of Mrs. Gurubai Kamarkar, of Bombay," *Life and Light for Women XXIX:*3 *(*March 1899), 119–121.

84 A Year's Work in India, 12.

85 V.R. Karmarkar, *The Story of One Famine Orphan* (New York: Women's Board of Foreign Missions of the Presbyterian Church, n.d.).

86 Abbott, "Gurubai Karmarkar, M.D.", 8; Report of the AMM for the Year 1918 (Barsi: AMM, 1919), 3.

87 "India's sacrifice to America," *Missionary Herald* CXV:1 *(*Jan 1919), 7.

88 Conroy-Krutz, *Christian Imperialism*, 88–89.

89 "My adopted sister," *India Alliance* XI:3 (Sep 1911), 636–37.

90 Ibid., 636.

91 Ibid., 637.

92 N. Maksudyan, *Orphans and Destitute Children in the Late Ottoman Empire* (Syracuse: Syracuse University Press, 2014), 117–19; N. Maksudyan, "'Being saved to serve': Armenian orphans of 1894–1896 and interested relief in missionary orphanages," *Turcica* 42 (2010), 47–88.

93 Hazen, *A Century in India*, 85.

94 Chapter VI of C. M. Pepper, *Life-Work of Louis Klopsch. Romance of a Modern Knight of Mercy* (New York: Christian Herald, 1910).

95 A Year's Work in India, 59.

96 H. Fairbank, "Industrial Work in the American Marathi Mission, Ahmednagar, Western India" (Sixth Indian Industrial Conference 1910, Amraoti, 1911), 261.

97 A Year's Work in India, 9.

98 Phipps would later fund the Imperial Agricultural Research Institute at Pusa.

99 Houghton Library, Boston, ABCFM Papers film A467 (Robert A. Hume), "Robert A. Hume to James Barton" (2 July 1902); "Robert A. Hume to Henry Phipps," undated. The following letters received or written by Hume, cited in this chapter, are from the same film reel.

100 A. Zimmerman, "A German Alabama in Africa: the Tuskegee expedition to German Togo and the transnational origins of West African cotton growers," *The American Historical Review* 110:5 (Dec 2005), 1362–98.

101 "Robert Allen Hume to Henry Phipps" undated.

102 Ibid.

103 "Robert A. Hume to R.A. Lamb," (10 Feb 1902).

104 K.L. Butterfield, *The Christian Mission in Rural India. Report and Recommendations* (New York: International Missionary Council, 1930), 40.

105 "Hume of Ahmednagar – His story," 2.

106 Hazen, *A Century in India*, 84; Twentieth Annual Report of the Department of Land Records and Agricultural, Bombay Presidency for the Year 1902–1903 (Bombay: Government Central Press, 1904).

107 Report of the AMM 1902 (Madras, 1903), 37.

108 Ibid., 38.

109 A Year's Work in India, 7.

110 Report of the AMM 1902, 38–39.

111 After their return to North America, he and his wife applied their knowledge of Indian cotton designs, and became successful proprietors of Churchill Weavers in Kentucky. For a detailed discussion of Churchill's further career in India see A. McGowan, *Crafting the Nation in Colonial India* (New York: Palgrave Macmillan, 2009), 170–81; Haynes, *Small Town Capitalism in Western India*, 207–10.

112 A. Kumar Bagchi, "Indian economy and society during World War One," *Social Scientist* 42:7/8 (2014), 5–27.

113 D. Arnold, "Death and the modern empire: the 1918–19 Influenza epidemic in India," *Transactions of the Royal Historical Society* 29 (2019), 191–92.

114 Charlesworth, *Peasants and Imperial Rule*, 204.

115 S. Ansari, "The Bombay Presidency's 'Home Front', 1914–1918" in R.D. Long and I. Talbot (eds), *India and World War I. A Centennial Assessment* (New York: Routledge, 2018), 60–62.

116 Chhabria, *Making the Modern Slum*, 185.

117 "Report of the Committee of the Bombay Presidency Famine Relief Fund "(Appendix 2) in Review of the Famine Relief Administration in the Bombay Presidency during the year 1918–1919 (Bombay Presidency, 1920), 5–14; For a discussion of the philanthropy of Indian traders and industrialists see A. Kumar, *Philanthropy and the Development of Modern India: In the Name of Nation* (New York: Oxford University Press, 2021).

118 Review of the Famine Relief Administration in the Bombay Presidency during the Year 1918–1919, 1–2.

119 Appendix 3 to the Review of the Famine Relief Administration in the Bombay Presidency during the Year 1918–1919, 15–19.

120 Budget of the Government of Bombay for 1921–22 (Bombay: Government of Bombay Financial Department, 1921), 8.

121 I.D. Mills, "The 1918–1919 Influenza pandemic: the Indian experience," *The Indian Economic & Social History Review* 23:1 (1986), 35–36; Also see Arnold, "Death and the modern empire. the 1918–19 influenza epidemic in India," 192.

122 D. Hardiman, "The Influenza epidemic of 1918 and the Adivasis of Western India," *Social History of Medicine* 25:3 (Aug 2012), 644–64; Arnold, "Death and the modern empire: the 1918–19 Influenza epidemic in India", 181–200; M. Ramanna, "Coping with the Influenza Pandemic: the Bombay Experience" in H. Phililips and D. Killingray (eds), *The Spanish Influenza Pandemic of 1918–19. New Perspectives* (London: Routledge, 2003), 86–98.

123 Ramanna, "Coping with the Influenza Pandemic: the Bombay Experience," 89.

124 Mills, "The 1918–1919 Influenza pandemic: the Indian experience," 34.

125 Review of the Famine Relief Administration in the Bombay Presidency during the year 1918–1919, 4.

126 Report of the AMM for the Year 1918 (Barsi: AMM, 1919), 4.

127 Report of the AMM for the Year 1919 (Barsi: AMM, 1920), 5.

128 Andover Harvard Theological Library. ABCFM, India. Records, 1813–1966, bMS 1264, box 2, folder 1, A.H. Clark, "War Strain in India" (AMM, 1918). This was a reprint from the *Congregationalist* edition of 12 Dec 1918.

129 Report of the AMM for the Year 1920 (Ahmednagar: AMM, 1921), 52.

130 Report of the AMM for the Year 1918, 19; Report of the AMM for the Year 1919, 35.

131 Report of the AMM for the Year 1920, unpag (Fundamental Facts and Figures).

132 Report of the AMM for the Year 1919, 51.

133 Report of the AMM. In Western India, 1923 (Vadala: AMM, 1924), 22.

134 Biographical information has been derived from a combination of the following sources: Report of the AMM 1909 (Ahmednagar: AMM, 1910), 5; "Anand Sidoba Hiwale", *Bowdoin Orient* XXXIX:20 (10 Dec 1909), 167; "Notes on the Rochester Convention," *Bowdoin Orient* XXXIX:23 (14 Jan 1910), 187; "Bowdoin in the East," *Bowdoin Orient* XXXIX:26 (18 Feb 1910), 213.

135 Report of the AMM for the Year 1919, 6.

136 Report of the AMM for the Year 1919, 13, 30.

137 Bombay Representative Council of Missions Bombay. Annual Report for 1917–1918 (Bombay, 1918), 13; Bombay Representative Council of Missions Bombay. Annual Report for 1919–1920 (Bombay: Council of Missions, 1920), 9.

138 Proceedings of the Sixth Meeting of the National Missionary Council, Lahore, 13–18 Nov 1919 (Calcutta: National Missionary Council, 1920), 54.

139 W.R. Hogg, *Ecumenical Foundations. A History of the International Missionary Council and Its Nineteenth-Century Background* (New York: Harper and Brothers Publishers, 1952), 152–53; B. Stanley, *The World Missionary Conference, Edinburgh 1910* (Grand Rapids: William B. Eerdmans Pub. Co, 2009).

140 Little, "An explosion of new endeavours," 2.

141 Whereas historians have long regarded the secularisation of philanthropy as all-encompassing, recent studies aptly demonstrate the continued importance of American religious and in particular missionary organisations in US-sponsored relief during World War I and in its aftermath. J. Beer, *The Philanthropic Revolution: An Alternative History of American Charity* (Philadelphia: University of Pennsylvania Press, 2015), 1–15; On the proposed secularisation of aid during World War I, see Cabanes, *The Great War and the Origins of Humanitarianism*, 4.

142 J.L. Barton, "The effect of the war on Protestant missions," *Harvard Theological Review* 12:1 (Jan 1919), 1–35; R.A. Wright, *A World Mission: Canadian Protestantism and the Quest for a New International Order, 1918–1939* (Montreal: McGill-Queen's University Press, 1991), 114–17.

143 "India's 'National' and Christian progress," *Missionary Review of the World* 42 (Jan 1919), 21–22.

144 H Fischer-Tiné, "'Unparalleled opportunities': the Indian Y.M.C.A.'s army work schemes for imperial troops during the Great War (1914–1920)," *The Journal of Imperial and Commonwealth History* 47:1 (2019), 100–137; J. Lovell Murray, *The Call of a World Task in War Time* (New York: Student Volunteer Movement, 1918), 86–87.

145 Murray, *The Call of a World Task in War Time*, 70.

146 J.M. Selles, *The World Student Christian Federation, 1895–1925: Motives, Methods, and Influential Women* (Eugene: Pickwick Publications, 2011), 102–3; N.D. Showalter, "Crusade or catastrophe? the student missions movement and the First World War," *International Bulletin of Missionary Research* 17:1 (Jan 1993), 13–17.

147 Murray, *The Call of a World Task in War Time*, 106; For a discussion of the book and Murray's role in the missionary movement see Wright, *A World Mission*, 114–21.

148 Tyrrell, *Reforming the World*, 67; B. Harder, "The student volunteer movement for foreign missions and its contribution to 20th century missions," *Missiology: An International Review* 8:2 (April 1980): 141–54; D.L. Robert, "The origin of the student volunteer watchword: 'the evangelization of the world in this generation'," *International Bulletin of Missionary Research* 10:4 (Oct 1986), 146–49.

149 Tyrrell, "Vectors of Practicality," 39–60.

150 C. Hodge Evans, *The Social Gospel in American Religion: A History* (New York: New York University Press, 2017); J.F. Fishburn, "The Social Gospel as Missionary Ideology" in W.R. Shenk (ed.), *North American Foreign Missions, 1810–1914: Theology, Theory, and Policy* (Grand Rapids: William B. Eerdmans Pub, 2004), 218–42.

151 Report of the AMM for the Year 1918, 2.

152 "India's famine spreading," *Christian Herald* 42:4 (1 March 1919), 248; Details on the work of Methodists during the war can be found in M. Renton, *War-Time Agencies of the Churches. Directory and Handbook* (New York: General War-Time Commission of the Churches, 1919), 66–71.

153 Report of the Foreign Missions Conference of North America (New York, 1920), 43.

154 "India's famine spreading," *Christian Herald* 42:4 (22 Jan 1919), 90; *India Alliance* XIX:4, 5, 6 (Oct, Nov & Dec 1918), 338–340.

155 Curtis, *Holy Humanitarians*, 265.

156 "Famine death roll grows in India," *Christian Herald* 42:12 (March 1919), 304.

157 "Death's harvest in India," *Christian Herald* 42:19 (10 May 1919), 553.

158 P. Harnetty, "The famine that never was: Christian Missionaries in India, 1918–1919," *The Historian* 6:3 (March 2001), 555–75.

159 Proceedings of the Sixth Meeting of the National Missionary Council, Lahore, 13–18 Nov 1919, 55.

160 "Missions and famine. India's debt to the West," *Times of India* (1 Dec 1919), 6.

161 Annual Report of the ABCFM 1921 (Boston, 1921), 84.

162 R.A. Hume, Report of the American Marathi Mission for the Year 1920 (Ahmednagar: ABCFM Marathi Mission, 1921), 52.

163 A Year's Work in India, 62.

Chapter 4. Promising Freedom from Famine: American Missionary Rural Reform, 1910s–1940s

1 E. Sherwood, *India Awakening* (New York: Missionary Education Movement of the US and Canada, 1912), 25–26.

2 M. Zanasi, "Western utopias, missionary economics, and the Chinese village," *The Journal of World History* 24:2 (2013), 362; On famine relief in China, see P. Fuller, *Famine Relief in Warlord China* (Cambridge: Harvard University Press, 2019).

3 P. Kumar, "'Modernization' and agrarian development in India, 1912–52," *The Journal of Asian Studies* 79:3 (2020), 638.

4 D. Naoroji, *Poverty and Un-British Rule in India* (London: Swan Sonnenschein, 1901); R. C. Dutt, *Indian Famines. Their Causes and Prevention* (London: P.S. King & Son, 1901).

5 Mukherjee, *Natural Disasters and Victorian Empire*, 29–60.

6 On the YMCA, see H. Fischer-Tiné, *The YMCA in Colonial India: Modernization, Philanthropy and American Soft Power in South Asia* (London: Bloomsbury Academic, 2022); H. Fischer-Tiné, S. Huebner, and I. Tyrrell (eds), *Spreading Protestant Modernity: Global Perspectives on the Social Work of the YMCA and YWCA* (Honolulu: University of Hawai'i Press, 2021).

7 Tyrrell, *Spreading Protestant Modernity*, 3; On the Social Gospel Movement, see Fishburn, "The Social Gospel as Missionary Ideology," 218–242; Evans, *The Social Gospel in American Religion*; W. J. Deichmann Edwards and C. De Swarte Gifford (eds), *Gender and the Social Gospel* (Urbana: University of Illinois Press, 2003).

8 Tyrrell, "Vectors of Practicality," 42–43.

9 Amrith and Clavin, "Feeding the world," 31.

10 Tyrrell, "Vectors of Practicality," 43–44.

11 S. Sohi, *Echoes of Mutiny: Race, Surveillance, and Indian Anticolonialism in North America* (New York: Oxford University Press, 2014), 148.

12 G.R. Hess, *Sam Higginbottom of Allahabad, Pioneer of Point Four to India* (Charlottesville: University Press of Virginia, 1967), 26.

13 "Death's harvest in India," *Christian Herald* 42:19 (10 May 1919), 552.

14 S. Higginbottom, *The Gospel and the Plow or the Old Gospel and Modern Farming in Ancient India* (New York: MacMillan Company, 1921), 50.

15 The procurement of funds of R.A. Hume from the Americo-India Famine Committee in 1901 is one such example (see the previous chapter for details). For other examples from China see R.E. Stross, *The Stubborn Earth: American Agriculturalists on Chinese Soil, 1898–1937* (Berkeley: University of California Press, 1989), 141; J.E. Baker, "Is it Missionary Work?" *The Chinese Recorder* (1 Sep 1924), 583; L. Impey, "The theory and practice of famine prevention," *The China Weekly Review* (30 Aug 1924), 441.

16 Stross, *The Stubborn Earth*, 141.

17 G.R. Hess, "American agricultural missionaries and efforts at economic improvement in India," *Agricultural History* 42:1 (1968), 23; Tyrrell, "Vectors of Practicality," 42.

18 M.P. Brunner, "Teaching development: debates on 'scientific agriculture' and 'rural reconstruction' at Khalsa College, Amritsar, c. 1915–47," *The Indian Economic & Social History Review* 55:1 (Jan 2018), 77–132.

19 "Sam Higginbottom," *Rotarian* (Oct 1944), 10–12; Kumar, "'Modernization' and agrarian development in India, 1912–52," 633–58; Chatterjee, *The Making of Indian Secularism*, 147–50; Hess, *Sam Higginbottom of Allahabad, Pioneer of Point Four to India.*

20 Fischer-Tiné, "The YMCA and low-modernist rural development in South Asia, c.1922–1957," 193–234.

21 H. Büschel, "Eine Brücke am Mount Meru: Zur Globalgeschichte von Hilfe zur Selbsthilfe und Gewalt in Tanganjika" in H. Büschel and D. Speich (eds), *Entwicklungswelten: Globalgeschichte der Entwicklungszusammenarbeit* (Frankfurt am Main: Campus, 2009), 175–206.

22 Fischer-Tiné, "The YMCA and low-modernist rural development in South Asia, c.1922–1957," 210; Kumar, "'Modernization' and agrarian development in India, 1912–52," 639.

23 "Pretty Ohio girl goes to famine-stricken India as farm missionary," *Day Book* 1:111 (3 Feb 1912), 8; *The Ohio State Bulletin, Alumni and Former Student Register (1878–1912)* (Nov 1912), 31; *The Cincinnatian 1902* (University of Cincinnati), 43; "A record graduating class," *Chicago Livestock World* (14 May 1912), 2; "Weekly Columbus letter," *Ohio Farmer* (22 June 1912), 7; On the history of the YWCA, see K.E. Phoenix, "Not by Might, nor by Power, but by Spirit': The Global Reform Efforts of the Young Women's Christian Association of the United States, 1895–1939" (Unpublished dissertation, University of Illinois, 2010).

24 B. Kim, "Lakshmi and the scientific housewife: a transnational account of Indian women's development and production of an Indian modernity," *Economic and Political Weekly* 38:11 (2003), 1057–58.

25 Hess, *Sam Higginbottom of Allahabad, Pioneer of Point Four to India*, 17–20.

26 Mrs. Sam Higginbottom, "Rural home section," *Allahabad Farmer* IX:3 (May 1935), 147.

27 Mrs. Sam Higginbottom, "An experiment in rural education: a review", *Allahabad Farmer* IX:3 (May 1935), 105–9.

28 Mrs. Ethel C. Higginbottom, "Course in home-making," *Allahabad Farmer* X:5 (Sep 1936), 213–5.

29 Matriculation was usually given after the completion of the tenth grade and qualified students for higher education.

30 Mrs. E.F. Vestal, "Short course for women," *Allahabad Farmer* X:6 (Nov 1936), 256–7.

31 E.L. Fleischmann, "At home in the world: globalizing domesticity through home economics in the interwar years," *Modern Asian Studies* 35:4 (2001), 886.

32 M. Hancock, "Home science and the nationalization of domesticity in Colonial India," *Modern Asian Studies* 35 no. 4 (Oct 2001), 871–903 in particular page 887.

33 Ibid., 895.

34 D.S. Hatch, *Further Upward in Rural India* (London: Oxford University Press, 1938), 145. Emphasis added.

35 Ibid., 146–47.

36 Fischer-Tiné, "The YMCA and low-modernist rural development in South Asia," 204.

37 E.G. Hatch, *Mar-Tan-Dam: Not on the Map?*, The International Committee of the Young Men's Christian Association of the United States and Canada, n.d [1933].

38 Hatch, *Mar-Tan-Dam: Not on the Map?*; For a discussion of her plays see Fischer-Tiné, "The YMCA and low-modernist rural development in South Asia," 215.

39 M.G. Thompson, *For God and Globe: Christian Internationalism in the United States between the Great War and the Cold War* (Ithaca: Cornell University Press, 2016).

40 A.E. Cornebise, *Soldier-Scholars: Higher Education in the AEF, 1917–1919* (Philadelphia: American Philosophical Society, 1997).

41 Jachertz and Nützenadel, "Coping with hunger?," 99–119; A. Nützenadel, "'A world without famine'? Internationale Ernährungspolitik im Zeitalter der Weltkriege," *Comparativ* 17:3 (2007), 12–27; Amrith and Clavin, "Feeding the world," 29–50.

42 *World Agriculture* V:3 (Oct 1925), 400. Emphasis added.

43 "China's place in world agriculture," *World Agriculture* II:1 (July 1921), 94–96.

44 "China's contributions to the world's food," *World Agriculture* II:1 (July 1921), 102–103.

45 Stross, *The Stubborn Earth*, 101.

46 Tyrrell, "Vectors of Practicality," 46; Stross, *The Stubborn Earth*, 101.

47 Tyrrell, "Vectors of Practicality," 46.

48 Butterfield, *The Christian Mission in Rural India. Report and Recommendations*; Duane Spencer Hatch, *Up From Poverty in Rural India* (Bombay: Oxford University Press, 1932), 228.

49 J.H. Reisner, *Reforesting China. Permanent Famine Prevention versus Famine Relief* (Nanking: Nanking University, 1921).

50 Andover Theological Library, Harvard University, Boston, ABCFM India, box 20, folder 1, "Reisner to the Christian Rural Fellowship, 30 June 1939," 2.

51 B.H. Hunnicutt and W.W. Ried, *The Story of Agricultural Missions* (New York: Missionary Education Movement of the US and Canada, 1931), v.

52 "The International Association of Agricultural Missions," *World Agriculture* 1:2 (June 1920), 27; 30. Tyrrell, "Vectors of Practicality," 52.

53 S. Billington Harper, *In the Shadow of the Mahatma: Bishop V.S. Azariah and the Travails of Christianity in British India* (Grand Rapids: W.B. Eerdmans Pub. Co, 2000).

54 Andover Theological Library, ABCFM, India, box 1, folder 10, "W. Hazen, *Missions and the Government in India*, 1943", 28.

55 C.M. Bauman, "Postcolonial anxiety and anti-conversion sentiment in the report of the Christian missionary activities enquiry committee," *The International Journal of Hindu Studies* 12:2 (Aug 2008), 181–213; C.M. Bauman, "Hindu-Christian conflict in India: globalization, conversion, and the coterminal castes and tribes," *The Journal of Asian Studies* 72:3 (Aug 2013), 633–53.

56 Andover Theological Library, ABCFM, India, box 20, folder 2, E.C. Bhatty and I.N. Singh, "Towards Richer Life. A Report on the Economic Life of the Church" (Mysore: National Christian Council, 1950).

57 Ibid., 14.

58 J. Bathgate, "Presbyterians and rural development in India," *The Journal of Presbyterian History* 62:3 (1984), 243; Kumar, "'Modernization' and agrarian development in India, 1912–52," 633–58.

59 Bhatty and Singh, "Towards richer life," 21; Fischer-Tiné, "The YMCA and low-modernist rural development in South Asia, c.1922–1957," 193–234.

60 J. Fischer, "India's Insoluble Hunger," *Harper's Magazine* (1 April 1945), 438–45.

61 J. Fischer, *India's Insoluble Hunger* (Bombay: Vora & Co., 1947).

62 Biographical information on John Fischer is taken from Yale University Library, Manuscripts and Archives, John Sylvester Fischer Papers (MS 850), "Biographical/Historical"; D.J. Lawler and E. R. Mahan (eds), *Foreign Relations of the United States: Diplomatic Papers, 1943, The Near East and Africa, vol. IV, India* (Washington: Government Printing Office, 2010), (document) 281; "Harper's editor to quit, successor named," *New York Times* (29 Jan 1953), 23.

63 M.S. Venkataramani, "The Roosevelt administration and the Great Indian Famine," *International Studies* 4:3 (July 1962), 241–64.

64 Fischer, "India's Insoluble Hunger," 445.

65 Fischer, "India's Insoluble Hunger," 440.

66 Ibid., 440.

67 Nair, "The construction of a 'population problem' in colonial India 1919–1947," 227–47.

68 A. Bashford, *Global Population: History, Geopolitics, and Life on Earth* (New York: Columbia University Press, 2014); M. Connelly, "Population control in India: prologue to the emergency period," *Population and Development Review* 32:4 (Dec 2006), 629–67.

69 S. Neale Alter, "The Future of Rural Christian Missions in India" (New York: Agricultural Missions, 1945).

70 I.W. Moomaw, "Missions and human need," *International Review of Mission* 39:156 (1950), 419.

71 Bathgate, "Presbyterians and Rural Development in India," 242.

72 W.H. Wiser and C.V. Wiser, *Behind Mud Walls, 1930–1960* (Berkeley: University of California Press, 1971).

73 Yale Divinity Library, Special Collections, William Hendricks Wiser and Charlotte Viall Wiser Papers (RG 128) (hereafter Wiser Papers), box 1, folder 3, "India Village Service" (pamphlet).

74 Sackley, "Village models:," 749–78.

75 India Village Service (pamphlet), 4; D. M. Ment, "The American role in education in the Middle East: ideology and experiment, 1920–1940," *Paedagogica Historica* 47:1–2 (Feb 2011), 179–81.

76 India Village Service (pamphlet), 5.

77 S. Daniel, "Working amongst women in Uttar Pradesh," *Community Development Bulletin* 8 no. 4 (1957), 79.

78 C.R. Unger, *Entwicklungspfade in Indien: eine internationale Geschichte 1947–1980* (Göttingen: Wallstein Verlag, 2015), 68–71; P. Das, "Women under India's community development programme," *International Labour Review* 80 (1959), 26–45.

79 Yale Divinity Library, Special Collections, Wiser Papers, box 1, folder 5, "Personnel for work among rural women," *India Village Service Chronicle* 184 (22 Sept 1956).

80 Unger, *Entwicklungspfade in Indien*, 70.

81 Training Within Industry Report, 1940–1945. A Record of the Development of Management Techniques for Improvement of Supervision. Their Use and the Results (Washington: War Manpower Commission, 1945), ix; S.A. Carpenter, "'Regular farm girl': the women's land army in World War II," *Agricultural History* 71:2 (1997), 163–85.

82 W.D. Rasmussen, *A History of the Emergency of Farm Labour Supply Program, 1943–47*, vol. 13 (Washington: US Department of Agriculture. Bureau of Agricultural Economics, 1951), 158–60.

83 C.R. Dooley, "Training Within Industry in the United States," *International Labour Review* 54:3–4 (1946), 161.

84 Yale Divinity Library, Special Collections, Wiser Papers, box 1, folder 3, "Nineteenth Quarterly Report, India Village Service (March, April and May 1950), part I," 7.

85 Ibid., 1.

86 Yale Divinity Library, Special Collections, Wiser Papers, box 1, folder 3, W.H. Wiser, "Key Point Lesson" (12 May 1950).

87 "Key point lessons," *Allahabad Farmer* XXV:4 (July 1951), 135; A. Mayer, *Pilot Project, India: The Story of Rural Development at Etawah, Uttar Pradesh* (Berkeley: University of California Press, 1958), 165.

88 "Nineteenth Quarterly Report," 2.

89 Yale Divinity Library, Special Collections, Wiser Papers, box 1, folder 3, "India Village Service after Four Years" (pamphlet), 9.

90 "India village service after four years", 10.

91 Visual Aids: A Guide for Peace Corps Volunteers (Washington: Peace Corps Information Collection and Exchange, 1976).

Chapter 5. Famine Amid Swadeshi and Swaraj, 1900s–1920s

1 "The Famine Union," *Times of India* (11 July 1901), 4.

2 N. Owen, *The British Left and India: Metropolitan Anti-Imperialism, 1885–1947* (Oxford: Oxford University Press, 2007), 35–48.

3 "The Indian Famine Union," *India* (14 June 1901), 282–84; "Indian Famine Union," *The Speaker* (6 July 1901), 387–88.

4 S.K. Ratcliffe, *Sir William Wedderburn and the Indian Reform Movement* (London: George Allen & Unwin Ltd., 1923), 110–19.

5 S. Bose, "Pondering Poverty, Fighting Famines: Towards a New History of Economic Ideas" in K. Basu and R. Kanbur (eds), *Arguments for a Better World: Essays in Honor of Amartya Sen*, vol. 2 (Oxford: Oxford University Press, 2009), 425–28.

6 D. Patel, *Naoroji: Pioneer of Indian Nationalism* (Cambridge: Harvard University Press, 2020), 46–86.; A.K. Dasgupta, *A History of Indian Economic Thought* (Routledge, 2002), 74–86; V. Visana, "Vernacular liberalism, capitalism, and anti-imperialism in the political thought of Dadabhai Naoroji," *The Historical Journal* 59:3 (Sep 2016), 775–97.

7 M. Goswami, *Producing India: From Colonial Economy to National Space* (Chicago: University of Chicago Press, 2004), 225–31.

8 W. Digby, *The Famine Campaign in Southern India, 1876–1878* (London: Longmans, Green, and Co., 1878); W. Digby, *India for the Indians-and for England* (London: Talbot Brothers, 1885).

9 "Causes of poverty in India. Letter from Sir A. MacDonnell," *Times of India* (27 July 1901), 8; "An open letter to Sir Antony MacDonnell," *Times of India* (19 March 1901), 5.

10 W. Digby, *Prosperous India. A Revelation from Official Records* (London: Unwin, 1901).

11 D. Naoroji, *Poverty and Un-British Rule in India* (London: Swan Sonnenschein, 1901); R.C. Dutt, *Indian Famines. Their Causes and Prevention* (London: P.S. King & Son, 1901); See also R.C. Dutt, *England and India: A Record of Progress during a Hundred Years, 1785–1885* (London: Chatto & Windus, 1897).

12 R.C. Dutt, *Open Letters to Lord Curzon on Famines and Land Assessments in India* (London: Kegan Paul, Trench, Truebner & Co., 1900).

13 W. Wedderburn, "The reorganisation of agricultural credit in India," *The Journal of the Society of Arts* 40 (Nov 1891), 646–54.

14 W. Wedderburn, "The Skeleton and the (Jubilee) Feast" in *Speeches and Writings of William Wedderburn* (Madras: G.A. Natesan & Co., 1918), 284–322.

15 Digby, *Prosperous India. A Revelation from Official Records*, xix.

16 "The Government of India and the Famine Union," *Homeward Mail* (15 Sep 1902), 1281–82.

17 Report of the 17th INC held at Calcutta in Dec 1901 (Calcutta, INC, 1902), 4; "The National Congress: President's Address. Bombay Famine Policy," *Times of India* (27 Dec 1901), 5.

18 Report of the 20th INC held at Bombay in Dec 1904 (Bombay: INC, 1905), 118–29.

19 P.R. Shankar, "Some reflections on the correspondence between G.K. Gokhale and William Wedderburn" *Proceedings of the Indian History Congress* 63 (2002), 606; E. Valdameri, *Indian Liberalism between Nation and Empire: The Political Life of Gopal Krishna Gokhale* (London: Routledge, 2022), 133–34.

20 Fourth Annual Report of the Indian Famine Union (Liverpool: Indian Famine Union Liverpool Branch, Dec 1905), 10.

21 On the Rai's political ideology, see V. Bhargav, "Lala Lajpat Rai's ideas on caste: conservative or radical?," *Studies in Indian Politics* 6:1 (1 June 2018), 15–26; For an analysis of the first two decades of the INC see J.R. McLane, *Indian Nationalism and the Early Congress* (Princeton: Princeton University Press, 1977).

22 S. Sarkar, *The Swadeshi Movement in Bengal, 1903–1908* (New Delhi: People's Publishing House, 1973); D. Menon, "The many spaces and times of Swadeshi', *Economic and Political Weekly* 47:42 (2012), 44–52; K. Manjapra, "Knowledgeable internationalism and the Swadeshi Movement, 1903–1921," *Economic and Political Weekly* 47:42 (2012), 53–92; M. Goswami, "From swadeshi to swaraj: nation, economy, territory in colonial South Asia, 1870 to 1907," *Comparative Studies in Society and History* 40:4 (1998), 609–36.

23 Sarkar, *The Swadeshi Movement in Bengal*, 31–63

24 Patel, *Naoroji* here in particular pages 5–8.

25 Report of the Proceedings of the 22nd Indian National Congress held at Calcutta in Dec 1906 (Calcutta, 1907), 106–13.

26 F. Osella, "Charity and philanthropy in South Asia: an introduction," 4–34; D.E. Haynes, "From tribute to philanthropy: the politics of gift giving in a western Indian city," *The Journal of Asian Studies* 46:2 (May 1987), 339–60; M. Kasturi, "'All gifting is sacred'," 107–39.

27 Brewis, "'Fill full the mouth of famine'," 887–918.

28 Watt, *Serving the Nation*, here in particular pages 83–85; P. Kidambi, *The Making of an Indian Metropolis: Colonial Governance and Public Culture in Bombay, 1890–1920* (Aldershot: Ashgate, 2007), 203–33; Brewis, "Education for service: social service and higher education in India and Britain," 119–36; G. Beckerlegge, "Swami Vivekananda and Seva: Taking 'Social Service' Seriously," in W. Radice (ed.), *Swami Vivekananda and the Modernisation of Hinduism* (New Delhi: Oxford Univ. Press, 1999), 158–93.

29 Watt, *Serving the Nation* here in particular page 7–8; C.A. Watt, "Education for national efficiency: constructive nationalism in North India, 1909–1916," *Modern Asian Studies* 31:2 (1997), 339–74.

30 U. Stark,"Lucknow's Jalsa-i Tahzib: Urbane Elite, Organisierte Handlungskompetenz und Frühe 'Associational Culture'," in H. Fischer-Tiné (ed.), *Handeln und Verhandeln. Kolonialismus, Transkulturelle Prozesse und Handlungskompetenz* (Münster: LIT, 2002), 51–73.

31 D. Kopf, *The Brahmo Samaj and the Shaping of the Modern Indian Mind* (Princeton: Princeton University Press, 1979), 129–54, here in particular 143; K.W. Jones, *Socio-Religious Reform Movements in British India* (Cambridge: Cambridge University Press, 1989), 36–37; Brewis, "'Fill full the mouth of famine'," 902; 906.

32 C. Bhatt, *Hindu Nationalism: Origins, Ideologies, and Modern Myths* (Oxford: Berg, 2001), 16–22 and 41–44; On the role of the Arya Samaj in fuelling communal violence, see K.W. Jones, "Communalism in the Punjab: the Arya Samaj contribution," *The Journal of Asian Studies* 28:1 (1968), 39–54; D. Hardiman, "Purifying the nation: the Arya Samaj in Gujarat 1895–1930," *The Indian Economic & Social History Review* 44:1 (March 2007), 41–65; C.S. Adcock, "Debating conversion, silencing caste: the limited scope of religious freedom," *The Journal of Law and Religion* 29:3 (2014), 363–77.

33 "A nation which cannot protect its own orphans and waifs [...] which lets them fall into the hand of its antagonist without raising a finger for their rescue, [...] can have no claim to command respect at the hand of other people." Cf. "Lajpat Raj to the editor of *The Tribune* of 24 Oct 1899" reprinted in B.R. Nanda, *The Collected Works of Lala Lajpat Rai [CWLLR]*, vol. 1 (New Delhi: Manohar, 2008), 123.

34 K.W. Jones, *Socio-Religious Reform Movements in British India*, 94–103; Soni, "Learning to labour," 15–42; R. Raj, "Of Swadeshi, self-reliance and self-help: a study of the Arya Samaj in Colonial Punjab, 1890–1920s," *Moving the Social* 62 (2019), 51–72.

35 Report of the Indian Famine Commission (1901), 80.

36 Some of Rai's statements on famine, including his appeals for famine relief, are reprinted in Nanda, *CWLLR*, vol. 1, 121–40 and B.R. Nanda, *CWLLR*, vol. 3 (New Delhi: Manohar, 2008), 104–5; Rai gives an account of the famine relief of the Arya Samaj in the 1890s in L. Rai, *The Arya Samaj: An Account of Its Origin, Doctrines, and Activities* (London: Longmans, Green, 1915), 211–18.

37 Brewis, "Education for service," 126.

38 "The famine relief of the Brahmo Samaj," *Indian Messenger* (9 Feb 1908), 39; "To save the starving," *Indian Messenger* (8 March 1908), 91; "Brahmo Samaj famine relief work: an appeal," *Indian Messenger* (15 March 1908), 145; "Famine relief work of the theistic churches," *Indian Messenger* (1 Sep 1908), 484.

39 Kopf, *The Brahmo Samaj and the Shaping of the Modern Indian Mind*, 149.

40 "To save the starving," *Indian Messenger* (8 March 1908), 91.

41 N. Nair, *Changing Homelands: Hindu Politics and the Partition of India* (Harvard University Press, 2011), 13–30; S. Pal, "Agrarian unrest in Punjab and Sardar Ajit Singh," *Proceedings of the Indian History Congress* 70 (2009), 451–64.

42 Nair, *Changing Homelands*, 27.

43 Watt, *Serving the Nation*, 83; F. Chand, *Lajpat Rai: Life and Work* (New Delhi: Government of India Publications Division, 1978), 221–22; L. Rai, *The Arya Samaj*, 218–19.

44 Rai's statements on famine relief in 1908 are reprinted in Nanda, *CWLLR*, vol. 3, 93–127. Cf. page 106.

45 "Famine Relief Work in U.P. and other Provinces," in Nanda, *CWLLR*, vol. 3, 102–26 Cf. page 112.

46 Watt, *Serving the Nation*, 17–19.

47 G.K. Devadhar, "The famine of 1908 in India and the work done by non-official agencies," *Modern Review* 6:3 (Sep 1909), 250–264; G.K. Devadhar, "The present famine in the United Provinces," *Indian Social Reformer* XIX:4 (27 Sep 1908), 40–41.

48 Brewis, "Education for service," 119–36.

49 Gokhale had only recently become the subject of critical historical research. See Valdameri. *Indian Liberalism between Nation and Empire.*

50 Devadhar, "The present famine in the United Provinces (continued)," *Indian Social Reformer* XIX:5 (4 Oct 1908), 52.

51 R. Srivatsan, "Concept of 'seva' and the 'sevak' in the Freedom Movement," *Economic and Political Weekly* 41:5 (2006), 427–38.

52 Devadhar, "The famine of 1908," 255.

53 Watt, 103; P. Anagol, *The Emergence of Feminism in India, 1850–1920* (Burlington: Ashgate, 2005), 67–72.

54 G.K. Devadhar (ed.), *Poona Seva Sadan Society. Review & Report with an Album* (Poona: Seva Sadan Society, 1934), 36; 48; G.K. Devadhar, *The Servants of India Society. A Lecture delivered by Mr. Gopal Krishna Devadhar* (Poona: Aryabhushan Press, 1914), 9.

55 A Brief Account of the Work of the Servants of India Society. From 1st Jan 1917 to 30th June 1923 (Poona, 1924), 38–41.

56 "Famine Relief Work in U.P. and other Provinces" in Nanda, *CWLLR*, vol. 3, 104.

57 G.K. Devadhar, "The present famine in the United Provinces," *Indian Social Reformer* XIX:4 (27 Sep 1908), 41.

58 Devadhar, "The famine of 1908," 250.

59 P.E. Roy, *A White Man's Province: British Columbia Politicians and Chinese and Japanese Immigrants, 1858–1914* (Vancouver: University of British Columbia Press, 1990); S. Sohi, "Race, surveillance, and Indian anticolonialism in the transnational Western U.S.-Canadian Borderlands," *The Journal of American History* 98:2 (1 Sep 2011), 420–36.

60 "Gift of famine: invasion of Sikhs from the Punjab," *International Wood-Worker* 17:10 (Oct 1907), 4–5.

61 D. Hoerder, *Cultures in Contact: World Migrations in the Second Millennium*, Comparative and International Working-Class History (Durham: Duke University Press, 2002), 373–76; V. Bald, *Bengali Harlem and the Lost Histories of South Asian America* (Cambridge: Harvard University Press, 2013), 101–2; M. Ramnath, *Haj to Utopia*, 18.

62 H.J.M. Johnston, *The Voyage of the Komagata Maru: The Sikh Challenge to Canada's Colour Bar* (Vancouver Toronto: UBC Press, 2014), 5–6; J.M. Brown, *Global South Asians. Introducing the Modern Diaspora* (Cambridge: Cambridge University Press, 2006), 29–58.

63 Numbers are given in J. Brown, *Global South Asians*, 38; For an overview of South Asian immigration into North America see V. Lal, *The Other Indians: A Political and Cultural History of South Asians in America* (New Delhi: HarperCollins Publishers India, 2008), 12–22.

64 "Gift of famine: invasion of Sikhs from the Punjab," 4–5.

65 J. Price, "'Orienting' the Empire: Mackenzie King and the aftermath of the 1907 Race Riots," *BC Studies: The British Columbian Quarterly* 156–157 (May 2008), 53–81.

66 Sohi, *Echoes of Mutiny*, 25–41; Gould, *Sikhs, Swamis, Students, and Spies*, 87–114; A. Gera Roy, *Imperialism and Sikh Migration: The Komagata Maru Incident* (London: Routledge, 2018), 69–110.

67 This argument has been made by Seema Sohi in *Echoes of Mutiny*, summarised in the introduction to her monograph, and subsequently fleshed out in the chapters.

68 S. Bonsal, "Aroused India faces mutiny and invasion," *New York Times* (10 May 1908), 2.

69 Michael Silvestri, *Policing 'Bengali Terrorism' in India and the World: Imperial Intelligence and Revolutionary Nationalism, 1905–1939* (Cham: Palgrave Macmillan, 2019), 33.

70 Bonsal, "Aroused India faces mutiny and invasion," *New York Times* (10 May 1908), 2.

71 H. Fischer-Tiné, *Shyamji Krishnavarma. Sanskrit, Sociology and Anti-Imperialism* (New Delhi: Routledge India, 2014); H. Fischer-Tiné, "Indian nationalism and the 'world forces': transnational and diasporic dimensions of the Indian freedom movement on the eve of the First World War," *The Journal of Global History* 2:3 (Nov 2007), 333–35.

72 On the global itinerary of Taraknath Das and his activities in North America see M. Framke, "Shopping Ideologies for independent India? Taraknath Das's engagement with Italian fascism and German national socialism," *Itinerario* 40:1 (April 2016), 55–81; N. Bose, "Taraknath Das (1884–1958): British Columbia, and the anti-colonial borderlands", *BC Studies: The British Columbian Quarterly* 204 (2020), 67–88; N. Bose, "Taraknath Das: A Global Biography," in *South Asian Migrations in Global History: Labour, Law and Wayward Lives* (Bloomsbury, 2020), 157–178.

73 Sohi, *Echoes of Mutiny*, 34–36; Slate, *Lord Cornwallis is Dead*, 121–23.

74 On the importance of the relative freedom to participate in anticolonial activism in Europe as a factor in the movements of Indians to and across Europe, see D. Brückenhaus, *Policing Transnational Protest. Liberal Imperialism and the Surveillance of Anticolonialists in Europe, 1905–1945* (New York: Oxford University Press, 2017), 13.

75 "Famine in Hindusthan and the only remedy," *Free Hindusthan* I:1 (April 1908), 2–3; "Victims of British rule," *Free Hindusthan* I:8 (Nov/Dec 1908).

76 "Famine in Hindusthan and the only Remedy,"2–3.

77 Bose, "Taraknath Das (1884–1958)," 75,

78 *Free Hindusthan* I:8 (Nov 1908), 1.

79 Robert Ward, *Some Fruits of the Great India Famine* (Madras: The M.E. Publishing House, 1902).

80 "Famine in Hindusthan and the only remedy," *Free Hindusthan* I:I (April 1908), 3.

81 Bose, "Pondering Poverty, Fighting Famine," 429.

82 Library and Archives Canada, RG7-G-21, vol. 200, File 332.3(b), "William C. Hopkinson, Immigration Inspector, to William W. Cory, Deputy Minister of the Interior" (25 Nov 1909), 1; Library and Archives Canada, RG211, vol. 703, file 21, "W. L. Mackenzie King, Deputy Minister of Labour, to Rudolphe Lemieux, Minister of Labour, for transmission to the Governor-General" (Aug 1908), 1–6.

83 Siegel, *Hungry Nation*, 10; S. Kamra, "Law and radical rhetoric in British India: the 1897 trial of Bal Gangadhar Tilak," *South Asia: Journal of South Asian Studies* 39:3 (July 2016), 546–59.

84 Valdameri, *Indian Liberalism between Nation and Empire*, 41–44.

85 Goswami, *Producing India*, 211–12.

86 Hall-Matthews, *Peasants, Famine and the State in Colonial Western India*, 14.

87 "The Brahmo Samaj and famine relief," *Indian Social Reformer* XXV:11 (15 Nov 1915), 122.

88 Resolution on the Administration of Famine Relief in the United Provinces of Agra and Oudh during the Years 1913 and 1914 (Allahabad: United Provinces, 1918); "Famine in the United Provinces," *Indian Social Reformer* XXIV:23 (8 Feb 1914), 271–72; "The famine in the United Provinces," XXVI:27 (8 March 1914), 318–19; "Famine in the United Provinces," *Indian Social Reformer* XXIV:32 (12 April 1914), 377.

89 Resolution on the Administration of Famine Relief in the United Provinces of Agra and Oudh, 22.

90 "An appeal to Hindustanees in Canada for U.P. Famine Relief," *Hindustanee* I:IV (1 April 1914), 1–3.

91 "A brief account of the work of the Servants of India Society (12 June 1905 to 31 Dec 1916)" (Poona, 1924), 9–10.

92 For biographical information on Kunzru, see M.S. Rajan, "Pandit Hriday Nath Kunzru: a memoir," *India Quarterly: A Journal of International Affairs* 34:4 (Oct 1978), 441–56.

93 "An appeal to Hindustanees in Canada for U.P. Famine Relief," *Hindustanee* I:IV (1 May 1914), 2.

94 Sohi, *Echoes of Mutiny*, 50–53.

95 A. Gera Roy and A.K. Sahoo, "The journey of the *Komagata Maru*: national, transnational, diasporic," *South Asian Diaspora* 8:2 (2016), 85–97; S. Chattopadhyay, *Voices of Komagata Maru: Imperial Surveillance and Workers from Punjab in Bengal* (New Delhi: Tulika, 2018); A. Gera Roy, *Imperialism and Sikh Migration: The Komagata Maru Incident* (London: Routledge, 2018).

96 Gandhi's time in South Africa is examined in A. Desai and G.H. Vahed, *The South African Gandhi: Stretcher-Bearer of Empire* (Stanford, California: Stanford University Press, 2015).

97 Sohi, *Echoes of Mutiny*, 112–17.

98 "An appeal to Hindustanees in Canada for U.P. Famine Relief," *Hindustanee* I:IV (1 May 1914), 2.

99 "U.P. Famine Relief Fund," *Hindustanee* I:IV (1 May 1914), 2–3 [note that the edition was published erroneously as vol. IV, and was followed by vol. V in June]; "A Lesson from the Famine in the United Provinces," *Hindustanee* I:V (1 June 1914), 4–6.

100 "A lesson from the famine in the United Provinces," *Hindustanee* I:V (1 June 1914), 4.

[101] The socialist Henry Mayers Hyndman (1842–1921), a vocal critic of colonialism, published his first study on India in 1875, followed by *The Indian Famine and the Crisis in India* two years later. H.M. Hyndman, *The Indian Famine and the Crisis in India* (London: Edward Stanford, 1877); M. Morris, "From anti-colonialism to anti-imperialism: the evolution of H.M. Hyndman's critique of empire, c.1875–1905," *Historical Research* 87:236 (May 2014), 293–314 here in particular page 297; P.R. Shanker, "Henry Mayers Hyndman (1842–1921) and the radicalization of the Indian National Congress," *Proceedings of the Indian History Congress* 66 (2005), 1041–49.

[102] Sohi, *Echoes of Mutiny*, 152–64; Slate, *Lord Cornwallis is Dead*, 119–27.

[103] The pro-German *Vital Issue* from New York published a two-page summary of his lecture in "Results of British Rule in India," *Vital Issue* 3:5 (1915), 12–13 Cf. page 12; See also "Million facing starvation in India," *Issues and Events* 5:4 (1916), 15.

[104] Ram Chandra, *A Few Facts about British Rule in India* (Bulletin of Hindustan Gadar, June 1915).

[105] C. Moffat, *India's Revolutionary Inheritance: Politics and the Promise of Bhagat Singh* (Cambridge: Cambridge University Press, 2019), 38–41; C. Moffat, "The itinerant library of Lala Lajpat Rai," *Th History Workshop Journal* 89 (1 April 2020), 121–39; D. Brundage, "Lala Lajpat Rai, Indian Nationalism, and the Irish Revolution, The View from New York, 1914–1920" in E. Dal Lago, R. Healy, and G. Barry (eds), *1916 in Global Context* (London: Routledge, 2017), 78–91.

[106] M. Bevir, "The formation of the All-India Home Rule League," *The Indian Journal of Political Science* 52:3 (1991), 341–56.

[107] A. Raucher, "American anti-imperialists and the pro-India movement, 1900–1932," *Pacific Historical Review* 43:1 (1974), 83–110; P.E. Teed, "Interfaith encounter and religious pluralism: J.T. Sunderland's mission to Brahmo Samajes of India, 1895–96," *American Studies* 50:1/2 (2009), 51–69.

[108] J.T. Sunderland, "The Cause of Indian Famines", *New England Magazine* 23:1 (1900), 58–60; J.T. Sunderland, *The Causes of Famines in India* (Toronto, 1904).

[109] E. Manela, *Wilsonian Moment: Self-Determination and the International Origins of Anticolonial Nationalism* (Cary: Oxford University Press, 2007), in particular 77–97; D. Ghosh, "Whither India? 1919 and the aftermath of the First World War," *The Journal of Asian Studies* 78:2 (2019), 389–97.

[110] D. Arnold, *Gandhi* (Harlow: Longman, 2001), 109–18.

[111] "Famine and grip sweeping India," *Young India* II:2 (Feb 1919), 41–43; "We want no charity," *Young India* II:6 (June 1919), 122–23; "India hungry and discontented," *Young India* II:6 (June 1919), 143; *Young India* II: 7 (July 1919), 150–51; "Famine increasing," *Young India* II:8 (Aug 1919), 178; "India-a factory of skeletons," *Young India* III:8 (Aug 1920), 187–89.

[112] See Chapter 3 for a discussion of famine conditions in India at the time.

[113] Hardiman, "The Influenza epidemic of 1918 and the Adivasis of Western India," 644–64; Arnold, "Death and the Modern Empire. The 1918–19 Influenza Epidemic in India," 181–200; Ramanna, "Coping with the Influenza Pandemic. The Bombay Experience," 86–98.

[114] Harnetty, "The famine that never was," 555–75; "The famine in India," *Toronto Globe* (12 April 1919), 6; "Appalling conditions prevail in India," *Toronto Globe* (24 April 1919), 6.

[115] "We want no charity", *Young India* II:6 (June 1919), 122.

[116] S. Sarkar, "Nationalism and poverty: discourses of development and culture in 20th century India," *Third World Quarterly* 29:3 (April 2008), 433–35; For a discussion on the power of nationalism to exclude and overwrite difference see A. Loomba, *Colonialism/Postcolonialism* (London: Routledge, 2015), 165–70.

[117] "Famine," *Young India* II:7 (July 1919), 150–51.

[118] On M.N. Roy see K. Manjapra, *M.N. Roy: Marxism and Colonial Cosmopolitanism* (New York: Routledge, 2010); M. Goebel, "Geopolitics, transnational solidarity or diaspora nationalism? The

global career of M.N. Roy, 1915–1930," *European Review of History: Revue Européenne d'histoire* 21:4 (4 July 2014), 485–99.

[119] "Hunger and revolution in India" was a serial article published in two editions of *The Call* (18 and 25 Sep 1919). It is also available in S. Ray *Selected Works of M.N. Roy*, vol. 1 (Oxford: Oxford University Press, 1987), 155–58; Also see I. Huacuja Alonso, "M.N. Roy and the Mexican Revolution: how a militant Indian nationalist became an international communist," *South Asia: Journal of South Asian Studies* 40:3 (3 July 2017), 524.

[120] British Library [BL], Asia, Pacific & Africa Collections [APAC], India Office Record [IOR]/ L/PJ/6/1787, File 310.

[121] BL, APAC, IOR/L/PJ/6/1882, File 5200.

[122] Slate, *Lord Cornwallis is Dead*, 127–38; Manela, *Wilsonian Moment*, 172; A. Chatfield, "The anti-imperialist moment," *The Australasian Journal of American Studies* 39:1 (2020), 81–100.

[123] Extracts of the article "Aid to starving India" from the *Boston American* (22 Sep 1922) are included in BL, APAC, IOR/L/PJ/6/1882, File 5200.

[124] Arnold, *Gandhi*, 114–26.

[125] *The Collected Works of Mahatma Gandhi* [CWMG], vol. 20 (Delhi: Ministry of Information and Broadcasting, 1958), 293–94; *CWMG*, vol. 21 (Delhi: Ministry of Information and Broadcasting, 1958), 85.

[126] I am referring to *Young India*, a publication started by Gandhi in 1919, not the eponymous journal of the India Home Rule League in New York. A summary of his writings on famine can also be found in V.B. Kher (ed.), *M.K. Gandhi, Social Service, Work and Reform* (Ahmedabad: Navajivan Publishing House, 1976), 159–60.

[127] "India-a factory of skeletons," *Young India* III:8 (Aug 1920), 187–89.

[128] *CWMG*, vol. 20, 380.

[129] *CWMG*, vol. 21, 286; *CWMG*, vol. 20, 304.

[130] L.N. Trivedi, "Visually mapping the 'nation': swadeshi politics in nationalist India, 1920–1930," *The Journal of Asian Studies* 62:1 (Feb 2003), 11–41; L. Trivedi, *Clothing Gandhi's Nation: Homespun and Modern India* (Bloomington: Indiana University Press, 2007); P. Gonsalves, *Khadi: Gandhi's Mega Symbol of Subversion* (Thousand Oaks: SAGE Publications, 2012).

[131] *Young India* (6 Oct 1921), 314 reprinted in Kher, *M.K. Gandhi. Social Service, Work and Reform*, vol. 1, 165.

[132] Ibid., 165.

[133] M. Gandhi, *The Wheel of Fortune* (Madras: Ganesh & Co., 1922), 97–106.

[134] *Young India* (6 Oct 1921), 314.

[135] Arnold, *Gandhi*, 114–26.

[136] L. Trivedi, *Clothing Gandhi's Nation: Homespun and Modern India* (Bloomington: Indiana University Press, 2007), 58.

[137] "Rural reconstruction," *Servant of India* (6 Oct 1927), 417–19.

[138] Report of the 17th INC held at Calcutta on the 26th, 27th & 28th Dec 1901 (Calcutta: Indian National Congress, 1902), 20.

[139] J. Vernon, *Modern Britain: 1750 to the Present* (Cambridge: Cambridge University Press, 2017), 418.

[140] Menon, "The many spaces and times of Swadeshi," 44–52; Fischer-Tiné, "Indian nationalism and the 'world forces'," 325–44.

[141] P. Roy, "Women, Hunger, and Famine: Bengal, 1350/1943" in B. Ray (ed.), *Women of India: Colonial and Post-Colonial Periods*, vol. 3 (New Delhi: Pauls Press, 2005), 397.

Chapter 6. Famine Relief and Nationalist Politics on the Eve of Independence: The Bengal Famine of 1942–44

1 M.B. Tauger, "The Indian famine crises of World War II," *British Scholar* 1:2 (March 2009), 186–95.

2 M.M. Khan, "Scurvy in the famine areas of Hissar District, Punjab," *The Indian Medical Gazette* 77:1 (Jan 1942), 6–12.

3 K.G. Sivaswamy et al., *Food Famine and Nutritional Diseases in Travancore (1943–44)* (Coimbatore: Sevindia Kerala Relief Centre, 1945).

4 The exodus formed the beginning of a larger pattern of displacement and migration that indicated the extent of economic hardship in Travancore. M.A. Oommen, "Famine and Elephants: Remembering Place-Making along Travancore's Forest Fringe," in K. Holmes and H. Goodall (eds), *Telling Environmental Histories: Intersections of Memory, Narrative and Environment* (Basingstoke: Palgrave Macmillan, 2017), 241–65.

5 Report of the Bijapur Famine Relief Committee (Bombay: Servants of India Society, 1944); S. Raghavan, *India's War: World War II and the Making of Modern South Asia* (New York: Basic Books, 2016), 352–55.

6 Tauger, "The Indian famine crises of World War II," 190.

7 Cf. S. Tomara, "Villages in India lack food," *New York Herald Tribune* (7 Oct 1942) copy included in BL, APAC, IOR/ L/I/1257, File 478/A.

8 BL, APAC, IOR/ L/I/1257. File 478/A, "B. Leech to A.P. Morley (10 Oct 1942)".

9 C.A. Bayly, "'The nation within': British India at war 1939–1947," *Proceedings of the British Academy* 125 (2004), 265–85; S. Raghavan, *India's War: World War II and the Making of Modern South Asia* (New York: Basic Books, 2016), 256–75.

10 BL, APAC, IOR/L/I/651, File 442/B.

11 Y. Khan, *The Raj at War: A People's History of India's Second World War* (London: Vintage, 2016), 93–95; Mukherjee, *Hungry Bengal*, 7–10.

12 B. Prasad and B.L. Raina (eds), *Official History of the Indian Armed Forces in the Second World War, 1939–45. Medical Services. Administration* (Delhi: Combined Inter-Services Historical Section, India & Pakistan, 1953), 564.

13 Tauger, "The Indian famine crises of World War II," 186–95.

14 Khan, *The Raj at War*, 142–57.

15 S. Hitesranjan, "The Quit India Movement in Medinipur District" in G. Pandey (ed.), *The Indian Nation in 1942* (Calcutta: Centre for Studies in Social Sciences, 1988), 19–76; B. Chakrabarty, "Political mobilization in the localities: the 1942 Quit India Movement in Midnapur," *Modern Asian Studies* 26:4 (Oct 1992), 791–814; P.R. Greenough, "Political mobilization and the underground literature of the Quit India Movement, 1942–44," *Modern Asian Studies* 17:3 (1983), 353.

16 L. Brennan, "Government famine relief in Bengal, 1943," *The Journal of Asian Studies* 47:3 (Aug 1988), 549–52.

17 "Handcuffs on relief workers," *People's War* (13 Dec 1942), 3; West Bengal State Archives, Home Department Files, Visit of Congress Workers to the Cyclone affected areas of Midnapore, "Office of the District Magistrate in Midnapore to Blair (27 Dec 1942)"; "Cowgill to Khan" (27 Dec 1942).

18 Ramakrishna Mission Bengal & Orissa Cyclone Relief: 1942–1944" (Howrah: The Secretary from Belur Math, July 1945).

19 "Brahmo Samajes relief mission report," *Indian Messenger* (7 March 1943), 52–54.

20 A letter of Amrita Lal Chatterjee dated 20 Feb 1943 was published in *Indian Messenger* (7 March 1943), 55–56.

21 BL, APAC, IOR/L/I/1/651, File 442/B, "Press clipping".

22 BL, APAC, IOR/L/I/1/651: File 442/B, "Correspondence between the High Commissioner, L.S. Amery, the Mayor of London and John Anderson".

23 BL, APAC, IOR/L/I/1/651, File 442/B, "Amery to the Governor of Bengal" (31 Dec 1942).

24 Mukherjee, *Hungry Bengal*, see pages 7– 16 for an overview of the causes of the famine that are explored in detail throughout the book.

25 On colonial censorship in India during World War II, see S. Bhattacharya, "Wartime policies of state censorship and the civilian population: Eastern India, 1939–45," *South Asia Research* 17:2 (Oct 1997), 140–77; S. Bhattacharya, *Propaganda and Information in Eastern India, 1939–45: A Necessary Weapon of War* (Richmond, Surrey: Curzon, 2001). On the censorship of the press, see in particular pages 125–26.

26 NMML, AC 3432 and AC 3433, "Indian Freedom Struggle documents at political activities. Papers regarding Second visit of S.C. Bose to Germany 1939–1944".

27 S. Bose, *His Majesty's Opponent: Subhas Chandra Bose and India's Struggle against Empire* (Cambridge: Harvard University Press, 2011), 250; R. Hayes, *Subhas Chandra Bose in Nazi Germany: Politics, Intelligence and Propaganda 1941–1943* (Oxford: Oxford University Press, 2011).

28 I. Stephens, *Monsoon Morning* (London: Ernest Benn, 1966), 182.

29 Vernon, *Hunger*, 148.

30 NMML, SPM Private Papers, Subject File 108, Letter to SPM (14 Oct 1943).

31 *ABP* (1 Sep 1943), 1.

32 A. Sen, *Development as Freedom* (Oxford: Oxford University Press, 2001); D. Banik, *Starvation and India's Democracy* (London: Routledge, 2007); O. Rubin, *Democracy and Famine* (Milton Park: Routledge, 2011).

33 The relief and rehabilitation work of non-official organisations is, amongst others, described by Freda Bedi, who noted the notable scope of the Bengal Relief Committee that worked under the auspices of S.P. Mukherjee. Freda Bedi, *Bengal Lamenting* (Lahore: Lion Press, 1944), 89.

34 N. Nakazato, "The Role of Colonial Administration, 'Riot Systems' and Local Networks during the Calcutta Disturbances of August 1946" in T. Sarkar and S. Bandyopadhyay (eds), *Calcutta: The Stormy Decades* (Routledge, 2017), 267–319.

35 A. Sarkar, "Fed by famine," 2022–86; Roy, *Making Peace, Making Riots*, 68–103.

36 For a recent in-depth analysis of Sarvarkar's writings, in particular his use of history and history-writing see V. Chaturvedi, *Hindutva and Violence: VD Savarkar and the Politics of History* (Albany: State University of New York Press, 2022).

37 *Akhand Hindustan* (or Hindusthan) was a term coined by Kanialal M. Munshi, a Congressman with Hindu right-wing leanings, who called for the Akhand Hindustan Conference in February 1942. While the idea of a united Hindu India had undergirded Hindu communitarian movements since the nineteenth century, in 1942 *Akhand Hindustan* was associated with the role of princely states in India, which the Mahasabha saw as purely and originally Hindu. They were to take the lead in the movement towards a united Hindu nation. M. Bhagavan, "Princely states and the Hindu imaginary: exploring the cartography of Hindu nationalism in Colonial India," *The Journal of Asian Studies* 67:3 (Aug 2008), 881–915.

38 J. Chatterji, *Bengal Divided: Hindu Communalism and Partition, 1932–1947* (Cambridge, New York: Cambridge University Press, 1994), 191–98; P. Bapu, *Hindu Mahasabha in Colonial North India, 1915–1930: Constructing Nation and History* (London: Routledge, 2013); N. Gondhalekar and S. Bhattacharya, "The all India Hindu Mahasabha and the end of British rule in India, 1939–1947," *Social Scientist* 27:7/8 (July 1999), 48.

39 S. Aiyar, "Fazlul Huq, region and religion in Bengal: the forgotten alternative of 1940–43," *Modern Asian Studies* 42:6 (Nov 2008), 1217.

40 P. Bapu, *Hindu Mahasabha in Colonial North India, 1915–1930: Constructing Nation and History* (London: Routledge, 2013), 93–104.

41 W. Gould, "Hindu militarism and partition in 1940s United Provinces: rethinking the politics of violence and scale," *South Asia: Journal of South Asian Studies* 42:1 (Feb 2019), 134–51; B. Kumar Jha, "Militarizing the Community: Hindu Mahasabha's initiative (1915–1940)," *Studies in History* 29:1 (Feb 2013), 119–46.

42 Gondhalekar and Bhattacharya, "The All India Hindu Mahasabha," 48–74.

43 K. Meadowcroft, "The Emergence, Crystallization and Shattering of Right-Wing Alternative to Congress Nationalism: The All-India Hindu Mahasabha, 1937–52" (Unpublished dissertation, Montreal, Concordia University, 2003), 179–83.

44 On the relationship between Hindu Mahasabha and the Congress, see Chatterjee, *Bengal Divided*, 130–149.

45 Aiyar, "Fazlul Huq, Region and Religion in Bengal," 1213–49.

46 Gondhalekar and Bhattacharya, "The All India Hindu Mahasabha," 48–74.

47 R. Batabyal, *Communalism in Bengal: From Famine to Noakhali, 1943–47* (New Delhi: Sage Publications, 2005), 106–7.

48 NMML, SPM Private Papers, Subject File 110, "Bengal Relief Committee Minutes of the First and Second Committee Meetings held on 30 July 1943 and 31 July 1943".

49 O. Goswami, "Then came the Marwaris: some aspects of the changes in the pattern of industrial control in Eastern India," *The Indian Economic & Social History Review* 22:3 (Sep 1985), 225–49; O. Goswami, "Sahibs, Babus, and Banias: changes in industrial control in Eastern India, 1918–50," *The Journal of Asian Studies* 48:2 (May 1989), 289–309.

50 T. Omar Ali, *A Local History of Global Capital: Jute and Peasant Life in the Bengal Delta* (Princeton: Princeton University Press, 2018).

51 Mukherjee, *Hungry Bengal*, 230–31.

52 Its predecessor, the Marwari *Sahayak Samity* (1913) had been charged with supporting anti-governmental action. As it became the Marwari Relief Society, the former Samity declared itself a social service organisation, thereby emphasising its non-political nature. After the cyclone hit Midnapore in 1942, the Marwari Relief Society was seen doling out aid along with Congress workers, leading to a government raid on the Society's offices in Calcutta in October. Though anti-government leaflets were confiscated during the raid, the Society's members remained free. West Bengal State Archives, Home Department Files, Marwari Relief Society, 391/43, "Deputy Secretary to the Govt. of India in the External Affairs Department to the Chief Secretary to the Government of Bengal," dated 2 June 1943.

53 Mukherjee, *Hungry Bengal*, 177.

54 The Muslim Chamber of Commerce closed down its relief committee at the end of the year; it was substituted by a Muslim League Relief Committee. "Muslims and Bengal Famine Relief," *Dawn* (30 Nov 1943), 4.

55 Mukherjee, *Hungry Bengal*, 176–77.

56 NMML, SPM Private Papers, "How the Hindu Mahasabha fought Bengal famine. Report of Relief Work by Bengal Provincial Hindu Mahasabha."

57 "Mahasava relief work,"*ABP* (2 Dec 1943), 5.

58 "Starving destitutes in streets of Calcutta," *ABP* (5 Sep 1943), 6; "Home for Bengal orphans," *ABP* (12 Sep 1943), 5.

59 "Bengal orphan boys," *ABP* (7 Sept 1943), 1.

60 NMML, AIWC Papers, Correspondence relating to Relief Work done by AIWC in Bengal Famine of 1943 (F. No. 311 Reel 19), "AIWC Calcutta Branch. An Appeal."

61 A. Raza and F. Roy, "Paramilitary organisations in interwar India," *South Asia: Journal of South Asian Studies* 38:4 (2015), 671–89; D. Reetz, *Islam in the Public Sphere. Religious Groups in India, 1900–1947* (Oxford: Oxford University Press, 2006), 79–81, 222–30.

62 Raza and Roy, "Paramilitary Organisations in Interwar India," 671–89; D. Panigrahi, *India's Partition: The Story of Imperialism in Retreat* (London: Routledge, 2004), 165.

63 Y. Bahadur Mathur, *Muslims and Changing India* (Delhi: Trimurti Publications, 1972), 208.

64 "Mahasava relief work," *ABP* (2 Dec 1943), 5.

65 In 1944, the government estimated that *Khaksars* comprised of 1.6 million members. Reetz, *Islam in the Public Sphere*, 80.

66 "Maintain 6 lakhs of destitutes," *ABP* (20 Sep 1943), 3.

67 "Mahasava relief work," *ABP* (2 Dec 1943), 5; Mathur, *Muslims and Changing India*, 217.

68 "Famine, disease and devastation," *ABP* (10 Dec 1943), 3.

69 "Maintain 6 lakhs of destitutes," *ABP* (20 Sep 1943), 3.

70 Reetz, *Islam in the Public Sphere*, 79.

71 Mathur, *Muslims and Changing India*, 217.

72 "Home for Bengal orphans," *ABP* (12 Sep 1943), 5.

73 "Khaksars and destitutes," *ABP* (28 Dec 1943), 3.

74 Mathur, *Muslims and Changing India*, 218; On Mashriqi, see also M. Daechsel, "Scientism and its discontents: the Indo-Muslim 'facism' Od Inayatullah Khan Al-Mashriqi," *Modern Intellectual History* 3:3 (Nov 2006), 443–72.

75 The efforts of the provincial Mahasabha to project itself on a national political landscape through seeking alliances with the Congress have also been described by Anwesha Roy, see Roy, *Making Peace, Making Riots*, 87–88.

76 Relief Organisations Fight Bengal Famine (Calcutta: Relief Co-ordination Committee, 1943).

77 Bedi, *Bengal Lamenting*, 90–91.

78 Chatterji, *Bengal Divided*, 48; N. Sengupta, *Bidhan Chandra Roy* (New Delhi: Ministry of Information and Broadcasting, 2002).

79 Sengupta, *Bidhan Chandra Roy*, 67–71.

80 H. Tinker, "A Forgotten Long March: The Indian exodus from Burma, 1942," *The Journal of Southeast Asian Studies* 6:1 (March 1975), 1–15.

81 G. Carnall, *Gandhi's Interpreter: A Life of Horace Alexander* (Edinburgh: Edinburgh University Press, 2010), 162–63; the Bengal Provincial Congress Committee formed the Bengal Civil Protection Medical Committee in February 1942 at the request from the then-President of the Working Committee of the INC, Abul Kalam Azad. In addition to B.C. Roy, the medical committee acted under the auspices of the physician K.S. Ray. The Quaker J.R. Symonds later became a member of the Committee as well. "Civil Protection Committee," *ABP* (31 Dec 1943), 5.

82 "Maha Bodhi Bengal Relief Work. Organising Secretary's Second Report (n.d)"

83 B. Schedneck, "Buddhist International Organizations," in M.K. Jerryson (ed.), *Oxford Handbook of Contemporary Buddhism* (New York: Oxford University Press, 2017), 398–416; A. Trevithick, *The Revival of Buddhist Pilgrimage at Bodh Gaya (1811–1949): Anagarika Dharmapala and the Mahabodhi Temple* (Delhi: Motilal Banarsidass Publishers, 2006).

84 "Rescue Buddhagaya," *The Maha Bodhi and the United Buddhist World* 31 (1923), 39; *Maha Bodhi Society of India. Diamond Jubilee Souvenir. 1891–1951* (Calcutta: Sri Gouranga Press, 1951), 75.

85 Stolte and Fischer-Tiné, "Imagining Asia in India," 82–88; H. Fischer-Tiné, "'The Cult of Asianism': Asiendiskurse in Indien zwischen Nationalismus und Internationalismus (ca. 1885–1955)," *Comparativ* 18:6 (2008), 16–33; C. Stolte, "Compass Points: Four Indian Cartographies of Asia, c. 1930–55" in M. Frey and N. Spakowski (eds), *Asianisms. Regionalist Interactions & Asian Integration*

(Singapore: National University of Singapore Press, 2016), 49–74; M. Moritz, "'The Empire of Righteousness': Anagarika Dharmapala and his Vision of Buddhist Asianism (c. 1900)" in *Asianisms: Regionalist Interactions & Asian Integration*, 19–48.

86 V. Verma, "Reinterpreting Buddhism: Ambedkar on the Politics of Social Action," *Economic and Political Weekly* 45:49 (Dec 2010), 56–65; S. Jondhale and J. Beltz (eds), *Reconstructing the World: B.R. Ambedkar and Buddhism in India* (New Delhi: Oxford University Press, 2004); B.S. Jürgens, *B.R. Ambedkar: Religionsphilosophie eines Unberührbaren* (Frankfurt am Main: P. Lang, 1994).

87 D.F. Ober, "From Buddha bones to bo trees: Nehruvian India, Buddhism, and the poetics of power, 1947–1956," *Modern Asian Studies* 53:4 (July 2019), 1312–50; C. Jaffrelot, "India's look east policy: an Asianist strategy in perspective," *India Review* 2:2 (April 2003), 35–68.

88 "Maha Bodhi Bengal Relief Work. Organising Secretary's Second Report" (Calcutta, 1944).

89 Framke, "'We must send a gift worthy of India and the Congress!'," 1973; B. Tsui, "World Revolution: China, Pan-Asianism and India" in *China's Conservative Revolution: The Quest for a New Order, 1927–1949* (Cambridge: Cambridge University Press, 2018), 195–228; T. Sen, *India, China, and the World: A Connected History* (Lanham: Rowman & Littlefield, 2017), 293–377.

90 "Chinese urged to help India," *Dawn* (1 Nov 1943), 4; "Famine in India. China's concern," *ABP* (1 Nov 1943), 3.

91 "Maha Bodhi Bengal Relief Work. Organising Secretary's Second Report."

92 A.M. Saklani, "Nehru, Chiang Kai-Shek, and the Second World War" in M. Thampi (ed.), *India and China in the Colonial World* (New Delhi: Routledge, 2005), 167–90.

93 "China's part in India's relief," *ABP* (16 Dec 1943), 2.

94 D.N. Gupta, *Communism and Nationalism in Colonial India, 1939–45* (Delhi: Sage, 2008); S. Rai Chowdhuri, *Leftism in India, 1917–47* (London: Palgrave Macmillan, 2007); Batabyal, *Communalism in Bengal*, 121–26.

95 B. Chandra, "P.C. Joshi: A Political Journey" (P.C. Joshi Birth Centenary Memorial Lecture, Delhi, 2007); M.R. Masani, "The Communist Party in India," *Pacific Affairs* 24:1 (March 1951), 18–38.

96 People's Relief Committee Bengal. Annual Report. 1943–44 (Calcutta, n.d.), 3.

97 "For Merger of All Relief," *People's War* (3 Oct 1943), 2; S. Rahman and C. Baxter, *Historical Dictionary of Bangladesh* (Lanham: Scarecrow Press, 2010), 66 and 218.

98 People's Relief Committee Bengal. Annual Report. 1943–44, 3.

99 Ibid., 9–10.

100 Framke, "Political humanitarianism in the 1930s," 63–81.

101 In February 1944, it united the Bengal Civil Protection Committee, Bengal Relief Committee, Bengal Muslim League Relief Committee, Bengal Provincial Hindu Mahasabha Relief Committee, Bengal Provincial Trade Union Congress Relief Committee, Bengal Medial Relief Committee, Friends Ambulance Unit, AIWC Relief Committee, All Bengal Mahila Rakhsa Samiti, Bengal Women's Food Committee, Mayor's Relief Fund, People's Relief Committee, Indian Medical Association, Servants of India Society, United Teachers and Students' Relief Committee, Calcutta Relief Committee, Marwari Relief Society. See Library of the Society of Friends [LSF], Horace Alexander Papers: TEMP MSS 577 12 114, "Note on the activities of the Bengal Medical Relief Co-ordination Committee."

102 Framke, "'We must send a gift worthy of India and the Congress!'," 1969–98.

103 Ibid., 1969–98.

104 "China's attitude to Congress and the Indian People," *People's War* (15 Aug 1943), concluding page; Zhu De had been the Commander in chief of the eight-route army, which was assisted by the Indian Medical Mission.

105 "Congress Medical Mission in China," *People's War* (29 Aug 1943), 1; "All Lahore greets Dr. Basu" *People's War* (24 Oct 1943), 5.

106 BL, APAC, IOR/L/PS/12/854.

107 People's History Museum [PHM], CP/IND/DUTT/11/10, "Indian People's Theatre Association: Children of the Earth (Dharti ke lal). The Story".

108 S. Sunderason, "As 'Agitator and Organizer': Chittaprosad and Art for the Communist Party of India, 1941–48," *Object* 13 (2011), 76–95; Chittaprosad and Delhi Art Gallery, *Hungry Bengal: A Tour through Midnapur District, by Chittaprosad, in November, 1943* (New Delhi: Delhi Art Gallery, 2011); K. Singh and S.K. Mallik, *Chittaprosad, 1915–1978: A Retrospective* (New Delhi: Delhi Art Gallery, 2011); S. Janah, *Photographing India* (New Delhi: Oxford University Press, 2013); N. Sarkar, *Matter of Conscience: Artists Bear Witness to The Great Bengal Famine of 1943* (Calcutta: Punasha, 1998); E. Sen and Z. Abedin, *Darkening Days, Being a Narrative of Famine-Stricken Bengal* (Calcutta: Susil Gupta, 1944). For a discussion of Sunil Janah's photography placed in a longer historical trajectory of famine photography see E. Terracciano, "Beyond or in Emergency? The Emergence of Photography during the 1943 Bengal Famine: Sunil Janah" in *Art and Emergency: Modernism in Twentieth-Century India* (London: I.B. Tauris & Co. Ltd, 2018), 69–124.

109 S. Sunderason, "Shadow-lines: Zainul Abedin and the afterlives of the Bengal Famine of 1943," *Third Text* 31:2–3 (April 2017), 239–59.

110 R. Dasgupta, "'The People' in People's Art and People's War", in G. Chakravartty (ed.), *People's 'Warrior': Words and Worlds of P.C. Joshi*, 443–56 (New Delhi: Tulika Books, 2014); S. Mitra, "Progressive cultural movement in Bengal," *Social Scientist* 8:5/6 (Dec 1979), 115; V. Dalmia, *Poetics, Plays, and Performances: The Politics of Modern Indian Theatre* (New Delhi: Oxford University Press, 2006); M. Bhattacharya, "IPTA in Bengal," *The Journal of Arts and Ideas* 2 (March 1983), 5–22.

111 *India People's Theatre Association Bulletin* 1 (July 1943).

112 N. Bhatia, *Acts of Authority, Acts of Resistance: Theatre and Politics in Colonial and Postcolonial India* (New Delhi: Oxford University Press, 2004), 76–94; B. Bhattacharya, "People's art or performance of the elites?: Debating the history of IPTA in Bengal," *Rupkatha Journal on Interdisciplinary Studies in Humanities* V:2 (2013), 178–88; Bhattacharya, "IPTA in Bengal," 5–22.

113 *India People's Theatre Association Bulletin* 1 (July 1943), 2–3.

114 S. Saha, "The Commune-Ist Air: The Case of the IPTA Central Squad" in *Theatre and National Identity in Colonial India*, by S. Saha (Singapore: Springer Singapore, 2018), 119–52.

115 B. Dutt and U. Sarkar Munsi, *Engendering Performance: Indian Women Performers in Search of an Identity* (New Delhi: SAGE Publications, 2010), 86–94. For a discussion of *Nabanna* see P. Roy, "Women, Hunger, and Famine: Bengal, 1350/1943" in B. Ray (ed.), *Women of India: Colonial and Post-Colonial Periods* (New Delhi: Pauls Press, 2005), 403–7.

116 Cf. documents relating to IPTA's production of Dharti Ke Lal that are part of Rajani Palme Dutt Papers. See PHM, CP/IND/DUTT/11/10, "Souvenirs of the 1946 visit – 'Peoples Theatre' performance".

117 For a brief discussion of the reception of the movie see Siegel, *Hungry Nation*, 40–41.

118 "Bengal bhooka nahin marega," *People's War* (5 Dec 1943), concluding page.

119 "'Bhooka Bengal' exhibition," *People's War* (9 Jan 1944), 3.

120 "'Bhooka Bengal' exhibition," *People's War* (9 Jan 1944), 3.

121 "Must Kerala go to the Bengal war?" *People's War* (26 Dec 1943), concluding page.

122 People's Relief Committee Bengal. Annual Report. 1943–44, 11–12.

123 Ibid., 12.

124 "Starvation in India," *Daily Worker* (14 Sep 1943), 2; "'Help tackle food hoarders' Call to Indian people," *Daily Worker* (28 Sep 1943), 4; "Workers demand relief for starving India," *Daily Worker* (27 Sept 1943), 3; "NCOs-Army and ATS-Tell Govt. 'Starving people of India are our fellow citizens'," *Daily Worker* (1 Oct 1943), 3.

125 "I came to know the R.A.F.," *People's War* (23 Jan 1944), 2.

[126] Ibid., 2. The picture is printed on page 3.

[127] C. Branson, *British Soldier in India: The Letters of Clive Branson.* (London: The Communist Party, 1944).

[128] Khan, *The Raj at War*, 144–46.

[129] "Starvation in India," *Daily Worker* (14 Sep 1943), 2.

[130] Khan, *The Raj at War*, 152.

[131] Ibid., 162–64.

[132] Y. Khan, "Sex in an imperial war zone: transnational encounters in Second World War India," *The History Workshop Journal* 73:1 (April 2012), 240–58.

[133] S. Howe, *Anticolonialism in British Politics: The Left and the End of Empire, 1918–1964* (Oxford: Oxford University Press, 1993), 107; Owen, *The British Left*, 265.

[134] PHM, LP/ID/IND/1/103, "India Famine Emergency Assembly. Invitation sent by the India Famine Relief Committee"; D.R. Prem, "India Relief Committee Birmingham. Annual Report 1943–1944" (n.d), 3; I discuss the work of the Committee and other Indian nationalist groups in Simonow, "The Great Bengal Famine in Britain," 168–97.

[135] "The India Relief Committee," *NewsIndia* 6 (Oct 1943), 2; PHM, LP/ID/IND/1/86, "Appeal for Immediate Relief for the Victims of Famine, Pestilence and Distress in India".

[136] BL, APAC, IOR/L/PJ/12/455, File 721/32, Extract from Scotland Yard Report no. 248 (23 June 1943), 2.

[137] C. Chambers, *The Story of Unity Theatre* (New York: St. Martin's Press, 1989), 236; D. Watson, "The People's War and the People's Theatre: British Socialist Theatre 1939–45" in C. Creighton and M. Shaw (eds), *The Sociology of War and Peace* (Basingstoke: Macmillan, 1987), 168–70; A. Davies, "The War Years" in M. Balfour (ed.), *Theatre and War, 1933–1945: Performance in Extremis,* 61 (New York: Berghahn Books, 2001).

[138] For an analysis of the effect of the Bengal famine on the relationship of Australia and India also see K. Maclean, *British India, White Australia. Overseas Indians, Intercolonial Relations and the Empire* (Sydney: University of New South Wales Press, 2020), 200–25.

[139] A. Jackson, *The British Empire and the Second World War* (London: Hambledon Continuum, 2006), 463–95.

[140] H. Goodall, "Shared Hopes, New Worlds. Indians, Australians, and Indonesians in the Boycott of Dutch Shipping, 1945–1949" in S. Moorthy and A. Jamal (eds), *Indian Ocean Studies: Cultural, Social, and Political Perspectives* (New York: Routledge, 2010), 159.

[141] "Australian communists pledge to help us," *People's War* (12 Sep 1943), concluding page.

[142] See the following editions of the *Tribune:* 30 March 1943, 7; 13 April 1944, 1; 22 June 1944, 2–3; 27 July 1944, 4; 31 Aug 1944, 2; 14 Sep 1945, 4; 30 Nov 1945, 4; 5 Nov 1946, 6; 27 Aug 1946, 4.

[143] Goodall, "Shared Hopes, New Worlds. Indians, Australians, and Indonesians in the Boycott of Dutch Shipping, 1945–1949," 165–67.

[144] "Indian student tells miners about famine," *Tribune* (30 March 1944), 2.

[145] "Australians demand Indian settlement," *People's War* (2 Jan 1944), 1.

[146] Maclean, *British India, White Australia*, 208–12.

[147] J.M. Ward, "Stevens, Sir Bertram Sydney (1889–1973)" in J. Ritchie (ed.) *Australian Dictionary of Biography*, vol 12 (Melbourne: Melbourne University Press, 1990).

[148] "Australia for Indian Famine Relief," *People's War* (30 April 1944), 4.

[149] "Sydney Button Day for Indian famine," *Tribune* (22 June 1944), 3.

[150] R. Lockwood, *Black Armada: Australia and the Struggle for Indonesian Independence, 1942–49* (Sydney: Australasian Book Society, 1975); H. Goodall, *Beyond Borders: Indians, Australians and the Indonesian Revolution, 1939 to 1950* (Amsterdam: Amsterdam University Press, 2019); D. Cottle and A. Keys, "Transnational Struggle: Asian Seafarers and the Struggle for Indonesian Independence

in Australia" paper presented at the 17th Biennial Conference of the Asian Studies Association of Australia, Melbourne, 2008); N. Manjrekar, "'To Shake Hands Across the Ocean': The Political Worlds of South Asian Seamen, c. 1918–1946" (Unpublished dissertation, London, University of London. SOAS, 2017).

151 R. Visram, *Asians in Britain: 400 Years of History* (London: Pluto Press, 2002), 257.

152 Howe, *Anticolonialism in British Politics*, 112–13; Visram, *Asians in Britain*, 225–53.

153 "Amery gets the bird," *Socialist Appeal* (Nov 1943), 3; "Police protection for Amery," *Town Crier* (18 Dec 1943), 1.

154 Devenish, *Debating Women's Citizenship in India*, 158–79; G. Chakravartty, "Emergence of Mahila Atma Raksha Samiti in the Forties. Calcutta Chapter" in T. Sarkar and S. Bandyopadhyaya (eds), *Calcutta: The Stormy Decades* (New Delhi: Social Science Press, 2015), 177–203.

155 Regarding the participation of Indian women in WIDF see M.S. Roy, "'The call of the world': women's memories of global socialist feminism in India," *International Review of Social History* 67:S30 (2022), 237–62.

156 M. Sinha, "Gender and the Critiques of Colonialism and Nationalism. Locating the 'Indian Woman'," in S. Sarkar and T. Sarkar (eds), *Woman and Social Reform in Modern India. A Reader* (Bloomington: Indiana University Press, 2008), 452–72.

157 A. Loomba, *Revolutionary Desires: Women, Communism, and Feminism in India* (London: Routledge, 2018), 206; Chakravartty, "Emergence of Mahila Atma Raksha Samiti," 177–203, here in particular 199–201.

158 M. Sen, *In Search of Freedom: An Unfinished Story* (Calcutta: Stree, 2001), 74–77; Chakravarrty, "Emergence of Mahila Atma Raksha Samiti," 180; Ela Sen authored a chronicle of the famine that used the illustrations of Zainul Abedin, see E. Sen and Z. Abedin, *Darkening Days, Being a Narrative of Famine-Stricken Bengal* (Calcutta: Susil Gupta, 1944).

159 Loomba, *Revolutionary Desires*, 195–240.

160 Chakravartty, "Emergence of Mahila Atma Raksha Samiti," 179.

161 "To defend motherland, homes & honour. Bengal's women organise," *People's War (23* May 1943), 3; The songbook of IPTA, hence, included, amidst 'Food' and 'Lenin' also a song named 'Zoya'. *India People's Theatre Association Bulletin* 1 (July 1943), 27; "Andhra kisan women admire picture of Russian heroine Zoya," *People's War* (9 April 1944), 8.

162 P. Roy, "Women, Hunger, and Famine: Bengal, 1350/1943" in B. Ray (ed.), *Women of India: Colonial and Post-Colonial Periods* (New Delhi: Pauls Press, 2005), 392–98.

163 The efforts of MARS to organise and train women to be fit for defence purposes seem to parallel the militarisation of Muslim and Hindu communal organisations, but these have generally focused on men. B. Kumar Jha, "Militarizing the community: Hindu Mahasabha's initiative (1915–1940)," *Studies in History* 29:1 (Feb 2013), 119–46; A. Raza and F. Roy, "Paramilitary organisations in Interwar India," *South Asia: Journal of South Asian Studies* 38:4 (2015), 671–89; P. Bapu, "The Militarisation of Hindu Society" in *Hindu Mahasabha in Colonial North India, 1915–1930: Constructing Nation and History* (London: Routledge, 2013), 93–104.

164 Call to women. Unite to fight death, to save Bengal (Calcutta: MARS, n.d. [1943]). The rise of prostitution in Bengal during the famine see Y. Khan, "Sex in an imperial war zone," 240–58.

165 Call to women. Unite to fight death, to save Bengal (Calcutta: MARS, n.d. [1943]); A. Roy, *Making Peace, Making Riots: Communalism and Communal Violence, Bengal 1940–1947* (New Delhi: Cambridge University Press, 2018), 69–103.

166 The AIWC 17th Session (7–10 April 1944) (Bombay: AIWC, 1944), 78.

167 Call to Women. Unite to fight death, to save Bengal (Calcutta: MARS, n.d. [1943]).

[168] Devenish, *Debating Women's Citizenship*, 162; Loomba, *Revolutionary Desires*, 206–8; Chakravarrty, "Emergency of Mahila Atma Raksha Samiti," 201.

[169] "Self-Reliant Andhra Women will resist the Japs!" *People's War* (18 April 1943), concluding page.

[170] "Memorandum submitted by the All Bengal Mahila Atma Raksha Samiti," 1; R. Chakravarty "Bustee Women Flock to Food Rally," *People's War* (n.d) [Jan 1943], unpag.

[171] Chakravartty, "Emergence of Mahila Atma Raksha Samiti," 201–3.

[172] R. Ray, *My Reminiscences: Social Development During Gandhian Era and After* (New Delhi: Allied Publishers, 1982), 107; D. Rama Rau, *An Inheritance: The Memoirs of Dhanvanthi Rama Rau* (Bombay: Allied Publishers, 1960), 211.

[173] The AIWC 17th Session (7–10 April 1944) (Bombay: AIWC, 1944), 8.

[174] "Tried to bury starving child," *Daily Record* (26 Oct 1943), 3; "Human beings become wild beasts," *Birmingham Daily Post* (26 Oct 1943), 3.

[175] NMML, AIWC Papers: Correspondence relating to Relief Work done by AIWC in Bengal Famine of 1943, "The Bengal & Other Afflicted Areas Relief Committee's Report" (Oct 1943).

[176] NMML, AIWC Papers: Correspondence relating to Relief Work done by AIWC in Bengal Famine of 1943, "N.M. Joshi to Pandit" (22 Nov 1943).

[177] The committee was initially named Save the Children Fund, but its name was changed to All-India Save the Children Committee in February 1944. This was because it was frequently confused with the homonymous committee in Geneva. The AIWC 17th Session (7–10 April 1944) (AIWC: Bombay, 1944), 10.

[178] Devenish, *Debating Women's Citizenship*, 100–22.

[179] NMML, AIWC Papers: Correspondence relating to Relief Work done by AIWC in Bengal Famine of 1943, "Pandit to the AIWC Standing Committee" (9 Oct 1943).

[180] BL, APAC, IOR/ L/PJ/12/455, Extract from New Scotland Yard Report no. 248 (23 June 1943).

[181] At other times referred to as well as Jai Kishore or Jaykishori Handoo.

[182] BL, APAC, IOR/ L/PJ/12/455, Extract from New Scotland Yard Report no. 257 (27 Oct 1943).

[183] Extract from New Scotland Yard Report no. 257 (27 Oct 1943).

[184] The AIWC 17th Session, 50.

[185] L. Stec, "Pacifism, Vera Brittain, and India," *Peace Review* 13:2 (June 2001), 237–44.

[186] R. Ankit, "Quaker pacifist and Indian politics: Horace Gundry Alexander and India, 1947–77," *Quaker Studies* 18:2 (2014), 191–210; On the India conciliation group, see Carnall, *Gandhi's Interpreter*, 96–102; H. Tinker, "The India Conciliation Group, 1931–50: dilemmas of the mediator," *The Journal of Commonwealth & Comparative Politics* 14:3 (Nov 1976), 224–41.

[187] NMML, AIWC Papers: Correspondence relating to Relief Work done by AIWC in Bengal Famine of 1943, "N.M. Joshi to V.L. Pandit" (22 Nov 1943).

[188] "Participation in international conferences," *India Quarterly* 2:1 (Jan 1946), 63.

[189] M. Framke and E. Möller, "From local philanthropy to political humanitarianism: South Asian and Egyptian humanitarian aid during the period of decolonization," *ZMO Working Papers* 22 (2019), 1–17.

[190] On the importance of social work in the women's movement in the 1940s: E. Rook-Koepsel, *Democracy and Unity in India: Understanding the All India Phenomenon, 1940–1960* (London: Routledge, 2019), 73–101; Devenish, *Debating Women's Citizenship in India*, 100–28; U. Sen, "Social work, refugees and national belonging: evaluating the 'Lady Social Workers' of West Bengal," *South Asia: Journal of South Asian Studies*, 44:2 (2021), 344–61.

Chapter 7. American Food Aid for Independent India

1 J.F. Muehl, *American Sahib* (New York: The John Day Company, 1946), 213.

2 "J.F. Muehl. Memoir" in *Proceedings of the University of Michigan. May Meeting, 1965* (Ann Arbor: University of Michigan, 1965), 800.

3 Muehl, *American Sahib*, 207–20.

4 BL, APAC, IOR/ L/PJ/12/487, File 262/35, "The Indian Problem in American Politics" (5 Sep 1944), 1.

5 On the India League's activities during the Second World War, see also D. Patel, "'One man lobby'? Propaganda, nationalism in the diaspora, and the India League of America during the Second World War," *The Journal of Imperial and Commonwealth History* 49:6 (Nov 2021), 1110–40.

6 R. Shaffer, "J.J. Singh and the India League of America, 1945–1959: pressing at the margins of the Cold War consensus," *The Journal of American Ethnic History* 31:2 (2012), 68–103.

7 "A voice for the 400 million," *Silhouette* (9 Aug 1943), 27.

8 NYPL, Manuscripts and Archives Division, SH Papers, box 25, "Tentative Draft for National Campaign" authored by Sidney Hertzberg.

9 B. Kampmark, "Fearing union and imperialism: American publicists and the campaign against aiding Britain, 1939–1941," *The Australasian Journal of American Studies* 29:1 (July 2010), 1–14; R. Sarles and B. Kauffman, *A Story of America First: The Men and Women Who Opposed U.S. Intervention in World War II* (Westport, Conn: Praeger, 2003).

10 "India. The time for mediation is Now," *New York Times* (28 Sep 1942), 9.

11 NYPL, Manuscripts and Archives Division, SH Papers, box 25, "Press release of the India League of America."

12 N. Slate, *Colored Cosmopolitanism: The Shared Struggle for Freedom in the United States and India* (Cambridge: Harvard University Press, 2012); G. Horne, *The End of Empires: African Americans and India* (Philadelphia: Temple University Press, 2008).

13 NYPL, Manuscripts and Archives Division, Georgia Lloyd papers, box 69, "India League Mass Meeting at Town Hall" (9 Aug 1943).

14 "Speakers demand freedom for India," *New York Times* (10 Aug 1943), 7.

15 G. Hess, "The 'Hindu' in America: immigration and naturalization policies and India, 1917–1946," *Pacific Historical Review* 38 (1969), 59–79.

16 R. Nutt, "G. Sherwood Eddy and the attitudes of Protestants in the United States toward global mission," *Church History* 66:3 (Sep 1997), 502; M.G. Thompson, "Sherwood Eddy, the missionary enterprise, and the rise of Christian internationalism in 1920s America," *Modern Intellectual History* 12:1 (April 2015), 65–93; Thompson, *For God and Globe*, 27–46.

17 K.J. Leong, *The China Mystique: Pearl S. Buck, Anna May Wong, Mayling Soong, and the Transformation of American Orientalism* (Berkeley: University of California Press, 2005); P. Conn, *Pearl S. Buck: a cultural biography* (Cambridge: Cambridge University Press, 1998); D.A. Hollinger, *Protestants Abroad: How Missionaries Tried to Change the World but Changed America* (Princeton: Princeton University Press, 2017), 24–58.

18 BL, APAC, IOR/L/P&J/12/487, "India League of America" (14 June 1943); "Pearl Buck in new post," *New York Times* (5 Sep 1944), 15; This is also discussed in Patel, "'One Man Lobby'?," 1118.

19 On postal and telegraphic censorship in India during World War II, see Bhattacharya, *Propaganda and Information in Eastern India*, 122–25.

20 M.S. Venkataramani, *Bengal Famine of 1943. The American Response* (Delhi: Vikas, 1974), 27–28.

21 "The Bengal Famine," *Life* (20 Nov 1943), 16; 19–21.

22 "'India League head asks aid in famine," *New York Times* (2 Nov 1943), 8; "U.S. seeks to aid India," *New York Times* (30 Oct 1943), 7; "India League of America's appeal to President Roosevelt," *ABP* (4 Oct. 1943), 2.

23 BL, APAC, IOR/L/PJ/12/487, "The Indian Problem with American Politics" (5 Sep 1943), 5; See also H.A. Gould, *Sikhs, Swamis, Students, and Spies*, 320.

24 M.N. Barnett, *Empire of Humanity: A History of Humanitarianism* (Ithaca: Cornell University Press, 2011), 111.

25 J. Reinisch, "'Auntie UNRRA' at the Crossroads." *Past & Present* 218:8 (2013), 70–97; J. Reinisch, "Internationalism in Relief: The Birth (and Death) of UNRRA," *Past & Present* 210:6 (Jan 2011), 258–89; B. Riley, *The Political History of American Food Aid: An Uneasy Benevolence* (New York: Oxford University Press, 2017), 90–92.

26 M. Sinha, "The United Nations Relief and Rehabilitation Administration (UNRRA) and the Question of Food Aid during the Bengal Famine of 1943," *Proceedings of the Indian History Congress* 73 (2012), 1042–52.

27 Venkataramani, "The Roosevelt administration and the Great Indian Famine," 241–64.

28 S. Chary, *The Eagle and the Peacock: U.S. Foreign Policy Toward India Since Independence* (Connecticut: Greenwood Press, 1995), 23–37; Venkataramani, *Bengal Famine of 1943*, 40–59; M. Mukerjee, *Churchill's Secret War*, 207–13.

29 J.M. Rothgeb, *U.S. Trade Policy: Balancing Economic Dreams and Political Realities* (Washington D.C.: CQ Press, 2001), 65–68.

30 Chary, *The Eagle and the Peacock*, 25.

31 Library of Congress [LoC], Henry Luce Papers, box 1, folder 13, "Buck to Luce" (8 Oct 1943); Although no formal ties existed between the India League and the India Famine Relief Committee, many of the Committee's members had either already joined the League or would do so in late 1944, when the India Famine Relief Committee was dissolved.

32 T.C. Jespersen, "'Spreading the American dream' of China: United China Relief, the Luce family, and the creation of American conceptions of China before Pearl Harbor," *The Journal of American-East Asian Relations* 1:3 (Fall 1992), 269–94.

33 LoC, Henry Luce Papers, box 24, folder 2, "Bailey to Walsh" (2 Dec 1943).

34 Leong, *The China Mystique*, 43.

35 W. Vandivert, "Famine in India," *Life* (20 Dec 1943), 38–44; A. Müller, *Indien im Sucher Fotografien und Bilder von Südasien in der deutschsprachigen Öffentlichkeit, 1920–1980.* (Göttingen: Vandenhoeck & Ruprecht, 2019), 318.

36 His images again fuelled debates in Britain, where they were published in the British socialist press in December. This was at a time when the British government attempted to declare the famine as over, even as Indian nationalists and the British left strove to prove the continuing havoc caused by famine in India. "'India facing worse famine," *Socialist Appeal* (March 1944), 1.

37 Hollinger, *Protestants Abroad*, 26; Henry Luce, "The American Century," *Life* (17 Feb 1941), 61–65; I. Tyrrell, *Transnational Nation: United States History in Global Perspective since 1789* (London: Palgrave Macmillan, 2015), 187–88; A. Brinkley, *The Publisher Henry Luce and His American Century* (New York: Vintage Books, 2011).

38 J.E. Davies, C.P. Taft, and C. Warren, *Voluntary War Relief during World War II. A Report to the President.* (Washington: United States Government Printing Office, 1946); R.M. McCleary, *Global Compassion: Private Voluntary Organizations and U.S. Foreign Policy since 1939* (Oxford: Oxford University Press, 2009), 40–59.

39 Barnett, *Empire of Humanity*, 108–9.

40 USNA, American Relief for India, Inc, folder 2, Record Group 220, General Record Files, 1942–46, "Report to Members and Contributors" (Oct 1944), 6.

41 LoC, Henry Luce Papers, box 24, folder 2, "Smith to Fox" (27 Jan 1944).

42 S.M. Cutlip, *Fund Raising in the United States: Its Role in America's Philanthropy.* (New Brunswick: Transaction Publishers, 1990), 406–7.

43 S.L. Ashenberg and N. Kolko Philips, "The relationship between social work and labor unions: a history of strife and cooperation," *The Journal of Sociology and Social Welfare* 15 (1988), 105–18.

44 USNA, American Relief for India, Inc., folder 1, Record Group 220, General Record Files, 1942–46, "Report to the Board of Directors of the Indian Famine Relief Committee, Inc., 21 April 1944."

45 USNA, American Relief for India, Inc., folder 2, "Memorandum. United States Government. Allen to Bagby" (16 Sep 1944).

46 USNA, American Relief for India, Inc., folder 1, "Report to the Board of Directors of the Indian Famine Relief Committee, Inc." (21 April 1944); "Excerpt from Minutes of Board Meeting" (20 April 1944).

47 USNA, American Relief for India, Inc., folder 2, "Report to Members and Contributors. India's Relief Dilemma" (14 Oct 1944).

48 Haverford College Quaker and Special Collection, John Rich Papers, box 1, folder 14, A report on Friends' Famine Relief in India 1945. Together with a report of American Relief for India, Inc. (Philadelphia: AFSC, 1945).

49 LoC, Henry Luce Papers, box 24, folder 2, "Report to Members and Contributors. We are Re-organizing" (Oct 1944).

50 USNA, American Relief for India, Inc., folder 2, "Memorandum. United States Government. Allen to Bagby" (16 Sep 1944).

51 American Relief for India, Inc. consisted of the general secretary of the ASFC Rufus Jones who acted as honorary chairman, Philipp Murray of the CIO, William Green of the AFL, William Philips, former personal representative of the US President to India, who had flagged the need to assist Bengal as early as 1943, Henry Luce and Henry F. Grady, among others. A Report on Friends' Famine Relief in India 1945 (Philadelphia: AFSC, 1946).

52 D.C. Engerman, *The Price of Aid: The Economic Cold War in India* (Cambridge: Harvard University Press, 2018), 45.

53 "Our stake in India," *New York Times* (27 May 1945), 54.

54 On rationing R. De, "'Commodities must be controlled': economic crimes and market discipline in India (1939–1955)," *The International Journal of Law in Context* 10:3 (Sep 2014), 277–94; On the deteriorating food situation, see Siegel, *Hungry Nation*, 53.

55 "India's food crisis" (Washington, D.C.: Government of India Information Services, 1946), 6.

56 S. Chary, *The Eagle and the Peacock. U.S. Foreign Policy Toward India Since Independence* (Connecticut: Greenwood Press, 1995), 23–33.

57 Cf. Collingham, *The Taste of War*, 68; R. Jachertz and A. Nützenadel, "Coping with hunger?," 105–6.

58 Collingham, *The Taste of War*, 68–69; 124–26.

59 Jachertz and Nützenadel, "Coping with hunger?," 111.

60 "India Favours World Food Board" (Washington: Government of India Information Services, 1946).

61 Jachertz and Nützenadel, "Coping with hunger?," 111.

62 N. Cullather, "Hunger and containment: how India became 'important' in US Cold War strategy," *India Review* 6:2 (14 May 2007), 59–90; A.J. Rotter, *Comrades at Odds: The United States and India, 1947–1964* (Ithaca: Cornell University Press, 2000); S. Chary, *The Eagle and the Peacock: U.S. Foreign Policy Toward India Since Independence* (Connecticut: Greenwood Press, 1995); R.J. McMahon, *The Cold War on the Periphery: The United States, India and Pakistan* (New York: Columbia University Press, 1996); D. Merrill, *Bread and the Ballot: The United States and India's Economic Development, 1947–1963* (Chapel Hill: University of North Carolina Press, 1990).

63 Venkataramani, *Bengal Famine of 1943*; Venkataramani, "The Roosevelt administration and the Great Indian Famine," 241–64.

64 Jachertz and Nützenadel, "Coping with hunger?," 105–6.

65 Collingham, *The Taste of War*, 476–78; A.L. Bentley, "Uneasy sacrifice: the politics of United States famine relief, 1945–48," *Agriculture and Human Values* 11:4 (Sep 1994), 4–18.

66 Reprinted in S. Sarkar and S. Bhattacharya (eds), *Towards Freedom: Documents on the Movement for Independence in India 1946. Part I* (New Delhi: Oxford University Press, 2007), 855.

67 Sarkar and Bhattacharya (eds), *Towards Freedom*, 853–58.

68 PHM, CP/IND/DUTT/11/11: "Rajani Palme Dutt's visit to India", "India at the end of world food queue" (newsclipping).

69 "Travel Notes No. 3", *Labour Monthly* (July 1946), 218.

70 Cf. "The Commissioner in India (Merrell) to the Secretary of State. New Delhi, April 15, 1946" in N.O. Sappington et al. (eds), *Foreign Relations of the United States, 1946, The British Commonwealth, Western and Central Europe*, vol. V (Washington: United States Government Printing Office, 1969), 90–91.

71 "They face death by famine," *Tribune* (5 Nov 1946), 6.

72 Both Bourke-White and Janah took photographs of the famine. Hers were reproduced in *Interview with India* whereas Janah's images appeared in the *People's Age* throughout the year 1946. See S. Liegmann, "Inszenierungsformen durch die Linse: Leid Abbildende Pressefotografie während der Umbrüche 1946/47 in Indien," *Interdisziplinäre Zeitschrift für Südasienforschung* 2 (2017), 41–69; Sunil Janah, *Photographing India* (New Delhi: Oxford University Press, 2013), 30.

73 "The Hoover trip, it seems to me, was primarily a public relations job on all fronts." Cited from "Oral History Interview with Dennis A. Fitzgerald" (Washington, 21 June 1971) [https://www.trumanlibrary.org/oralhist/fitz.html].

74 BL, APAC, IOR/L/E/8/3406: E 15/46 Part 47, "Mr. Hoover's Statement on the Food Situation in India."

75 "The Commissioner in India (Merrell) to the Secretary of State. New Delhi, April 15, 1946," 90–91.

76 "India's Hunger: Report of the American Famine Mission to India" (New York: American Relief for India, 1946).

77 SH Papers, box 24, "India Famine Emergency Committee. Draft Report" (Dec 1946).

78 "Studying India's food needs with the U.S. Famine Mission," *Illustrated Weekly Bombay* (28 July 1946) (an unpaginated copy is kept in SH Papers box 43).

79 SH Papers, box 23, "Buck to Abbas" (12 June 1946); "Chaudhury to Buck" (14 June 1946).

80 SH Papers, box 23, "Maidment to Secretary of US Famine Mission" (1 July 1946); "Maidment to Hertzberg" (3 July 1946).

81 For a photograph of the meeting see SH Papers, box 43.

82 SH Papers, box 24, "Draft Report of the India Famine Emergency Committee" (Dec 1946), 11.

83 T.W. Schultz Cf. "American Famine Mission arrives in Karachi," *Pioneer* (26 June 1946); "Famine mission from America: arrival in Karachi," *Pioneer* (25 June 1946), 5.

84 SH Papers, box 24, "Summary of publicity received by American Famine Mission to India" (7 Oct 1946).

85 "India's Hunger: Report of the American Famine Mission to India" (New York: American Relief for India, 1946).

86 SH Papers, box 24, "India's food situation following the Indian food delegation."

87 R. Prasad, "Food position not free from anxiety," *Indian Information* (15 May 1947), 445 (citation); R. Prasad, "India's food front", *The March of India* 1:1 (1947), 9–11.

88 M. Kudaisya, "'A mighty adventure': institutionalising the idea of planning in Post-Colonial India, 1947–60," *Modern Asian Studies* 43:4 (July 2009), 939–78.

89 B. Zachariah, *Developing India: An Intellectual and Social History, c. 1930–50* (New Delhi: Oxford University Press, 2005), 216–20.

90 Amrith, "Food and welfare in India," 1010–35; R.Mukherjee, *Food Planning for Four Hundred Millions* (London: Macmillan and Co., 1938); P. Thakurdas et al., *A Plan of Economic Development For India* (London: Penguin Books, 1944).

91 Siegel, *Hungry Nation*, 21–49.

92 Zachariah, *Developing India*, 220–22.

93 K.A. Abbas and N.G. Jog, *A Report to Gandhiji: A Survey of Indian and World Events During the 21 Months of Gandhiji's Incarceration* (Bombay: Hind Kitabs, 1944), 60–70.

94 B.N. Banerjee, G.D. Parikh, and V.M. Tarkunde, 'People's Plan for Economic Development of India' (Delhi: Indian Federation of Labour, 1944).

95 S.N. Agarwal, *The Gandhian Plan of Economic Development for India* (Bombay: Padma Publications, 1944), 2.

96 Thakurdas et al., *A Plan of Economic Development for India*, 12–13.

97 Banerjee, Parikh, and Tarkunde, "People's Plan for Economic Development of India," 11.

98 Agarwal, *The Gandhian Plan of Economic Development for India*, 55–56.

99 Sherman, "From 'grow more food' to 'miss a meal'," 575–76.

100 Sherman, "From 'grow more food' to 'miss a meal'," 576; R. De, "'Commodities must be controlled'," 277–94; I. Kamtekar, "A different war dance: state and class in India 1939–1945," *Past & Present* 176 (Aug 2002), 187–221.

101 Sherman, "From 'grow more food' to 'miss a meal'," 583.

102 Ibid., 586.

103 *Unity* 1:12 (July 1952), 1.

104 Ibid., 1.

105 Sherman, "From 'grow more food' to 'miss a meal'," here in particular page 586; R. Trumbull, "Communists may profit by famine in India," *New York Times* (6 April 1952), 6.

106 C.M. Stolte, "Orienting India: Interwar Internationalism in an Asian Inflection, 1917–1937" (Leiden, University of Leiden, 2013), 159–95.

107 Engerman, *The Price of Aid, 52–57.*

108 R.J. McMahon, "Food as a diplomatic weapon: the India wheat loan of 1951," *Pacific Historical Review* 56:3 (Aug 1987), 349–77; Cullather, "Hunger and containment," 59–90; Chary, *The Eagle and the Peacock. U.S. Foreign Policy Toward India Since Independence.*

109 Sherman, "From 'grow more food' to 'miss a meal'," 583–4.

110 "Parliamentary Debates. Official Report, vol. VII, 1951. Third Session (Second Part) of the Parliament of India (7 April 1951)," 6306.

111 "Rice imports from China," *Times of India* (5 Dec 1952), 8; *India To-day* 1:2 (June 1951), 3–4.

112 A. Majumdar, *The Tebhaga Movement. Politics of Peasant Protest in Bengal 1946–1950* (Delhi: Aakar, 2011), 128–29; M.R. Masani, "The Communist Party in India," *Pacific Affairs* 2:1 (March 1951), 26–27; R, Chandavarkar, "From communism to 'social democracy': the rise and resilience of communist parties in India, 1920–1995," *Science & Society* 61:1 (Spring 1997), 99–106.

113 Masani, "The Communist Party in India," 32; R. Dave, *The Communist Culture Front* (Bombay: Democratic Research Service, 1951).

114 S. Saha *Theatre and National Identity in Colonial India* (Singapore: Springer Singapore, 2018), 148–49.

115 *Unity* 1:12 (July 1952), 2.

116 In the case of the art of Chittaprosad, Sanjoy Kumar Mallik attributed the shift to Chittaprosad's turn to the peace movement and his dissociation from communist politics. S. Kumar Mallik, "The Detached Line" in *Chittaprosad. A Retrospective. 1915–1978* (New Delhi: Delhi Art Gallery, 2011), 57–61; Also see S. Sunderason, "As 'agitator and organizer': Chittaprosad and art for the Communist Party of India, 1941–48," *Object* 13 (2011), 87.

117 For a discussion of Soviet aid, and its competition with American food aid, see B.R. Siegel, "'Fantastic Quantities of Food Grains': Cold War Visions and Agrarian Fantasies in Independent India" in Leslie James and Elisabeth Leake (eds), *Decolonization and the Cold War: Negotiating Independence* (London: Bloomsbury, 2015), 21–42.

118 "China's offer of food grains to India," *Times of India* (12 April 1952), 6.

119 McMahon, "Food as a diplomatic weapon," 349–77.

120 *India To-day* II:10 (May 1951), 2.

121 "Russia's desire to help India," *Times of India* (2 June 1951), 1.

122 "Food gift to India," *Times of India* (8 May 1951), 1.

123 *India Today* 1:2 (June 1951), 4.

124 Siegel, *Hungry Nation*, 1.

125 *India Today*, (April 1952), 4.

126 "India and the U.S.A," *March of India* 1 no. 1 (1947), 6.

127 T.J. Watts, "Watumull Foundation" in H. Ling and A.W. Austin (eds), *Asian American History and Culture: An Encyclopaedia* (London: Routledge, 2015), 363–64; BL, APAC, IOR/L/P&J/12/487, File 262/35, "The Indian Problem in American Politics" (5 Sep 1944), 4–7.

128 R. Ariyoshi, "Ellen Jensen Watumull. A Lady with a Fierce Grace" in B. Dye (ed.), *Hawai'i Chronicles II: Contemporary Island History from the Pages of Honolulu Magazine* (Honolulu: University of Hawai'i Press, 1998), 212.

129 "India's Minister of Health pushes family planning," *War on Hunger* II:2 (Feb 1968), 6; 12.

130 *Infant Mortality, Population Growth and Family Planning in India* (London: Allen & Unwin, 1972), 233. For more information on Chandrasekhar's international career and views on population control see Alison Bashford, *Global Population: History, Geopolitics, and Life on Earth* (New York: Columbia University Press, 2014), 341–4.

131 On the role of family planning and population control in Indian nation-building, as well as the transnational networks that informed them, see M. Sreenivas, *Reproductive Politics and the Making of Modern India* (Seattle: University of Washington Press, 2021).

132 The Watumull Foundation, American Wives of India, Church Federation of LA, Golden Rule Foundation, India Cultural Society, India Students Association of America, National Conference of Methodist Youth, New Century Foundation, Samaritan Society, UNESCO LA, World Students Association, The World Student Service Fund.

133 Rama Rau, *An Inheritance*, 241 (citation); M. Connelly, "Population control in India: prologue to the Emergency Period," *Population and Development Review* 32:4 (Dec 2006), 630–31.

134 A. Devenish, *Debating Women's Citizenship in India, 1930–1960* (Delhi: Bloomsbury, 2019), 204–10.

135 E. Jensen Watumull, "Margaret Sanger as I knew her," *The Journal of Family Welfare* 13:1 (Sept 1965), 7.

136 Watts, "Watumull Foundation," 364; Sanger's response to the meeting is recorded in "Sanger to Folsome" (23 Jan 1951) in E. Katz, P.C. Engelman and C. Moran Hajo (eds), *The Selected Papers of Margaret Sanger*, vol. 4 (Illinois: University of Illinois Press, 2016), 191.

137 "The Watumull Foundation Contribution to East–West Understanding," *Congressional Record: Proceedings and Debates of the 86th Congress, First Session*, 1.

138 "Watumull Foundation Report, 1951–1952" (California, 1952), 6.

139 A. Ram Mehta, "Population Control of India" in *Population Control in Relation to Food in India* (Los Angeles: The Watumull Foundation, 1952), 6.

140 Bashford, *Global Population*, 241–46.

141 N. Gangulee, *Health and Nutrition in India* (London: Faber and Faber, 1939), 130.

Conclusion

1 R. Dirks, "Social responses during severe food shortages and famine [and comments and reply]," *Current Anthropology* 21:1 (Feb 1980), 21–44; Arnold, "Social Crisis and Epidemic Disease," 385–404.

2 *GHI 2021. Hunger and Food Systems in Conflict Settings* (Bonn and Dublin, Oct 2021). The report can be accessed at https://www.globalhungerindex.org/india.html [date of access 19 April 2022].

3 The GHI applies five categories: low, moderate, serious, alarming, extremely alarming.

4 V. Paul, Anamika Singh, and Sneha Palit, "POSHAN Abhiyaan: Nourishing India: Making Nutrition a National Priority," *Proceedings of the Indian National Science Academy* 84 (July 2018), 835–41; R. Dasgupta et al., "An uphill task for POSHAN Abhiyan: examining the missing link of 'convergence'," *Indian Pediatrics* 57:2 (Feb 2020), 109–13.

5 "Statement of Ministry of Women and Child Development on Global Hunger Report 2021," Government of India, https://pib.gov.in/PressReleasePage.aspx?PRID=1764191 [date of access 19 April 2022].

6 Y. Seth Sharma, "India swings into action to climb up Global Hunger Index," The Economic Times India (17 Nov 2021), https://economictimes.indiatimes.com/news/economy/indicators/india-swings-into-action-to-climb-up-global-hunger-index/articleshow/87748226.cms [date of access 10 April 2022].

7 "'Government's response to Global Hunger Index is mischievous': Jean Drèze," *The Wire Science* (19 Oct 2021), https://science.thewire.in/external-affairs/world/global-hunger-index-india-rank-jean-dreze [date of access 20.04.2022].

8 A. VP, "Global Hunger Index: Govt Living in Denial," OxfamIndia (18 Oct 1021), https://www.oxfamindia.org/press-release/global-hunger-index-govt-living-denial [date of access 19 April 2022]; K. Iyer, "Hungry Nation: Govt Outrage About Global Ranking Cannot Hide Distress in Indian Homes" (article14, 23 Oct 2021), https://article-14.com/post/hungry-nation-govt-outrage-about-global-ranking-cannot-hide-distress-in-indian-homes--6173821b68de6 [date of access 19 April 2022].

9 H. Mander, *Locking Down the Poor* (New Delhi: Speaking Tiger, 2021), 1–17; On the use of history during the Covid-19 pandemic, see D. Arnold, "Pandemic India: Coronavirus and the uses of history," *The Journal of Asian Studies* 79:3 (Aug 2020), 569–77.

10 J. Drèze and A. Sen, *Hunger and Public Action* (Oxford: Oxford University Press, 1991), 126; A. Sen, *Development as Freedom* (Oxford: Oxford University Press, 2001),178–88; B. Currie, *Politics of Hunger in India: A Study of Democracy, Governance and Kalahandi's Poverty* (Basingstoke: Palgrave Macmillan, 2014), 124–25.

11 Amrith, "Food and welfare in India, 1010–35; Siegel, *Hungry Nation*, 220–33; J. Drèze, "Famine Prevention" in J. Drèze and A. Sen (eds), *The Political Economy of Hunger* (Oxford: Oxford University Press, 1991), 13.

12 S. Sen, "Between Homes; Without Homes. Migration, Circularity and Domesticity" in R. Bandyopadhyay, P. Banerjee and R. Samaddar (eds), *India's Migrant Workers and the Pandemic* (New York: Routledge, 2021),138–64; U. Sarmin, "Hunger, Humiliation, and Death: Perils of Migrant Worker in the Time of Covid-19" in *India's Migrant Workers and the Pandemic*, 167–76.

13 D. Sinha, "Hunger and food security in the times of Covid-19," *The Journal of Social and Economic Development* 23:2 (Sep 2021), 320–31; J. Drèze and A. Somanchi, "The Covid-19 Crisis and People's Right to Food" (SocArXiv, 1 June 2021), 1–22; J.Drèze, "What free food rations hide: a rollback of social security in India" (2 Jan 2023), https://scroll.in/article/1040987/what-free-food-rations-hide-a-rollback-of-social-security-in-india [date of access 20 March 2023].

Bibliography

A Brief Account of the Work of the Servants of India Society, Poona. From 1st of January 1917 to 30th June 1923 (Pune: Servants of India Society, April 1924).

A Report on Friends' Famine Relief in India 1945. Together with a Report of American Relief for India, Inc. (Philadelphia: AFSC, 1946).

A Year's Work in India. Report of the American Marathi Mission of Western India of 1900 (Bombay: ABCFM, 1901).

Abbas, Khwaja Ahmad, and Narayan Gopal Jog, *A Report to Gandhiji: A Survey of Indian and World Events During the 21 Months of Gandhiji's Incarceration* (Bombay: Hind Kitabs, 1944).

Adcock, Cassie S., "Debating conversion, silencing caste: the limited scope of religious freedom," *Journal of Law and Religion* 29:3 (2014), 363–77.

Agarwal, Shriman Narayan, *The Gandhian Plan of Economic Development for India* (Bombay: Padma Publications, 1944).

Ahuja, Ravi, "State formation and 'famine policy' in early colonial South India," *The Indian Economic & Social History Review* 39:4 (Dec 2002), 351–80.

———, "'The Bridge-Builders': Some Notes on Railways, Pilgrimage and the British 'Civilizing Mission' in Colonial India" in Harald Fischer-Tiné and Michael Mann (eds), *Colonialism as Civilizing Mission: Cultural Ideology in British India*, 95–116 (London: Anthem Press, 2004).

———, "Authoritarian shadows," *Social Scientist* 46: 5–6 (2018), 3–20.

———, "A Beveridge Plan for India? Social insurance and the making of the 'formal sector'," *International Review of Social History* 64:2 (2019), 207–48.

Aiyar, Sana, "Fazlul Huq, region and religion in Bengal: the forgotten alternative of 1940–43," *Modern Asian Studies* 42:6 (Nov 2008), 1213–49.

Aldous, Chris, "Contesting famine: hunger and nutrition in Occupied Japan, 1945–1952," *Journal of American-East Asian Relations* 17:3 (2010), 230–56.

Ali, Tariq Omar, *A Local History of Global Capital: Jute and Peasant Life in the Bengal Delta* (Princeton: Princeton University Press, 2018).

All-India Women's Conference 17th Session. Bombay, April 7 to 10, 1944 (Bombay: The All-India Women's Conference, 1944).

Alonso, Isabel H., "M.N. Roy and the Mexican Revolution: how a militant Indian nationalist became an international communist," *South Asia: Journal of South Asian Studies* 40:3 (July 2017), 517–30.

Alter, S. Neale, "The Future of Rural Christian Missions in India" (New York: Agricultural Missions, 1945).

Ambirajan, Srinivasa, "Malthusian population theory and Indian famine policy in the nineteenth century," *Population Studies* 30:1 (March 1976), 5–14.

American Relief for India (New York: American Relief for India, Inc., 1945).

Amrith, Sunil S., "Food and welfare in India, c. 1900–1950," *Comparative Studies in Society and History* 50:4 (Oct 2008), 1010–35.

———, *Unruly Waters: How Rains, Rivers, Coasts, and Seas Have Shaped Asia's History* (New York: Basic Books, 2018).

——— and Patricia Clavin, "Feeding the world: connecting Europe and Asia, 1930–1945," *Past & Present* 218:8 (Jan 2013), 29–50.

Anagol, Padma, *The Emergence of Feminism in India, 1850–1920* (Burlington: Ashgate, 2005).

Ankit, Rakesh. "Quaker pacifist and Indian politics: Horace Gundry Alexander and India, 1947–77," *Quaker Studies* 18:2 (2014), 191–210.

Annual Report 1951 (Philadelphia: American Friends Service Committee, 1951).

Annual Report of the ABCFM 1900 (Boston, 1900).

Annual Report of the ABCFM 1921 (Boston, 1921).

Ansari, Sarah, "The Bombay Presidency's 'Home Front', 1914–1918" in Roger D. Long and Ian Talbot (eds), *India and World War I: A Centennial Assessment* (New York: Routledge, 2018), 60–78.

Ariyoshi, Rita, "Ellen Jensen Watumull. A Lady with a Fierce Grace" in Bob Dye (ed.), *Hawai'i Chronicles II: Contemporary Island History from the Pages of Honolulu Magazine* (Honolulu: University of Hawai'i Press, 1998), 208–14.

Arnold, David, "Looting, grain riots and government policy in South India 1918," *Past and Present* 84:1 (1979), 111–45.

———, *Famine: Social Crisis and Historical Change* (Oxford: Basil Blackwell, 1988).

———, *Colonizing the Body: State Medicine and Epidemic Disease in Nineteenth-Century India* (Berkeley: University of California Press, 1993).

———, "Social crisis and epidemic disease in the famines of nineteenth-century India," *Social History of Medicine* 6:3 (1993b), 385–404.

———, "The 'discovery' of malnutrition and diet in Colonial India," *The Indian Economic & Social History Review* 31:1 (March 1994), 1–26.

———, *Gandhi*. Profiles in Power (Harlow: Longman, 2001).

———, "The Colonial Prison: Power, Knowledge and Penology in Nineteenth Century India" in David Hardiman and David Arnold (eds), *Subaltern Studies VIII. Essays in Honour of Ranajit Guha* (New Delhi: Oxford University Press, 2008), 148–87.

———, "The medicalization of poverty in colonial India," *Historical Research* 85:229 (Aug 2012), 488–504.

———, "India: The Contested Prison" in Frank Dikötter and Ian Brown (eds), *Cultures of Confinement: A History of the Prison in Africa, Asia, and Latin America* (Ithaca: Cornell University Press, 2018), 147–84.

———, "Death and the modern Empire: the 1918–19 influenza epidemic in India," *Transactions of the Royal Historical Society* 29 (2019), 181–200.

———, "Pandemic India: Coronavirus and the uses of history" *The Journal of Asian Studies* 79: 3 (Aug 2020), 569–77.

———, "The Good, the Bad, and the Toxic: Moral Foods in British India" in Angela Ki Che Leung, Melissa L. Caldwell, Robert Ji-Song Ku and Christine R. Yano (eds), *Moral Foods* (University of Hawaii Press, 2020), 111–29.

Ashenberg, Shulamith Lala, and Norma Kolko Philips, "The relationship between social work and labor unions: A History of strife and cooperation," *Journal of Sociology and Social Welfare* 15 (1988), 105–18.

Aykroyd, Wallace Ruddell, *The Nutritive Value of Indian Foods and the Planning of Satisfactory Diets*, Health Bulletin 23 (Delhi, 1959)

———, *The Conquest of Famine* (New York: Reader's Digest Press, 1975).

——— and Etienne Burnet, *Nutrition and Public Health*, Bulletin of the LNHO IV:2 (June 1935).

——— and B.G. Krishnan, "The carotene and Vitamin A requirements of children," *Indian Journal of Medical Research* 23 (1936), 741–45.

———, B.G. Krishnan and K.B. Madhava, "The effect of skimmed milk, soya bean, and other foods in supplementing typical Indian diets," *Indian Journal of Medical Research* 24 (1937), 1093–1206.

———, B.G. Krishnan, Reginald Passmore, and A.R. Sundarajan, *The rice problem in India* (Indian Medical Research Memoirs no. 32, 1940).

Bagchi, Amiya Kumar, "Indian economy and society during World War One," *Social Scientist* 42:7,8 (2014), 5–27.

Baker, Christopher John, *The Politics of South India. 1920–1937* (Cambridge: Cambridge University Press, 1976).

Bald, Vivek, *Bengali Harlem and the Lost Histories of South Asian America* (Cambridge: Harvard University Press, 2013).

Balfour, Michael (ed.), *Theatre and War, 1933–1945: Performance in Extremis* (New York: Berghahn Books, 2001).

Banerjee, Brajendra Nath, Govardhan Dhanaraj Parikh, and Vithal Mahadeo Tarkunde, *People's Plan for Economic Development of India* (Delhi: Indian Federation of Labour, 1944).

Banik, Dan, *Starvation and India's Democracy* (London: Routledge, 2007).

Bapu, Prabhu, *Hindu Mahasabha in Colonial North India, 1915-1930: Constructing Nation and History* (London: Routledge, 2013).

Barnett, Michael, *Empire of Humanity: A History of Humanitarianism* (Ithaca: Cornell University Press, 2011).

Barton, James L., "The effect of the War on Protestant missions," *Harvard Theological Review* 12:1 (Jan 1919), 1–35.

Bashford, Alison, *Global Population: History, Geopolitics, and Life on Earth* (New York: Columbia University Press, 2014).

Basu, Aparna, and Bharati Ray, *Women's Struggle. A History of the All India Women's Conference, 1927-2016* (New Delhi: Manohar, 2018).

Batabyal, Rakesh, *Communalism in Bengal: From Famine to Noakhali, 1943–47* (New Delhi: Sage Publications, 2005).

Bathgate, John, "Presbyterians and rural development in India," *Journal of Presbyterian History* 62:3 (1984), 237–45.

Bauman, Chad M., "Hindu-Christian conflict in India: globalization, conversion, and the coterminal castes and tribes," *The Journal of Asian Studies* 72:3 (Aug 2013), 633–53.

———, "Postcolonial anxiety and anti-conversion sentiment in the report of the Christian missionary activities enquiry committee," *International Journal of Hindu Studies* 12:2 (Aug 2008), 181–213.

Bayly, Christopher A., "'The nation within': British India at war 1939-1947," *Proceedings of the British Academy* 125 (2004), 265–85.

Beckerlegge, Gwilym, "Swami Vivekananda and Seva: Taking 'Social Service' Seriously" in William Radice (ed.), *Swami Vivekananda and the Modernisation of Hinduism* (New Delhi: Oxford Univ. Press, 1999), 158–93.

Bedi, Freda, *Bengal Lamenting* (Lahore: Lion Press, 1944).

Beer, Jeremy, *The Philanthropic Revolution: An Alternative History of American Charity* (Philadelphia: University of Pennsylvania Press, 2015).

Bender, Arnold E., "Problems of human protein nutrition," *Royal Society of Health Journal* 89:5 (Sept 1969), 221–26.

Bentley, Amy L., "Uneasy sacrifice: the politics of United States famine relief, 1945-48," *Agriculture and Human Values* 11:4 (Sep 1994), 4–18.

Berg, Alan, *The Nutrition Factor: Its Role in National Development* (Washington: Brookings Institution, 1973).

Berger, Rachel, "Alimentary affairs: historicizing food in modern India," *History Compass* 16:2 (Feb 2018), e12438.

Bevir, Mark, "The formation of the All-India Home Rule League," *The Indian Journal of Political Science* 52:3 (1991), 341–56.

Bhagavan, Manu, "Princely states and the Hindu imaginary: exploring the cartography of Hindu Nationalism in colonial India," *The Journal of Asian Studies* 67:3 (Aug 2008), 881–915.

Bhandari, B.N., "Œdema and scabies in a famine hospital," *The Indian Medical Gazette* 79:12 (Dec 1944), 574–78.

Bhargav, Vanya, "Lala Lajpat Rai's ideas on caste: conservative or radical?," *Studies in Indian Politics* 6:1 (June 2018), 15–26.

Bhatia, Balmokand M., *Famines in India: A Study of the Aspects of the Economic History of India (1860-1965)*. (New Delhi: Asia Publishing House, 1963).

Bhatia, Nandi, *Acts of Authority, Acts of Resistance: Theatre and Politics in Colonial and Postcolonial India* (New Delhi: Oxford University Press, 2004).

Bhatt, Chetan, *Hindu Nationalism: Origins, Ideologies, and Modern Myths* (Oxford: Berg, 2001).

Bhattacharya, Binayak, "People's art or performance of the elites? Debating the History of IPTA in Bengal," *Rupkatha Journal on Interdisciplinary Studies in Humanities* V:2 (2013), 178–88.

Bhattacharya, Malini, "IPTA in Bengal," *The Journal of Arts and Ideas* 2 (March 1983), 5–22.

Bhattacharya, Sanjoy, *Propaganda and Information in Eastern India, 1939-45: A Necessary Weapon of War* (Richmond, Surrey: Curzon, 2001).

———, "Wartime policies of state censorship and the civilian population: Eastern India, 1939-45," *South Asia Research* 17:2 (Oct 1997), 140–77.

Bhatty, E.C., and I.N. Singh, *Towards Richer Life: A Report on the Economic Life of the Church* (Mysore: National Christian Council, 1950).

Bihar and Orissa Famine Code (1930) (Calcutta: Jharna Supply Syndicate, 1950).

Bizzo, Maria Leticia Galluzzi, "Postponing Equality: From Colonial to International Nutritional Standards, 1932–1950" in Alexandra Widmer and Veronika Lipphardt (eds), *Health and Difference: Rendering Human Variation in Colonial Engagements* (New York: Berghahn Books, 2016), 129–48.

Black, Maggie, *Children First: The Story of UNICEF. Past and Present* (Oxford: Oxford University Press, 1996).

Bombay Representative Council of Missions Bombay. Annual Report for 1917–1918 (Bombay: Council of Missions, 1918).

Bombay Representative Council of Missions Bombay. Annual Report for 1919–1920 (Bombay: Council of Missions, 1920).

Borowy, Iris, *Coming to Terms with World Health: The League of Nations Health Organisation 1921-1946* (Frankfurt am Main: Peter Lang, 2009).

Bose, Neilesh, "Taraknath Das (1884–1958): British Columbia, and the anti-colonial borderlands," *BC Studies: The British Columbian Quarterly* 204 (2020), 67–88.

——— "Taraknath Das: A Global Biography" in Neilesh Bose (ed.), *South Asian Migrations in Global History: Labour, Law and Wayward Lives* (Bloomsbury, 2020), 157–78.

Bose, Sugata, *His Majesty's Opponent: Subhas Chandra Bose and India's Struggle Against Empire*. (Cambridge: Belknap Press of Harvard University Press, 2011).

———, "Pondering Poverty, Fighting Famines: Towards a New History of Economic Ideas" in Kaushik Basu and Ravi Kanbur (eds), *Arguments for a Better World: Essays in Honor of Amartya Sen* (Oxford: Oxford University Press, 2009), 425–35.

———, and Ayesha Jalal, *Modern South Asia: History, Culture, Political Economy* (New York: Routledge, 2004).

Branson, Clive, *British Soldier in India: The Letters of Clive Branson* (London: The Communist Party, 1944).

Brennan, Lance, "Government famine relief in Bengal, 1943," *The Journal of Asian Studies* 47:3 (Aug 1988), 541–66.

———, "The Development of the Indian Famine Codes. Personalities, Politics, and Policies" in Bruce Currey and Hugo Graeme (eds), *Famine as a Geographical Phenomenon* (Dordrecht: Springer, 2012), 91–111.

Brewis, Georgina, "The Making of an Imperial Ideal of Service: Britain and India before 1914" (Unpublished dissertation, University of East London, 2009).

———, "'Fill full the mouth of famine': Voluntary action in famine relief in India 1896–1901," *Modern Asian Studies* 44:4 (July 2010), 887–918.

———, "Education for service: Social service and higher education in India and Britain, 1905–1919," *History of Education Review* 42:2 (Oct 2013), 119–36.

Brinkley, Alan, *The Publisher Henry Luce and His American Century* (New York: Vintage Books, 2011).

Brooks, Jane, "'Uninterested in anything except food': The work of nurses feeding the liberated inmates of Bergen-Belsen. Nursing work and the liberation of Bergen-Belsen," *The Journal of Clinical Nursing* 21:19–20 (Oct 2012), 2958–65.

Brown, Judith M., *Global South Asians. Introducing the Modern Diaspora* (Cambridge: Cambridge University Press, 2006).

Brückenhaus, Daniel, *Policing Transnational Protest. Liberal Imperialism and the Surveillance of Anti-colonialists in Europe, 1905–1945* (New York: Oxford University Press, 2017).

Brundage, David, "Lala Lajpat Rai, Indian Nationalism, and the Irish Revolution, The View from New York, 1914-1920" in Enrico Dal Lago, Róisín Healy, and Gearóid Barry (eds), *1916 in Global Context* (London: Routledge, 2017), 78–91.

Brunner, Michael P., "Teaching development: debates on 'scientific agriculture' and 'rural reconstruction' at Khalsa College, Amritsar, c. 1915–47," *The Indian Economic & Social History Review* 55:1 (Jan 2018), 77–132.

Buchanan, W.J., "A note on the nutritive values of the dietaries of Bengal Jails," *The Indian Medical Gazette* 31:5 (May 1896), 153–58.

Budget of the Government of Bombay for 1921–22 (Bombay: Government of Bombay Financial Department, 1921).

Büschel, Hubertus, "Eine Brücke am Mount Meru: Zur Globalgeschichte von Hilfe zur Selbsthilfe und Gewalt in Tanganjika" in Hubertus Büschel and Daniel Speich (eds), *Entwicklungswelten: Globalgeschichte der Entwicklungszusammenarbeit* (Frankfurt am Main: Campus, 2009), 175–206.

Butterfield, Kenyon L., *The Christian Mission in Rural India. Report and Recommendations* (New York: International Missionary Council, 1930).

Cabanes, Bruno, *The Great War and the Origins of Humanitarianism, 1918–1924* (Cambridge: Cambridge University Press, 2014).

Caldwell, John C., "Malthus and the less developed world: the pivotal role of India," *Population and Development Review Population and Development Review* 24:4 (Dec 1998), 675–96.

Call to Women. Unite to Fight Death, to Save Bengal (Calcutta: MARS, n.d. [1943]).

Carnall, Geoffrey, *Gandhi's Interpreter: A Life of Horace Alexander* (Edinburgh: Edinburgh University Press, 2010).

Carpenter, Stephanie Ann, "'Regular Farm Girl': The women's land army in World War II," *Agricultural History* 71:2 (1997), 163–85.

Centennial Report of the AMM (Ahmednagar: AMM, 1913).

Chakrabarty, Bidyut, "Political mobilization in the localities: The 1942 Quit India Movement in Midnapur," *Modern Asian Studies* 26:4 (Oct 1992), 791–814.

Chakrabarty, M.L., "Blood-sugar levels in slow starvation," *The Lancet* 251:6503 (April 1948), 596–97.

Chakravartty, Gargi, "Emergence of Mahila Atma Raksha Samiti in the Forties. Calcutta Chapter," in Tanika Sarkar and Sekhara Bandyopadhyaya (eds), *Calcutta: The Stormy Decades* (New Delhi: Social Science Press, 2015), 177–203.

Chambers, Colin, *The Story of Unity Theatre* (New York: St. Martin's Press, 1989).

Chand, Feroz, *Lajpat Rai: Life and Work* (New Delhi: Government of India Publications Division, 1978).

Chandavarkar, Rajnarayan, "From communism to 'social democracy': The rise and resilience of communist parties in India, 1920-1995," *Science & Society* 61:1 (Spring 1997), 99–106.

Chandra, Bipan, *"P.C. Joshi: A Political Journey"* (Delhi, 2007).

Chari, M.V., "Inanition cases: A report," *The Indian Medical Gazette* 79:12 (Dec 1944), 578–82.

Charlesworth, Neil, *Peasants and Imperial Rule. Agriculture and Agrarian Society in the Bombay Presidency, 1850–1935* (Cambridge: Cambridge University Press, 1985).

Chary, Srinivas, *The Eagle and the Peacock: U.S. Foreign Policy Toward India Since Independence* (Connecticut: Greenwood Press, 1995).

Chatfield, Andrew, "The anti-imperialist moment," *Australasian Journal of American Studies* 39:1 (2020), 81–100.

Chatterjee, Nandini, *The Making of Indian Secularism: Empire, Law and Christianity, 1830-1960* (Basingstoke: Palgrave Macmillan, 2011).

Chatterji, Joya, *Bengal Divided: Hindu Communalism and Partition, 1932-1947* (Cambridge: Cambridge University Press, 1994).

———, *The Spoils of Partition: Bengal and India, 1947–1967* (Cambridge; New York: Cambridge University Press, 2007).

Chattopadhyay, Suchetana, *Voices of Komagata Maru: Imperial Surveillance and Workers from Punjab in Bengal* (New Delhi: Tulika, 2018).

Chaturvedi, Vinayak, *Hindutva and Violence: VD Savarkar and the Politics of History* (Albany: State University of New York Press, 2022).

Chhabria, Sheetal, *Making the Modern Slum: The Power of Capital in Colonial Bombay* (Seattle: University of Washington Press, 2019).

Chittaprosad, *Hungry Bengal: A Tour through Midnapur District, by Chittaprosad, in November, 1943* (New Delhi: Delhi Art Gallery, 2011).

Chowdhuri, Satyabrata Rai, *Leftism in India, 1917–47* (London: Palgrave Macmillan, 2007).

Collingham, Lizzie, *The Taste of War: World War II and the Battle for Food* (Penguin, 2012).

Conn, Peter, *Pearl S. Buck: A Cultural Biography* (Cambridge: Cambridge University Press, 1998).

Connelly, Matthew, "Population control in India: prologue to the emergency period," *Population and Development Review* 32:4 (Dec 2006), 629–67.

Conroy-Krutz, Emily, *Christian Imperialism: Converting the World in the Early American Republic* (Ithaca: Cornell University Press, 2015).

Cornebise, Alfred E., *Soldier-Scholars: Higher Education in the AEF, 1917–1919* (Philadelphia: American Philosophical Society, 1997).

Cornish, William R., *Reports on the Nature of the Food of the Inhabitants of the Madras Presidency and on the Dietaries of Prisoners in Zillah Jails* (Madras: United Scottish Press, 1863).

———, *Observations, on the Nature of the Food of the Inhabitants of Southern India; and, on Prison Dietaries in the Madras Presidency* (Madras: Adelphi Press, 1864).

Cottle, Drew, and Angela Keys, "Transnational Struggle: Asian Seafarers and the Struggle for Indonesian Independence in Australia," *Paper presented at the 17th Biennial Conference of the Asian Studies Association of Australia in Melbourne* (Melbourne, 2008), 1–3.

Cullather, Nick, "The foreign policy of the calorie," *The American Historical Review* 112: 2 (1 April 2007), 337–64.

———, "Hunger and containment: How India became 'important' in US Cold War Strategy," *India Review* 6:2 (14 May 2007), 59–90.

———, "When Is Famine Not a Famine? Gauging Indian Hunger in Imperial and Cold War Contexts" in Elizabeth Neswald, David F. Smith, and Ulrike Thoms (eds), *Setting Nutritional Standards: Theory, Policies, Practices* (Rochester: University of Rochester Press, 2017), 195–212.

Currie, Bob, *Politics of Hunger in India: A Study of Democracy, Governance and Kalahandi's Poverty.* (Basingstoke: Palgrave Macmillan, 2014).

Curtis, Heather, D. "Depicting distant suffering: evangelicals and the politics of pictorial humanitarianism in the age of American empire," *Material Religion* 8:2 (June 2012), 153–82.

———, *Holy Humanitarians: American Evangelicals and Global Aid* (Cambridge: Harvard University Press, 2018).

Cutlip, Scott M., *Fund Raising in the United States: Its Role in America's Philanthropy (*New Brunswick: Transaction Publishers, 1990).

Daechsel, Markus, "Scientism and its discontents: The Indo-Muslim 'facism' of Inayatullah Khan Al-Mashriqi," *Modern Intellectual History* 3:3 (Nov 2006), 443–72.

Dalmia, Vasudha, *Poetics, Plays, and Performances: The Politics of Modern Indian Theatre* (New Delhi: Oxford University Press, 2006).

Damodaran, Vinita, "Famine in a forest tract: Ecological change and the causes of the 1897 famine in Chotanagpur, Northern India," *Environment and History* 1:2 (1995), 129–58.

———, "Famine in Bengal: A comparison of the 1770 famine in Bengal and the 1897 famine in Chotanagpur," *The Medieval History Journal* 10:1–2 (Oct 2006), 143–81.

———, "The East India Company, Famine and Ecological Conditions in Eighteenth-Century Bengal" in Vinita Damodaran (ed.) *The East India Company and the Natural World* (Houndsmill: Palgrave Macmillan, 2015), 80–101.

Daniel, Shanti, "Working amongst Women in Uttar Pradesh," *Community Development Bulletin* 8:4 (1957), 79–85.

Das, Parimal, "Women under India's community development programme," *International Labour Review* 80 (1959), 26–45.

Dasgupta, Ajit K., *A History of Indian Economic Thought* (London: Routledge, 2002)

Dasgupta, Rajarshi, "'The People' in People's Art and People's War" in Gargi Chakravartty (ed.), *People's 'Warrior': Words and Worlds of P.C. Joshi* (New Delhi: Tulika Books, 2014), 443–56.

Dave, Rohit, *The Communist Culture Front* (Bombay: Democratic Research Service, 1951).

Davies, Joseph E., Charles P. Taft, and Charles Warren, *Voluntary War Relief during World War II. A Report to the President* (Washington: United States Government Printing Office, 1946).

Davies, Tegla, *Friends Ambulance Unit. The Story of the F.A.U. in the Second World War 1939–1946* (London: George Allen & Unwin Ltd., 1947).

Davis, Mike, *Late Victorian Holocausts: El Niño Famines and the Making of the Third World.* (London: Verso, 2007).

De, Rohit, "'Commodities must be controlled': economic crimes and market discipline in India (1939–1955)," *The International Journal of Law in Context* 10:3 (Sept 2014), 277–94.

De Waal, Alex, "Social contract and deterring famine: first thoughts," *Disasters* 20:3 (1996), 194–205.

———, *Mass Starvation: The History and Future of Famine* (Cambridge: Polity, 2018).

Desai, Ashwin, and Goolam H. Vahed, *The South African Gandhi: Stretcher-Bearer of Empire* (Stanford, California: Stanford University Press, 2015).

Devadhar, Gopal Krishna, *The Servants of India Society. A Lecture delivered by Mr. Gopal Krishna Devadhar* (Poona: Aryabhushan Press, 1914).

———, (ed.), *Poona Seva Sadan Society. Review & Report with an Album* (Poona: Seva Sadan Society, 1934).
Devenish, Annie, *Debating Women's Citizenship in India, 1930–1960* (Delhi: Bloomsbury, 2019).
Devereux, Stephen, *Theories of Famine* (New York: Harvester Wheatsheaf, 1993).
———, (ed.), *The New Famines: Why Famines Persist in an Era of Globalization* (London: Routledge, 2007).
Digby, William, *The Famine Campaign in Southern India, 1876–1878* (London: Longmans, Green, and Co., 1878).
———, *India for the Indians-and for England* (London: Talbot Brothers, 1885).
———, *Prosperous India: A Revelation from Official Records (*London: Unwin, 1901).
Dirks, Robert, "Social responses during severe food shortages and famine [and comments and reply]," *Current Anthropology* 21:1 (Feb 1980), 21–44.
"Discussion: the physiology and treatment of starvation," *Proceedings of the Royal Society of Medicine* 38:7 (1945), 395–97.
Dooley, C.R., "Training Within Industry in the United States," *International Labour Review* 54: 3–4 (1946), 160–78.
Drèze, Jean, and Amartya Sen, *Hunger and Public Action* (Oxford: Oxford University Press, 1991).
——— (eds), *The Political Economy of Hunger* (Oxford: Oxford University Press, 1991).
Drèze, Jean, and Anmol Somanchi, "The Covid-19 crisis and people's right to food," SocArXiv (1 June 2021).
Durbach, Nadja, *Many Mouths: The Politics of Food in Britain from the Workhouse to the Welfare State* (Cambridge: Cambridge University Press, 2020).
Dutt, Bishnupriya, and Urmimala Sarkar Munsi, *Engendering Performance: Indian Women Performers in Search of an Identity* (New Delhi: SAGE Publications, 2010).
Dutt, Romesh Chandra, *Open Letters to Lord Curzon on Famines and Land Assessments in India* (London: Kegan Paul, Trench, Truebner & Co., 1900).
———, *Indian Famines: Their Causes and Prevention* (London: P.S. King & Son, 1901).
Ebright, Donald F., *Free India. The First Five Years: An Account of the 1947 Riots, Refugees, Relief and Rehabilitation* (Nashville: Pantheon Press, 1954).
Edkins, Jenny, *Whose Hunger?: Concepts of Famine, Practices of Aid* (Minneapolis: University of Minnesota Press, 2000).
Edwards, Wendy J. Deichmann, and Carolyn De Swarte Gifford (eds), *Gender and the Social Gospel* (Urbana: University of Illinois Press, 2003).
Ellis, Catriona, "'If you cannot feed the body of a child you cannot feed the brain': education and nutrition in late colonial Madras," *South Asia: Journal of South Asian Studies* 44:1 (2021), 135–51.
Elman, Robert, "The intravenous use of protein and protein hydrolysates," *Annals of the New York Academy of Sciences* 47:3 (Oct 1946), 345–57.
Engerman, David C., *The Price of Aid: The Economic Cold War in India* (Cambridge: Harvard University Press, 2018).
Evans, Christopher Hodge, *The Social Gospel in American Religion: A History* (New York: New York University Press, 2017).
Ewart, Joseph, *The Sanitary Condition and Discipline of Indian Jails* (London: Smith, Elder and Co., 1860).
Fairbank, Henry, "Industrial Work in the American Marathi Mission, Ahmednagar, Western India" (Amraoti, 1911).
Famine Inquiry Commission. Final Report (Madras, 1945).
Famine Inquiry Commission. Report on Bengal (Madras, 1945).
Famine Relief Code. Bombay Presidency (Bombay: Government Central Press, 1904).
Famine Relief Code. Bombay Presidency (Bombay: Government of India, 1912).

Fildes, Valerie, Lara Marks, and Hilary Marland (eds), *Women and Children First: International Maternal and Infant Welfare, 1870–1945* (London: Routledge, 1992).

Fischer, John, *India's Insoluble Hunger* (Bombay: Vora & Co., 1947).

Fischer-Tiné, Harald, "Indian Nationalism and the 'world forces': transnational and diasporic dimensions of the Indian Freedom Movement on the eve of the First World War," *The Journal of Global History* 2:3 (Nov 2007), 325–44.

———, "'The Cult of Asianism': Asiendiskurse in Indien zwischen Nationalismus und Internationalismus (ca. 1885-1955)," *Comparativ: Zeitschrift für Globalgeschichte und Vergleichende Gesellschaftsforschung* 18:6 (2008), 16–33.

———, "Prarthana Samaj" in Denise Crush, Catherine A. Robinson, and Michael A. York (eds), *Encyclopedia of Hinduism* (New York: Routledge, 2008), 627–28.

———, *Shyamji Krishnavarma. Sanskrit, Sociology and Anti-Imperialism* (New Delhi: Routledge India, 2014).

———, "The YMCA and low-modernist rural development in South Asia, c.1922–1957," *Past & Present* 240:1 (Aug 2018), 193–234.

———, "'Unparalleled Opportunities': The Indian Y.M.C.A.'s Army Work schemes for imperial troops during the Great War (1914–1920)," *The Journal of Imperial and Commonwealth History* 47:1 (Jan 2019), 100–37.

———, *The YMCA in Colonial India: Modernization, Philanthropy and American Soft Power in South Asia* (London: Bloomsbury Academic, 2022).

——— and Nico Slate (eds), *The United States and South Asia from the Age of Empire to Decolonization. A History of Entanglements* (Leiden: Leiden University Press, 2022).

Fishburn, Janet F., "The Social Gospel as Missionary Ideology," in Wilbert R. Shenk (ed.), *North American Foreign Missions, 1810–1914: Theology, Theory, and Policy* (Grand Rapids: William B. Eerdmans Pub, 2004), 218–42.

Fleischmann, Ellen L., "At home in the world: globalizing domesticity through home economics in the interwar years," *Modern Asian Studies* 35:4 (2001), 886.

Forclaz, Amalia Ribi, "A new target for international social reform: The International Labour Organization and working and living conditions in Agriculture in the inter-war years, "*Contemporary European History* 20:3 (2011), 307–29.

Forth, Aidan, *Barbed-Wire Imperialism: Britain's Empire of Camps, 1876–1903* (Oakland: University of California Press, 2017).

Framke, Maria, "Political humanitarianism in the 1930s: Indian aid for republican Spain," *European Review of History: Revue Européenne d'histoire* 23:1–2 (Jan 2016), 63–81.

———, "Shopping ideologies for independent India? Taraknath Das's engagement with Italian fascism and German national socialism," *Itinerario* 40:1 (April 2016), 55–81.

———, "'We must send a gift worthy of India and the Congress!' War and political humanitarianism in late colonial South Asia," *Modern Asian Studies* 51:6 (Nov 2017), 1969–98.

——— and Esther Möller, "From local philanthropy to political humanitarianism: South Asian and Egyptian humanitarian aid during the period of decolonization," *ZMO Working Papers* 22 (2019), 1–17.

Fraser, Thomas W., *Culture & Change in India: The Barpali Experiment* (Massachusetts: University of Massachusetts Press, 1968).

Fuller, Pierre, *Famine Relief in Warlord China* (Cambridge: Harvard University Press, 2019).

Gandhi, Leela, *Affective Communities: Anticolonial Thought, Fin-De-Siècle Radicalism, and the Politics of Friendship* (Durham: Duke University Press, 2006).

Gandhi, Mohandas Karamchand, *The Wheel of Fortune* (Madras: Ganesh & Co., 1922).

———, *The Collected Works of Mahatma Gandhi*, vol. 20 (Delhi: Ministry of Information and Broadcasting, 1958).

———, *The Collected Works of Mahatma Gandhi*, vol. 21 (Delhi: Ministry of Information and Broadcasting, 1958).

Gangulee, Nargendranath, *Health and Nutrition in India* (London: Faber and Faber, 1939).

Ghosh, Durba, "Whither India? 1919 and the aftermath of the First World War," *The Journal of Asian Studies* 78:2 (2019), 389–97.

Glasman, Joël, "Measuring Malnutrition: The history of the MUAC tape and the commensurability of human needs," *Humanity: An International Journal of Human Rights, Humanitarianism, and Development* 9:1 (2018), 19–44.

Godbole, Narasinh Narayan, *Milk. The Most Perfect Food* (Benares: Benares Hindu University, 1936).

———, *Zur Kenntnis der höchstmolekularen gesättigten Säuren des Erdnussöles* (Berlin: Emil Dreyer's Buchdruckerei, 1925).

Goebel, Michael, "Geopolitics, transnational solidarity or diaspora nationalism? The global career of M.N. Roy, 1915–1930," *European Review of History: Revue Européenne d'histoire* 21:4 (July 2014), 485–99.

Gondhalekar, Nandini, and Sanjoy Bhattacharya, "The All India Hindu Mahasabha and the end of British rule in India, 1939–1947," *Social Scientist* 27:7/8 (July 1999), 48–74.

Gonsalves, Peter, *Khadi: Gandhi's Mega Symbol of Subversion* (Thousand Oaks: SAGE, 2012).

Goodall, Heather, "Shared Hopes, New Worlds: Indians, Australians, and Indonesians in the Boycott of Dutch Shipping, 1945–1949" in Shanti Moorthy and Ashraf Jamal (eds), *Indian Ocean Studies: Cultural, Social, and Political Perspectives* (New York: Routledge, 2010), 158–96.

———, *Beyond Borders: Indians, Australians and the Indonesian Revolution, 1939 to 1950* (Amsterdam: Amsterdam University Press, 2019).

Goswami, Manu, "From swadeshi to swaraj: nation, economy, territory in colonial South Asia, 1870 to 1907," *Comparative Studies in Society and History* 40:4 (1998), 609–36.

———, *Producing India: From Colonial Economy to National Space* (Chicago: University of Chicago Press, 2004).

Goswami, Omkar, "Then came the Marwaris: some aspects of the changes in the pattern of industrial control in Eastern India," *The Indian Economic & Social History Review* 22:3 (Sept 1985), 225–49.

———, "Sahibs, Babus, and Banias: Changes in industrial control in Eastern India, 1918–50," *The Journal of Asian Studies* 48:2 (May 1989), 289–309.

Götz, Norbert, Georgina Brewis, and Steffen Werther, *Humanitarianism in the Modern World: The Moral Economy of Famine Relief* (New York: Cambridge University Press, 2020).

Gould, Harold A., *Sikhs, Swamis, Students, and Spies: The India Lobby in the United States, 1900–1946* (New Delhi: Sage, 2006).

Gould, William, "Hindu militarism and partition in 1940s United Provinces: rethinking the politics of violence and scale," *South Asia: Journal of South Asian Studies* 42:1 (Feb 2019), 134–51.

Greenough, Paul R., *Prosperity and Misery in Modern Bengal: The Famine of 1943–1944* (New York: Oxford University Press, 1982).

———, "Political mobilization and the underground literature of the Quit India Movement, 1942–44," *Modern Asian Studies* 17:3 (1983), 11–47.

Grossmann, Atina, "Grams, calories, and food: languages of victimization, entitlement, and human rights in occupied Germany, 1945–1949," *Central European History* 44:1 (2011), 118–48.

Gupta, Devendra Nath, *Communism and Nationalism in Colonial India, 1939–45* (Delhi: Sage, 2008).

Hall-Matthews, David, "Historical roots of famine relief paradigms: ideas on dependency and free trade in India in the 1870s," *Disasters* 20:3 (Sep 1996), 216–30.

———, "Colonial ideologies of the market and famine policy in Ahmednagar District, Bombay Presidency, c. 1870–1884," *The Indian Economic & Social History Review* 36:3 (1999), 303–33.

———, *Peasants, Famine and the State in Colonial Western India* (Basingstoke: Palgrave Macmillan, 2005).

———, "Inaccurate conceptions: disputed measures of nutritional needs and famine deaths in Colonial India," *Modern Asian Studies* 42:6 (Nov 2008), 1189–1212.

Hancock, Mary, "Home science and the nationalization of domesticity in Colonial India," *Modern Asian Studies* 35:4 (Oct 2001), 871–903.

Harder, Ben, "The student volunteer movement for foreign missions and its contribution to 20th century missions," *Missiology: An International Review* 8:2 (April 1980), 141–54.

Hardiman, David, "Purifying the nation: the Arya Samaj in Gujarat 1895–1930," *The Indian Economic & Social History Review* 44:1 (March 2007), 41–65.

———, "The Influenza epidemic of 1918 and the Adivasis of Western India," *Social History of Medicine* 25:3 (Aug 2012), 644–64.

Harnetty, Peter, "The famine that never was: Christian missionaries in India, 1918–1919," *The Historian* 63:3 (March 2001), 555–75.

Harper, Susan Billington, *In the Shadow of the Mahatma: Bishop V.S. Azariah and the Travails of Christianity in British India* (Grand Rapids: W.B. Eerdmans Pub. Co, 2000).

Hatch, Duane Spencer, *Up from Poverty in Rural India* (Bombay: Oxford University Press, 1932).

———, *Further Upward in Rural India* (London: Oxford University Press, 1938).

Hauser, Julia, "Körper, Moral, Gesellschaft. Debatten über Vegetarismus zwischen Europa Und Indien, ca. 1850–1914" in Norman Aselmeyer and Veronika Settele (eds), *Geschichte des Nicht-Essens: Verzicht, Vermeidung und Verweigerung in der Moderne* (Berlin: De Gruyter Oldenbourg, 2018), 2599–87.

Hayes, Romain, *Subhas Chandra Bose in Nazi Germany: Politics, Intelligence and Propaganda 1941–1943* (Oxford: Oxford University Press, 2011).

Haynes, Douglas E., "From tribute to philanthropy: the politics of gift giving in a Western Indian city," *The Journal of Asian Studies* 46:2 (May 1987), 339–60.

———, *Small Town Capitalism in Western India: Artisans, Merchants and the Making of the Informal Economy, 1870–1960* (Cambridge: Cambridge University Press, 2012).

Hazen, William, *A Century in India. A Historical Sketch of the Marathi Mission of the ABCFM from 1813–1913* (Bombay: American Marathi Mission, 1913).

Heredia, Ruth, *The Amul India Story* (New Delhi: Tata McGraw-Hill Pub. Co, 1999).

Hess, Gary, *Sam Higginbottom of Allahabad, Pioneer of Point Four to India* (Charlottesville: University Press of Virginia, 1967).

———, "American agricultural missionaries and efforts at economic improvement in India," *Agricultural History* 42:1 (1968), 23–34.

———, "The 'Hindu' in America: immigration and naturalization policies and India, 1917–1946," *Pacific Historical Review* 38 (1969), 59–79.

Higginbottom, Sam, *The Gospel and the Plow or the Old Gospel and Modern Farming in Ancient India* (New York: MacMillan Company, 1921).

Hitesranjan, Sanyal, "The Quit India Movement in Medinipur District" in Gyanendra Pandey (ed.), *The Indian Nation in 1942* (Calcutta: Centre for Studies in Social Sciences, 1988), 19–76.

Hodges, Sarah, "Governmentality, population and reproductive family in Modern India," *Economic and Political Weekly* 39:11 (2004), 1157–63.

Hoerder, Dirk, *Cultures in Contact: World Migrations in the Second Millennium* (Durham: Duke University Press, 2002).

Hogg, William Richey, *Ecumenical Foundations. A History of the International Missionary Council and Its Nineteenth-Century Background* (New York: Harper and Brothers Publishers, 1952).

Hollinger, David A., *Protestants Abroad: How Missionaries Tried to Change the World but Changed America* (Princeton: Princeton University Press, 2017).

Horne, Gerald, *The End of Empires: African Americans and India* (Philadelphia: Temple University Press, 2008).

Howe, Paul and Stephen Devereux, "Famine intensity and magnitude scales: a proposal for an instrumental definition of famine," *Disasters* 28:4 (2004), 353–372.

Howe, Stephen, *Anticolonialism in British Politics: The Left and the End of Empire, 1918–1964* (Oxford: Oxford University Press, 1993).

Hume, Robert A., "Report of the American Marathi Mission for the Year 1920" (Ahmednagar: ABCFM Marathi Mission, 1921).

Hunnicutt, Benjamin H., and William Watkins Reid, *The Story of Agricultural Missions* (New York: Missionary Education Movement of the US and Canada, 1931).

Hunsberger, Alice C., "From Brimstone to the World's Fair: A Century of 'Modern Missions' as seen through the American Hume Missionary Family in Bombay" in Clifford Wallace Putney and Paul T. Burlin (eds), *The Role of the American Board in the World: Bicentennial Reflections on the Organization's Missionary Work, 1810–2010* (Eugene: Wipf & Stock, 2012), 102–41.

Hyndman, Henry Mayers, *The Indian Famine and the Crisis in India* (London: Edward Stanford, 1877).

India Favours World Food Board (Washington, D.C.: Government of India Information Services, 1946).

India's Food Crisis (Washington, D.C.: Government of India Information Services, 1946).

India's Hunger. Report of the American Famine Mission to India (New York: American Relief for India, 1946).

Interview with Henry Borsook, Transcript, 5 April 1978. The Caltech Institute Archives. https://oralhistories.library.caltech.edu/24/10/OH_Borsook_H.pdf.

Irwin, Julia, *Making the World Safe: The American Red Cross and a Nation's Humanitarian Awakening* (Oxford: Oxford University Press, 2013).

Jachertz, Ruth, and Alexander Nützenadel, "Coping with hunger? Visions of a global food system, 1930–1960," *The Journal of Global History* 6:1 (March 2011), 99–119.

Jackson, Ashley, *The British Empire and the Second World War* (London: Hambledon Continuum, 2006).

Jaffrelot, Christophe, "India's look east policy: an Asianist strategy in perspective," *India Review* 2:2 (April 2003), 35–68.

Jagadisan, T.N. and Shyamlal (eds), *Thakkar Bapa. Eightieth Birthday Commemoration Volume* (Madras: Diocesan Press, 1949).

Janah, Sunil, *Photographing India* (New Delhi: Oxford University Press, 2013).

Jespersen, T. Christopher, "'Spreading the American dream' of China: United China Relief, the Luce Family, and the creation of American conceptions of China before Pearl Harbor," *The Journal of American-East Asian Relations* 1:3 (Fall 1992), 269–94.

"J.F. Muehl. Memoir" in *Proceedings of the University of Michigan. May Meeting, 1965* (Ann Arbor: University of Michigan, 1965).

Jha, Bhuwan Kumar, "Militarizing the community: Hindu Mahasabha's initiative (1915–1940)," *Studies in History* 29:1 (Feb 2013), 119–46.

Jha, Madhavi, "'Men diggers and women carriers': gendered work on famine public works in colonial North India," *International Review of Social History* 65:1 (April 2020), 71–98.

Johnston, Hugh J. M., *The Voyage of the Komagata Maru: The Sikh Challenge to Canada's Colour Bar* (Vancouver Toronto: UBC Press, 2014).

Jondhale, Surendra, and Johannes Beltz (eds), *Reconstructing the World: B.R. Ambedkar and Buddhism in India* (New Delhi: Oxford University Press, 2004).

Jones, Kenneth W., "Communalism in the Punjab: the Arya Samaj contribution," *The Journal of Asian Studies* 28:1 (1968), 39–54.

———, *Socio-Religious Reform Movements in British India* (Cambridge: Cambridge University Press, 1989).

Jürgens, Bernd Sebastian, *B.R. Ambedkar: Religionsphilosophie eines Unberührbaren* (Frankfurt am Main: P. Lang, 1994).

Kalm, Leah M., and Richard D. Semba, "They starved so that others be better fed: remembering Ancel Keys and the Minnesota Experiment," *The Journal of Nutrition* 135:6 (June 2005). 1347–52.

Kalpagam, Uma, *Rule by Numbers: Governmentality in Colonial India* (Lanham: Lexington Books, 2014).

Kamat, Manjiri N., "'A relief camp in the Deccan countryside': the Sholapur cotton textile industry, Western India," *Proceedings of the Indian History Congress Proceedings of the Indian History Congress* 66 (2006/2005), 818–26.

Kampmark, Binoy, "Fearing union and imperialism: American publicists and the campaign against aiding Britain, 1939–1941," *The Australasian Journal of American Studies* 29:1 (July 2010), 1–14.

Kamra, Sukeshi, "Law and radical rhetoric in British India: The 1897 trial of Bal Gangadhar Tilak," *South Asia: Journal of South Asian Studies* 39:3 (2 July 2016), 546–59.

Kamtekar, Indivar, "A different war dance: state and class in India 1939–1945," *Past & Present* 176 (Aug 2002), 187–221.

Kannabiran, Vasantha, Rama Melkote, and Uma Maheshwari, *We Were Making History. Life Stories of Women in the Telangana People's Struggle* (New Delhi: Kali for Women, 1989).

Karmarkar, Vishvas Rao, *The Story of One Famine Orphan* (New York: Women's Board of Foreign Missions of the Presbyterian Church, n.d.).

Kasturi, Malavika, "'All gifting is sacred': the Sanatana Dharma Sabha Movement, the reform of *Dana* and civil society in Late Colonial India," *The Indian Economic & Social History Review* 47:1 (Jan 2010), 107–39.

———, "*Gurus* and gifting: *Dana*, the *Math* reform campaign, and competing visions of Hindu *Sangathan* in twentieth century India," *Modern Asian Studies* 52:1 (Jan 2018), 99–131.

Kaur, Rajender, and Anupama Arora, "India in the American Imaginary, 1780s–1880s" in Rajender Kaur and Anupama Arora (eds), *India in the American Imaginary, 1780s–1880s*, (New York: Palgrave Macmillan, 2017), 3–37.

Kent, Eliza F., "Tamil Bible women and the Zenana missions of colonial South India," *History of Religions* 39:2 (1999), 117–49.

———, *Converting Women: Gender and Protestant Christianity in Colonial South India.* (Oxford: Oxford University Press, 2004)

———, "'Mass Movements' in South India, 1877–1939" in Dennis C. Washburn and A. Kevin Reinhart (eds), *Converting Cultures: Religion, Ideology, and Transformations of Modernity* (Leiden: Brill, 2007), 367–94.

Khan, Yasmin, "Sex in an imperial war zone: transnational encounters in Second World War India," *The History Workshop Journal* 73:1 (April 2012), 240–58.

———, *The Great Partition: The Making of India and Pakistan* (New Haven: Yale University Press, 2017).

———, *The Raj at War: A People's History of India's Second World War* (London: Vintage, 2016).

Kher, V.B. (ed.), *M.K. Gandhi. Social Service, Work and Reform* (Ahmedabad: Navajivan Publishing House, 1976).

Kidambi, Prashant, *The Making of an Indian Metropolis: Colonial Governance and Public Culture in Bombay, 1890–1920* (Aldershot: Ashgate, 2007).

Kim, Berry, "Lakshmi and the scientific housewife: a transnational account of Indian women's development and production of an Indian modernity," *Economic and Political Weekly* 38:11 (2003), 1055–68.

Kopf, David, *The Brahmo Samaj and the Shaping of the Modern Indian Mind* (Princeton: Princeton University Press, 1979).

Kosambi, Meera, "A prismatic presence: the multiple iconisation of Dr Anandibai Joshee and the politics of life-writing," *Australian Feminist Studies* 16:35 (2001), 157–73.

Kramer, Paul A., "Empires, exceptions, and Anglo-Saxons: race and rule between the British and United States Empires, 1880–1910," *The Journal of American History* 88:4 (March 2002), 1315–53.

Krishnan, K.V., E.K. Narayanan, and G. Sankaran, "Protein hydrolysates in the treatment of inanition," *The Indian Medical Gazette* 79:4 (April 1944), 160–63.

Kudaisya, Medha, "'A mighty adventure': institutionalising the idea of planning in post-colonial India, 1947–60," *Modern Asian Studies* 43:4 (July 2009), 939–78.

Kumar, Arun, *Philanthropy and the Development of Modern India: In the Name of Nation* (New York: Oxford University Press, 2021).

Kumar, Prakash, "'Modernization' and agrarian development in India, 1912–52," *The Journal of Asian Studies* 79:3 (2020), 633–58.

Kupiec Cayton, Mary, "Canonizing Harriet Newell. Women, the Evangelical Press, and the Foreign Mission Movement in New England, 1800–1840" in Barbara Reeves-Ellington, Kathryn Kish Sklar, and Connie Anne Shemo (eds), *Competing Kingdoms: Women, Mission, Nation, and the American Protestant Empire, 1812-1960* (Durham: Duke University Press, 2010), 69–93.

Lal, Vinay, *The Other Indians: A Political and Cultural History of South Asians in America* (New Delhi: HarperCollins Publishers India, 2008).

Lall, Diwan Chaman, *Coolie: The Story of Labour and Capital in India* (Lahore: Oriental Publishing House, 1932).

Lambert, George, *India, the Horror-Stricken Empire* (Elkhart: Mennonite Publishing, 1898).

Leong, Karen J., *The China Mystique: Pearl S. Buck, Anna May Wong, Mayling Soong, and the Transformation of American Orientalism* (Berkeley: University of California Press, 2005).

Levinson, James, "Vital to the creation: interview with Alan Berg," *Development* 56:1 (March 2013), 24–36.

Liebau, Heike, "Country Priests, Catechists, and Schoolmasters as Cultural, Religious, and Social Middlemen in the Context of the Tranquebar Mission" in Robert Eric Frykenberg and Alaine M. Low (eds), *Christians and Missionaries in India: Cross-Cultural Communication since 1500, with Special Reference to Caste, Conversion, and Colonialism* (Grand Rapids: Routledge, 2003), 70–92.

Liegmann, Sarah, "Inszenierungsformen durch die Linse: Leid abbildende Pressefotografie während der Umbrüche 1946/47 in Indien" *Interdisziplinäre Zeitschrift Für Südasienforschung* 2 (2017), 41–69.

Lincoln, Murray D. and David Karp, *Vice President in Charge of Revolution* (New York: McGraw-Hill Book Company, 1960).

Little, Branden, "An explosion of new endeavours: global humanitarian responses to industrialized warfare in the First World War era," *First World War Studies* 5:1 (2 Jan 2014), 1–16.

Lockwood, Rupert, *Black Armada: Australia and the Struggle for Indonesian Independence, 1942–49* (Sydney: Australasian Book Society, 1975).

Loomba, Ania, *Colonialism/Postcolonialism* (London: Routledge, 2015).

———, *Revolutionary Desires: Women, Communism, and Feminism in India* (New York: Routledge, 2018).

Loveridge, Jack, "Between hunger and growth: pursuing rural development in partition's aftermath, 1947–1957," *Contemporary South Asia* 25:1 (2 Jan 2017), 56–69.

Maclean, Kama, "Hybrid Nationalist or Hindu Nationalist? The Life of Madan Mohan Malaviya (1861–1946)" in Kate Brittlebank (ed.), *Tall Tales and True: India, Historiography, and British Imperial Imaginings* (Clayton: Monash University Press, 2008), 107–24.

———, *British India, White Australia. Overseas Indians, Intercolonial Relations and the Empire* (Sydney: University of New South Wales Press, 2020).

Maha Bodhi Bengal Relief Work. Organising Secretary's Second Report (Calcutta, 1944).

Maha Bodhi Society of India. Diamond Jubilee Souvenir. 1891–1951 (Calcutta: Sri Gouranga Press, 1951).

Major, Andrea, "British humanitarian political economy and famine in India, 1838–1842," *The Journal of British Studies* 59:2 (April 2020), 221–44.

Majumdar, Ashok, *The Tebhaga Movement: Politics of Peasant Protest in Bengal 1946–1950* (Delhi: Aakar, 2011).

Malhotra, Ashok, "Race, diet, and class: Robert McCarrison's laboratory rat experiments in Coonoor, 1925–27," *Food Studies: An Interdisciplinary Journal* 9:1 (2019), 17–27.

———, "Cutting edge research in the contact zone? The establishment of the nutritional research laboratories in Coonoor (1925–27)," *South Asia: Journal of South Asian Studies* 44:1 (2 Jan 2021), 117–34.

Mallik, Sanjoy Kumar, "The Detached Line" in *Chittaprosad. A Retrospective. 1915-1978* (New Delhi: Delhi Art Gallery, 2011), 57–61.

Mander, Harsh, *Locking Down the Poor* (New Delhi: Speaking Tiger, 2021).

Manela, Erez, *Wilsonian Moment: Self-Determination and the International Origins of Anticolonial Nationalism* (Cary: Oxford University Press, 2007).

Manjapra, Kris, *M. N. Roy: Marxism and Colonial Cosmopolitanism* (New York: Routledge, 2010).

———, "Knowledgeable internationalism and the Swadeshi movement, 1903–1921," *Economic and Political Weekly* 47:42 (2012), 53–92.

Manjrekar, Naina, "'To Shake Hands Across the Ocean': The Political Worlds of South Asian Seamen, c. 1918–1946" (Unpublished dissertation, University of London. SOAS, 2017).

Masani, Minocher R., *Your Food: A Study of the Problem of Food and Nutrition in India* (Bombay: Padma Publications, 1944).

———, "The Communist Party in India," *Pacific Affairs* 24:1 (March 1951), 18–38.

Mathur, Yaduvansh Bahadur, *Muslims and Changing India* (Delhi: Trimurti Publications, 1972).

Mayer, Albert, *Pilot Project, India: The Story of Rural Development at Etawah, Uttar Pradesh* (Berkeley: University of California Press, 1958).

Michelle Burge McAlpin, *Subject to Famine: Food Crisis and Economic Change in Western India, 1860–1920* (Princeton: Princeton University Press, 1983).

McCleary, Rachel M., *Global Compassion: Private Voluntary Organizations and U.S. Foreign Policy since 1939* (Oxford: Oxford University Press, 2009).

McGowan, Abigail, *Crafting the Nation in Colonial India* (New York: Palgrave Macmillan, 2009).

McLane, John R., *Indian Nationalism and the Early Congress* (Princeton: Princeton University Press, 2015).

McLaren, Donald S., "The great protein fiasco," *Lancet* 304:7872 (1974), 93–96.

McMahon, Robert J., "Food as a diplomatic weapon: the India wheat loan of 1951," *Pacific Historical Review* 56:3 (Aug 1987), 349–77.

———, *The Cold War on the Periphery: The United States, India and Pakistan* (New York: Columbia University Press, 1996).

Meadowcroft, Keith, "The Emergence, Crystallization and Shattering of Right-Wing Alternative to Congress Nationalism: The All-India Hindu Mahasabha, 1937–52" (Unpublished dissertation, Concordia University, 2003).

Menon, Dilip, "The many spaces and times of Swadeshi," *Economic and Political Weekly* 47:42 (2012), 44–52.

Menon, Vengalil Krishnan Krishna, Ellen Wilkinson, Monica Whately and Leonard W. Matters, *The Condition of India. Being the Report of the Delegation sent to India by the India League, in 1932* (London: Essential News, 1934).

Ment, David M., "The American role in education in the Middle East: ideology and experiment, 1920–1940," *Paedagogica Historica* 47:1–2 (Feb 2011), 173–89.

Merrill, Dennis, *Bread and the Ballot: The United States and India's Economic Development, 1947–1963* (Chapel Hill: University of North Carolina Press, 1990).

Messer, Ellen, "The 'small but healthy' hypothesis: historical, political, and ecological influences on nutritional standards," *Human Ecology* 14:1 (1986), 57–75.

Miller, Ian, *Reforming Food in Post-Famine Ireland: Medicine, Science and Improvement, 1845–1922* (Manchester: Manchester University Press, 2014).

Mills, Ian D., "The 1918–1919 Influenza pandemic: the Indian experience," *The Indian Economic & Social History Review* 23:1 (1986), 1–40.

Misra, Jagannath P., *Madan Mohan Malaviya and the Indian Freedom Movement* (New Delhi: Oxford University Press, 2016).

Mitra, Sarojmohan, "Progressive cultural movement in Bengal," *Social Scientist* 8:5/6 (Dec 1979), 115–20.

Moffat, Chris, *India's Revolutionary Inheritance: Politics and the Promise of Bhagat Singh.* (Cambridge: Cambridge University Press, 2019).

———, "The itinerant library of Lala Lajpat Rai," *The History Workshop Journal* 89 (1 April 2020), 121–39.

Moomaw, Ira W., "Missions and human need," *International Review of Mission* 39:156 (1950), 418–24.

Moritz, Maria, "'The Empire of Righteousness': Anagarika Dharmapala and His Vision of Buddhist Asianism (c. 1900)" in Marc Frey and Nicola Spakowski (eds), *Asianisms. Regionalist Interactions & Asian Integration* (Singapore: National University of Singapore Press, 2016), 19–48.

Morris, David Morris, "Needed: a new famine policy," *Economic and Political Weekly* 10:5/7 (1975), 283–94.

Morris, Marcus, "From anti-colonialism to anti-imperialism: the evolution of H.M. Hyndman's Critique of Empire, c.1875–1905," *Historical Research* 87:236 (May 2014), 293–314.

Muehl, John Frederick, *American Sahib* (New York: The John Day Company, 1946).

Mukerjee, Madhusree, *Churchill's Secret War: The British Empire and the Ravaging of India during World War II* (New York: Basic Books, 2010).

Mukerjee, Radhakamal, *Food Planning for Four Hundred Millions* (London: Macmillan, 1938).

Mukherjee, Janam, *Hungry Bengal: War, Famine and the End of Empire* (Oxford: Oxford University Press, 2015).

Mukherjee, Upamanyu Pablo, *Natural Disasters and Victorian Empire: Famines, Fevers and the Literary Cultures of South Asia* (Basingstoke: Palgrave Macmillan, 2013).

Müller, Angela, *Indien im Sucher Fotografien und Bilder von Südasien in der deutschsprachigen Öffentlichkeit, 1920–1980* (Göttingen: Vandenhoeck & Ruprecht, 2019).

Murray, J. Lovell, *The Call of a World Task in War Time* (New York: Student Volunteer Movement, 1918).

Nag, Anindita, "Managing Hunger: Famine, Science and the Colonial State in India, 1860–1910" (Unpublished dissertation, UCLA, 2010).

Nair, Neeti, *Changing Homelands: Hindu Politics and the Partition of India* (Harvard University Press, 2011).

Nair, Rahul, "The construction of a 'population problem' in colonial India 1919–1947," *The Journal of Imperial and Commonwealth History* 39:2 (2011), 227–47.

Nakazato, Nariaki, "The Role of Colonial Administration, 'Riot Systems' and Local Networks during the Calcutta Disturbances of August 1946" in Tanika Sarkar and Sekhar Bandyopadhyay (eds), *Calcutta: The Stormy Decades* (New Delhi: Routledge, 2017), 267–319.

Nanda, Bal R. (ed.), *The Collected Works of Lala Lajpat Rai*, vol. 1 (New Delhi: Manohar, 2008).

———, *The Collected Works of Lala Lajpat Rai*, vol. 3 (New Delhi: Manohar, 2008).

Naoroji, Dadabhai, *Poverty and Un-British Rule in India* (London: Swan Sonnenschein, 1901).

Nash, Vaughan, *The Great Famine and Its Causes* (London: Longmans, Green, and Co., 1900).

Natarajan, C.P., and J.V. Shankar, "Food science and technology development in India: concepts and contributions of Dr. V. Subrahmanyan," *The Journal of Food Science and Technology* 17:1/ 2 (April 1980), 17–23.

Neswald, Elizabeth, "Nutritional Knowledge between the Lab and the Field. The Search for Dietary Norms in the Late Nineteenth and Early Twentieth Centuries" in Elizabeth Neswald, David F. Smith, and Ulrike Thoms (eds), *Setting Nutritional Standards: Theory, Policies, Practices* (Rochester: University of Rochester Press, 2017), 29–51.

Nevinson, Henry Woodd, *The New Spirit of India* (London: Harper & Brothers, 1908).

Nutt, Rick, "G. Sherwood Eddy and the attitudes of Protestants in the United States toward Global Mission," *Church History* 66:3 (Sept 1997), 502–21.

Nützenadel, Alexander, "'A world without famine'? Internationale Ernährungspolitik im Zeitalter der Weltkriege," *Comparativ* 17:3 (2007), 12–27.

Ó Gráda, Cormac, *Eating People Is Wrong, and Other Essays on Famine, Its Past, and Its Future* (Princeton: Princeton University Press, 2015), 38–91.

———, *Famine: A Short History* (Princeton: Princeton University Press, 2009).

Ober, Douglas F., "From Buddha bones to Bo trees: Nehruvian India, Buddhism, and the poetics of power, 1947–1956," *Modern Asian Studies* 53:4 (July 2019), 1312–50.

Oommen, Meera Anna, "Famine and Elephants: Remembering Place-Making along Travancore's Forest Fringe" in Katie Holmes and Heather Goodall (eds.), *Telling Environmental Histories: Intersections of Memory, Narrative and Environment* (Basingstoke: Palgrave Macmillan, 2017), 241–65.

Osella, Filippo, "Charity and philanthropy in South Asia: an introduction," *Modern Asian Studies* 52:1 (Jan 2018), 4–34.

Overstreet, Gene D., and Marshall Windmiller, *Communism in India* (Bombay: The Perennial Press, 1959).

Owen, Nicholas, *The British Left and India: Metropolitan Anti-Imperialism, 1885–1947* (Oxford: Oxford University Press, 2007).

Pal, R.K. and N. Singh, "The effect on rats of supplementing a North Indian diet with vegetable proteins (Sprouting Pulses) and calcium," *The Indian Medical Gazette* 73:8 (Aug 1938), 469–71.

Pal, Savinder, "Agrarian unrest in Punjab and Sardar Ajit Singh," *Proceedings of the Indian History Congress* 70 (2009), 451–64.

Palsetia, Jesse S., "Parsi and Hindu traditional and nontraditional responses to Christian conversion in Bombay, 1839–45," *The Journal of the American Academy of Religion* 74:3 (2006), 615–45.

———, "Parsis and Bombay City: Community Identity in the Nineteenth Century" in Prashant Kidambi, Manjiri Kamat and Rachel Dwyer (eds), *Bombay before Mumbai: Essays in Honour of Jim Masselos* (Oxford: Oxford University Press, 2019).

Panigrahi, Devendra, *India's Partition: The Story of Imperialism in Retreat* (London: Routledge, 2004).

Patel, Dinyar, *Naoroji: Pioneer of Indian Nationalism* (Cambridge: Harvard University Press, 2020).

———, "'One man lobby'? Propaganda, nationalism in the diaspora, and the India League of America during the Second World War," *The Journal of Imperial and Commonwealth History* 49:6 (2021), 1110–40.

Patel, Hitendra K., "Aspects of socio-political thoughts of Madan Mohan Malaviya, a liberal Hindu nationalist," *Proceedings of the Indian History Congress* 74 (2013), 581–86.

Patenaude, Bertrand M., *The Big Show in Bololand: The American Relief Expedition to Soviet Russia in the Famine of 1921* (Stanford: Stanford University Press, 2002).

Patwardhan, Nandini, *Radical Spirits: India's First Woman Doctor and Her American Champions* (Raleigh: Story Artisan Press, 2020).

Patwardhan, Vinayak Narayan, "The Problem of Nutrition in India & Asia," Paper presented at the Asian Relations Conference (New Delhi, 1944).

Pernau, Margrit, "Studying emotions in South Asia," *South Asian History and Culture* 12:2–3 (3 July 2021), 111–28.

Phoenix, Karen E., "'Not by Might, nor by Power, but by Spirit': The Global Reform Efforts of the Young Women's Christian Association of the United States, 1895–1939" (Unpublished dissertation, University of Illinois, 2010).

Popkin, Barry M., and Michael C. Latham, "The limitations and dangers of commerciogenic nutritious foods," *The American Journal of Clinical Nutrition* 26 (1973), 1015–23.

Porter, Stephen R., *Benevolent Empire: U.S. Power, Humanitarianism, and the World's Dispossessed* (Philadelphia: University of Pennsylvania Press, 2017).

Prakash, Gyan, *Another Reason: Science and the Imagination of Modern India* (Princeton: Princeton University Press, 1999).

Prasad, Bisheshwar, and B.L. Raina (eds), *Official History of the Indian Armed Forces in the Second World War, 1939–45. Medical Services. Administration* (Delhi: Combined Inter-Services Historical Section, India & Pakistan, 1953).

Prasad, Rajendra, "Food position not free from anxiety," *Indian Information* (15 May 1947).

———, "India's food front," *The March of India* 1:1 (1947).

Price, John, "'Orienting' the Empire: Mackenzie King and the aftermath of the 1907 Race Riots," *BC Studies: The British Columbian Quarterly* 156–157 (May 2008), 53–81.

Pripas-Kapit, Sarah Ross, "Educating Women Physicians of the World: International Students of the Woman's Medical College of Pennsylvania, 1883–1911" (Unpublished dissertation, UCLA, 2015).

Proceedings of the Sixth Meeting of the National Missionary Council, Lahore, November 13–18, 1919 (Calcutta: National Missionary Council, 1920).

Radhakrishna Rao, M.V., *Diet Manual: A Compilation of Diets Recommended in Health and Disease* (Bombay, 1959).

Raghavan, Srinath, *India's War: World War II and the Making of Modern South Asia* (New York: Basic Books, 2016).

Raheja, Gloria Goodwin, "'Hear the tale of the famine year': famine policy, oral traditions, and the recalcitrant voice of the colonized in nineteenth-century India," *Oral Tradition* 31:1 (2017), 155–98.

Rahman, Syedur, and Craig Baxter, *Historical Dictionary of Bangladesh* (Lanham: Scarecrow Press, 2010).

Raj, Richa, "Of Swadeshi, self-reliance and self-help: a study of the Arya Samaj in Colonial Punjab, 1890–1920s," *Moving the Social* 62 (2019), 51–72.

Rajan, M.S., "Pandit Hriday Nath Kunzru: a memoir," *India Quarterly: A Journal of International Affairs* 34:4 (Oct 1978), 441–56.

Rama Rau, Dhanvanthi, *An Inheritance: The Memoirs of Dhanvanthi Rama Rau* (Bombay: Allied Publishers, 1960).

Ramakrishna Mission Bengal & Orissa Cyclone Relief: 1942–1944 (Howrah: The Secretary from Belur Math, July 1945).

Ramanna, Mridula, "Coping with the Influenza Pandemic: The Bombay Experience," in Howard Phililips and David Killingray (eds), *The Spanish Influenza Pandemic of 1918–19. New Perspectives* (London: Routledge, 2003), 86–98.

Ramnath, Maia, *Haj to Utopia: How the Ghadar Movement Charted Global Radicalism and Attempted to Overthrow the British Empire* (Berkeley: University of California Press, 2011).

Rao, M.V.L., "Dr Vaidyanatha Subrahmanyan – a pioneering food scientist," *The Journal of Food Science and Technology* 17:1 & 2 (April 1980), 1–8.

Rao, N. Bhima, "Prof. V. Subrahmanyan – some personal reminiscences," *The Journal of Food Science and Technology* 17:1&2 (April 1980), 121–23.

Rashid, Salim, "The policy of laissez-faire during scarcities," *The Economic Journal* 90:359 (Sept 1980), 493–501.

Rasmussen, Wayne D., *A History of the Emergency of Farm Labour Supply Program, 1943–47* (Washington: US Department of Agriculture. Bureau of Agricultural Economics, 1951).

Ratcliffe, Samuel K., *Sir William Wedderburn and the Indian Reform Movement* (London: George Allen & Unwin Ltd., 1923).

Raucher, Alan, "American anti-imperialists and the pro-India movement, 1900–1932," *Pacific Historical Review* 43:1 (1974), 83–110.

Ray, Krishnendu, and Tulasi Srinivas (eds), *Curried Cultures: Globalization, Food, and South Asia* (Berkeley: University of California Press, 2012).

Ray, Renuka, *My Reminiscences: Social Development During Gandhian Era and After* (New Delhi: Allied Publishers, 1982).

Ray, Sibnarayan (ed.), *Selected Works of M.N. Roy*, vol 1 (1917–1922) (Oxford: Oxford University Press, 1987).

Ray, Utsa, "The body and its purity: dietary politics in Colonial Bengal," *The Indian Economic & Social History Review* 50:4 (Oct 2013), 395–421.

Raza, Ali, and Franziska Roy, "Paramilitary organisations in Interwar India," *South Asia: Journal of South Asian Studies* 38:4 (2015), 671–89.

Reetz, Dietrich, *Islam in the Public Sphere. Religious Groups in India, 1900-1947* (Oxford: Oxford University Press, 2006).

Reinisch, Jessica, "Internationalism in relief: the birth (and death) of UNRRA," *Past & Present* 210:6 (2011), 258–89.

———, "'Auntie UNRRA' at the crossroads," *Past & Present* 218:8 (2013), 70–97.

Reisner, John H., *Reforesting China: Permanent Famine Prevention versus Famine Relief* (Nanking: Nanking University, 1921).

Relief Organisations Fight Bengal Famine (Calcutta: Relief Co-ordination Committee, 1943).

Renton, Margaret, *War-Time Agencies of the Churches. Directory and Handbook* (New York: General War-Time Commission of the Churches, 1919).

Report of the AMM 1902 (Madras: AMM, 1903).

Report of the AMM 1909 (Ahmednagar: AMM, 1910).

Report of the AMM for the Year 1896 (Bombay: AMM, 1897).

Report of the AMM for the Year 1898 (Bombay: AMM, 1899).

Report of the AMM for the Year 1918 (Barsi: AMM, 1919).

Report of the AMM for the Year 1919 (Barsi: AMM, 1920).

Report of the AMM for the Year 1920 (Ahmednagar: AMM, 1921).

Report of the AMM. In Western India, 1923 (Vadala: AMM, 1924).

Report of the Bijapur Famine Relief Committee (Bombay: Servants of India Society, 1944).

Report of the Foreign Missions Conference of North America (New York, 1920).

Report of the Indian Famine Commission (London: Commission of Inquiry on Indian Famines, 1880).

Report of the Indian Famine Commission (London: Darling & Son, 1898).

Report of the Indian Famine Commission (Calcutta: Office of the Superintendent of Government Printing, 1901).

Report of the Madras Famine Code Revision Committee of 1938 (Madras, 1940).

Report of the 17th Indian National Congress held at Calcutta on the 26th, 27th & 28th December, 1901 (Calcutta: Indian National Congress, 1902).

Report on the Famine in the Bombay Presidency, 1899–1902 (Bombay: Government of Bombay, 1903).

Resolution on the Administration of Famine Relief in the United Provinces of Agra and Oudh during the Year 1907 and 1908 (Allahabad: Government Press, 1908).

Resolution on the Administration of Famine Relief in the United Provinces of Agra and Oudh during the Years 1913 and 1914 (Allahabad: United Provinces, 1918).

Review of the Famine Relief Administration in the Bombay Presidency during the year 1918–1919 (Bombay Presidency, 1920).

Riley, Barry, *The Political History of American Food Aid: An Uneasy Benevolence* (New York: Oxford University Press, 2017).

Robert, Dana L., "The origin of the student volunteer watchword: 'the evangelization of the world in this generation'," *International Bulletin of Missionary Research* 10:4 (Oct 1986), 146–49.

———, "The 'Christian Home' as a Cornerstone of Anglo-American Missionary Thought and Practice" in Dana Lee Robert (ed.), *Converting Colonialism: Visions and Realities in Mission History, 1706–1914* (Grand Rapids: William B. Eerdmans Pub. Co, 2008), 134–65.

Roberts, Hibbert R., "The Meals for Millions Foundation: A Study of a Non-Governmental Organization" (Unpublished dissertation, University of Washington, 1967. Ann Arbor, Michigan).

Robertson, Thomas, "New Frontiers: World War II Technologies and the Opening of Tropical Environments to Development" in Stephen J. Macekura and Erez Manela (eds), *The Development Century: A Global History* (Cambridge: Cambridge University Press, 2018), 107–29.

Rodogno, Davide, *Night on Earth: A History of International Humanitarianism in the Near East, 1918–1930* (New York: Cambridge University Press, 2022).

Rook-Koepsel, Emily, *Democracy and Unity in India: Understanding the All India Phenomenon, 1940–1960* (London: Routledge, 2019).

Rothgeb, John M., *U.S. Trade Policy: Balancing Economic Dreams and Political Realities* (Washington, D.C.: CQ Press, 2001).

Rotter, Andrew Jon, *Comrades at Odds: The United States and India, 1947–1964* (Ithaca: Cornell University Press, 2000).

Roy, Anjali Gera, *Imperialism and Sikh Migration: The Komagata Maru Incident* (London: Routledge, 2018).

——— and Ajaya K. Sahoo, "The journey of the *Komagata Maru*: national, transnational, diasporic," *South Asian Diaspora* 8:2 (2016), 85–97.

Roy, Anwesha, *Making Peace, Making Riots: Communalism and Communal Violence, Bengal 1940–1947* (New Delhi: Cambridge University Press, 2018).

Roy, Mallarika Sinha, "'The call of the world': women's memories of global socialist feminism in India," *International Review of Social History* 67:S30 (2022), 237– 62.

Roy, Parama, "Meat-eating, masculinity, and renunciation in India: a Gandhian grammar of diet," *Gender & History* 14:1 (April 2002), 62–91.

———, "Women, Hunger, and Famine: Bengal, 1350/1943" in Bharati Ray (ed.), *Women of India: Colonial and Post-Colonial Periods*, vol. 3 (New Delhi: Pauls Press, 2005), 392–423.

Roy, Patricia E., *A White Man's Province: British Columbia Politicians and Chinese and Japanese Immigrants, 1858–1914* (Vancouver: University of British Columbia Press, 1990).

Roy, Tirthankar, *Natural Disasters and Indian History* (Oxford: Oxford University Press, 2012).

Rubin, Olivier, "The merits of democracy in famine protection – fact or fallacy?" *The European Journal of Development Research* 21:5 (Dec 2009), 699–717.

———, *Democracy and Famine* (Milton Park: Routledge, 2011).

Ruswick, Brent, *Almost Worthy: The Poor, Paupers, and the Science of Charity in America, 1877–1917* (Bloomington: Indiana University Press, 2013).

Sackley, Nicole, "The village as Cold War site: experts, development, and the history of rural reconstruction," *The Journal of Global History* 6:3 (Nov 2011), 481–504.

———, "Village models: Etawah, India, and the making and remaking of development in the Early Cold War," *Diplomatic History* 37:4 (2013), 749–78.

Saha, Sharmistha, "The Commune-Ist Air: The Case of the IPTA Central Squad" in Sharmistha Saha (ed.) *Theatre and National Identity in Colonial India* (Singapore: Springer Singapore, 2018), 119–52.

Saklani, Avhinash Mohan, "Nehru, Chiang Kai-Shek, and the Second World War" in Madhavi Thampi (ed.), *India and China in the Colonial World* (New Delhi: Routledge, 2005), 167–90.

Sami, Leela, "Famine, Disease, Medicine and the State in Madras Presidency (1876–78)" (Unpublished dissertation, UCL, 2006).

Santhanam, K., *The Cry of Distress. A First-Hand Description and an Objective Study of the Indian Famine of 1943* (New Delhi: The Hindustan Times, 1943).

Sappington, N.O., John P. Glennon, Howard M. Smyth, and Frederick Aandahl (eds), *Foreign Relations of the United States, 1946. The British Commonwealth, Western and Central Europe*, vol. V (Washington: United States Government Printing Office, 1969).

Sarkar, Abhijit, "Fed by Famine: the Hindu Mahasabha's politics of religion, caste, and relief in response to the Great Bengal Famine, 1943–1944," *Modern Asian Studies* 54:6 (2020), 2022–86).

Sarkar, Nikhil, *Matter of Conscience: Artists Bear Witness to The Great Bengal Famine of 1943* (Calcutta: Punasha, 1998).

Sarkar, Sumit, *The Swadeshi Movement in Bengal, 1903–1908* (New Delhi: People's Publishing House, 1973).

———, *Modern India, 1885–1947* (Basingstoke: Macmillan, 1992).

———, "Nationalism and poverty: discourses of development and culture in 20th century India," *Third World Quarterly* 29:3 (April 2008), 429–45.

———, and Sabyasachi Bhattacharya (eds), *Towards Freedom: Documents on the Movement for Independence in India 1946. Part I* (New Delhi: Oxford University Press, 2007).

Sarles, Ruth, and Bill Kauffman, *A Story of America First: The Men and Women Who Opposed U.S. Intervention in World War II* (Westport, Conn: Praeger, 2003).

Sarmin, Utsa, "Hunger, Humiliation, and Death: Perils of Migrant Worker in the Time of Covid-19" in Ritajyoti Bandyopadhyay, Paula Banerjee, and Ranabir Samaddar (eds), *India's Migrant Workers and the Pandemic* (New York: Routledge, 2021), 167–76.

Sasson, Tehila, "From empire to humanity: the Russian famine and the imperial origins of international humanitarianism," *The Journal of British Studies* 55:3 (July 2016), 519–37.

———, and James Vernon, "Practising the British way of famine: technologies of relief, 1770–1985," *European Review of History: Revue Européenne d'histoire* 22:6 (2015), 860–72.

Sathyamala, Christina, "Sukathme's Legacy and the Indian Exceptionalism" in Murat Arsel, Anirban Dasgupta, and Servaas Storm (eds), *Reclaiming Development Studies: Essays for Ashwani Saith* (New York: Anthem Press, 2021), 179–96.

Scarfe, Wendy, and Allan Scarfe. *Tiger on a Rein: Report on the Bihar Famine* (London: G. Chapman, 1969).

Schedneck, Brooke, "Buddhist International Organizations" in Michael K. Jerryson (ed.), *Oxford Handbook of Contemporary Buddhism* (New York: Oxford University Press, 2017), 398–416.

Scholten, Bruce A., *India's White Revolution: Operation Flood, Food Aid and Development* (London: I.B. Tauris & Co., 2010).

Scott, Jefferson E., *In Famine Land: Observations and Experiences in India during the Great Drought of 1899–1900* (New York: Harper and Brothers Publishers, 1904).

Scott-Smith, Tom, "Beyond the 'raw' and the 'cooked': a history of fortified blended foods," *Disasters* 39:s2 (Oct 2015), 244–60.

———, *On an Empty Stomach: Two Hundred Years of Hunger Relief* (Ithaca: Cornell University Press, 2020).

Scrinis, Gyorgy, *Nutritionism: The Science and Politics of Dietary Advice* (New York: Columbia University Press, 2013).

Sebastian, Mrinalini, "Reading archives from a postcolonial feminist perspective: 'native' Bible women and the missionary ideal," *The Journal of Feminist Studies in Religion* 19:1 (2003), 5–25.

Seckler, David, "The 'small but healthy?' Hypothesis: a reply to critics," *Economic and Political Weekly* 19:44 (1984), 1886–88.

Selles, Johanna M., *The World Student Christian Federation, 1895–1925: Motives, Methods, and Influential Women* (Eugene: Pickwick Publications, 2011).

Sen, Amartya, "Starvation and exchange entitlements: a general approach and its application to the Great Bengal Famine," *The Cambridge Journal of Economics* 1:1 (March 1977), 33–59.

———, *Development as Freedom* (Oxford: Oxford University Press, 2001).

Sen, Ela and Zainul Abedin, *Darkening Days, Being a Narrative of Famine-Stricken Bengal* (Calcutta: Susil Gupta, 1944).

Sen, Manikuntala, *In Search of Freedom: An Unfinished Story* (Calcutta: Stree, 2001).

Sen, Samita, "Between Homes; Without Homes. Migration, Circularity and Domesticity" in Ritajyoti Bandyopadhyay, Paula Banerjee, and Ranabir Samaddar (eds), *India's Migrant Workers and the Pandemic* (New York: Routledge, 2021), 138–64.

Sen, Tansen, *India, China, and the World: A Connected History* (Lanham: Rowman & Littlefield, 2017).

Sen, Uditi, "Social work, refugees and national belonging: evaluating the 'Lady Social Workers' of West Bengal," *South Asia: Journal of South Asian Studies*, 44:2 (2021), 344–61.

Sengupta, Nitish, *Bidhan Chandra Roy* (New Delhi: Ministry of Information and Broadcasting, 2002).

Servants of India Society Report for 1939–40 (Poona, 1940).

Shaffer, Robert, "J. J. Singh and the India League of America, 1945–1959: pressing at the margins of the Cold War consensus," *The Journal of American Ethnic History* 31:2 (2012), 68–103.

Shahid, Amal, "Famine Labour and Coercion in Relief-based Public Works Construction in Colonial India in the Late Nineteenth Century" in Kate Ekama, Lisa Hellman and Matthias van Rossum (eds), *Slavery and Bondage in Asia, 1550–1850* (Berlin: De Gruyter, 2022), 231–50.

Shankar, Prabha Ravi, "Some reflections on the correspondence between G.K. Gokhale and William Wedderburn," *Proceedings of the Indian History Congress* 63 (2002), 604–14.

———, "Henry Mayers Hyndman (1842–1921) and the radicalization of the Indian National Congress," *Proceedings of the Indian History Congress* 66 (2005), 1041–49.

Sharma, Sanjay, *Famine, Philanthropy, and the Colonial State: North India in the Early Nineteenth Century* (New Delhi: Oxford University Press, 2001).

———, "Measuring Hunger. Debates on an 'Adequate' Diet in Colonial North India" in Kiranmayi Bhushi (ed.), *Farm to Fingers. The Culture and Politics of Food in Contemporary India* (New York: Cambridge University Press, 2018), 184–214.

———, "Poorhouses and Gratuitous Famine Relief in Colonial North India" in Ayesha Mukherjee (ed.), *A Cultural History of Famine: Food Security and the Environment in India and Britain* (London: Routledge, 2019), 129–48.

Sherman, Taylor C., "From 'grow more food' to 'miss a meal': hunger, development and the limits of post-colonial nationalism in India, 1947–1957," *South Asia: Journal of South Asian Studies* 36:4 (Dec 2013), 571–88.

Sherwood, Eddy, *India Awakening* (New York: Missionary Education Movement of the US and Canada, 1912).

Showalter, Nathan D., "Crusade or catastrophe? The student missions movement and the First World War," *International Bulletin of Missionary Research* 17:1 (Jan 1993), 13–17.

Shurtleff, William, and Akiko Aoyagi, *History of Meals for Millions, Soy, and Freedom from Hunger (1946–2011): Extensively Annotated Bibliography and Sourcebook* (Lafayette: Soyinfo Centre, 2011).

Siegel, Benjamin R., "'Fantastic Quantities of Food Grains': Cold War Visions and Agrarian Fantasies in Independent India," in Leslie James and Elisabeth Leake (eds), *Decolonization and the Cold War: Negotiating Independence* (London: Bloomsbury, 2015), 21–42.

———, "'Self-help which ennobles a nation': development, citizenship, and the obligations of eating in India's austerity years," *Modern Asian Studies* 50:3 (May 2016), 975–1018.

———, *Hungry Nation: Food, Famine, and the Making of Modern India* (Cambridge: Cambridge University Press, 2018).

———, "'The claims of Asia and the Far East': India and the FAO in the age of ambivalent internationalism," *The International History Review* 41:2 (4 March 2019), 427–50.

Silvestri, Michael, *Policing 'Bengali Terrorism' in India and the World: Imperial Intelligence and Revolutionary Nationalism, 1905–1939* (Cham: Palgrave Macmillan, 2019).

Simmons, Dana, "Starvation Science: From Colonies to Metropole" in Alexander Nützenadel and Frank Trentmann (eds), *Food and Globalization: Consumption, Markets and Politics in the Modern World* (Oxford: Oxford, 2008), 173–92.

Simonow, Joanna, "The Great Bengal Famine in Britain: metropolitan campaigning for food relief and the end of empire, 1943–44," *The Journal of Imperial and Commonwealth History* 48:1 (2020), 168–97.

Singh, Kishore, and Sanjoy Kumar Mallik, *Chittaprosad, 1915–1978: A Retrospective* (New Delhi: Delhi Art Gallery, 2011).

Singh, Maina Chawla, *Gender, Religion, and 'Heathen Lands': American Missionary Women in South Asia, 1860s–1940s* (New York: Garland Pub, 2000).

Sinha, Dipa, "Hunger and food security in the times of Covid-19," *The Journal of Social and Economic Development* 23:2 (1 Sept 2021), 320–31.

Sinha, Manish, "The United Nations Relief and Rehabilitation Administration (UNRRA) and the question of food aid during the Bengal Famine of 1943," *Proceedings of the Indian History Congress* 73 (2012), 1042–52.

Sinha, Mrinalini, "Gender and the Critiques of Colonialism and Nationalism. Locating the 'Indian Woman'" in Sumit Sarkar and Tanika Sarkar (eds), *Woman and Social Reform in Modern India. A Reader* (Bloomington: Indiana University Press, 2008), 452–72.

Sinha, Subir, "Lineages of the developmentalist state: transnationality and Village India, 1900–1965," *Comparative Studies in Society and History* 50:1 (Jan 2008), 57–90.

Sivaswamy, K.G., K.K. Chandy, T.S. Shastry, M.E. Naidu, and T.V.S. Shastry, *Food Famine and Nutritional Diseases in Travancore (1943–44)* (Coimbatore: Sevindia Kerala Relief Centre, 1945).

Slate, Nico, *Colored Cosmopolitanism: The Shared Struggle for Freedom in the United States and India* (Cambridge: Harvard University Press, 2012).

———, *Gandhi's Search for the Perfect Diet: Eating with the World in Mind* (Seattle: University of Washington Press, 2019a).

———, *Lord Cornwallis is Dead: The Struggle for Democracy in the United States and India* (Cambridge: Harvard University Press, 2019b).

Slobodkin, Yan, "Famine and the science of food in the French Empire, 1900–1939," *French Politics, Culture & Society* 36:1 (Jan 2018), 52–75.

Smith-Howard, Kendra, *Pure and Modern Milk: An Environmental History since 1900* (Oxford: Oxford University Press, 2014).

Social and Technical Assistance in India. 1951–1956. A Report to the International Cooperation Administration of the United States of America (Philadelphia: American Friends Service Committee, December 1956).

Sohi, Seema, "Race, surveillance, and Indian anticolonialism in the transnational Western U.S.-Canadian borderlands," *The Journal of American History* 98:2 (Sep 2011), 420–36.

———, *Echoes of Mutiny: Race, Surveillance, and Indian Anticolonialism in North America* (New York: Oxford University Press, 2014).

Soni, "Learning to Labour: 'Native' orphans in Colonial India, 1840s–1920s," *International Review of Social History* 65:1 (April 2020), 15–42.

Sreenivas, Mytheli, *Reproductive Politics and the Making of Modern India* (Seattle: University of Washington Press, 2021).

Srimanjari, *Through War and Famine. Bengal, 1939–45* (New Delhi: Orient BlackSwan, 2009).

Srivatsan, R., "Concept of 'Seva' and the 'Sevak' in the Freedom Movement," *Economic and Political Weekly* 41:5 (2006), 427–38.

Stahl, Rune Møller, "The Economics of Starvation: Laissez-Faire Ideology and Famine in Colonial India" in Mikkel Thorup (ed.), *Intellectual History of Economic Normativities* (New York: Palgrave Macmillan, 2016), 169–84.

Stanley, Brian, *The World Missionary Conference, Edinburgh 1910* (Grand Rapids: William B. Eerdmans Pub. Co, 2009).

Stanton, Jennifer., "Listening to the Ga: Cicely Williams' discovery of Kwashiorkor on the Gold Coast," *Clio Medica (Amsterdam, Netherlands)* 61 (2001), 149–71.

Stark, Ulrike, "Lucknow's Jalsa-i Tahzib: Urbane Elite, organisierte Handlungskompetenz und frühe 'Associational Culture'" in Harald Fischer-Tiné (ed.), *Handeln und Verhandeln. Kolonialismus, Transkulturelle Prozesse und Handlungskompetenz* (Münster: LIT, 2002), 51–73.

Stec, Loretta, "Pacifism, Vera Brittain, and India," *Peace Review* 13:2 (June 2001), 237–44.

Stephens, Ian, *Monsoon Morning* (London: Ernest Benn, 1966).

Stolte, Carolien, *Orienting India: Interwar Internationalism in an Asian Inflection, 1917–1937* (Leiden University, 2013).

———, "Compass Points: Four Indian Cartographies of Asia, c. 1930–55" in Marc Frey and Nicola Spakowski (eds), *Asianisms: Regionalist Interactions & Asian Integration* (Singapore: National University of Singapore Press, 2016), 49–74.

——— and Harald Fischer-Tiné, "Imagining Asia in India: nationalism and internationalism (ca. 1905–1940)," *Comparative Studies in Society and History* 54:1 (Jan 2012), 65–92.

Stross, Randall E., *The Stubborn Earth: American Agriculturalists on Chinese Soil, 1898–1937* (Berkeley: University of California Press, 1989).

Subrahmanyan, V., T.R. Doraiswamy, and R.K. Bhagavan, "The effect of supplementary processed protein foods fortified with essential vitamins and minerals on the growth and nutritional status of undernourished weaned infants," *The Indian Journal of Peadiatrics* 26:4 (1959), 406–13.

———, and Kantha Joseph, T.R. Doraiswamy, M. Narayanarao, A.N. Sankaran, and M. Swaminathan, "The effect of a supplementary multipurpose food on the growth and nutritional status of schoolchildren," *The British Journal of Nutrition* 11:4 (Dec 1957), 382.

———, and H.A.B. Parpia, "Some Aspects of the Utilization of Food Research in India," in R.C. Bhutiani, K.L. Raghakrishnan, and B.V. Subbarayappa (eds), *Some Aspects of Food Technology in India* (Mysore: CFTRI, 1959), 1–12.

———, H.A.B. Parpia and M. Swaminathan, "Plan for the Manufacture of Indian Multipurpose Food" in R.C. Bhutiani, K.L. Raghakrishnan, and B.V. Subbarayappa (eds), *Some Aspects of Food Technology in India* (Mysore: CFTRI, 1959), 96–99.

———, R. Rajagopalan, M. Narayana Rao, and M. Swaminathan, "Processed Protein Foods of Vegetable Origin Fortified with Vitamins and Minerals" in R.C. Bhutiani, K.L. Raghakrishnan, and B.V. Subbarayappa (eds), *Some Aspects of Food Technology in India*, (Mysore: CFTRI, 1959). 115–23.

Sukhatme, Pandurang V., "Protein strategy and agricultural development," *The Indian Journal of Agricultural Economics* XXVII:1 (1972), 1–24.

Sukthankar, S.E., *Report of the Fact-finding Committee for Survey of Scarcity Areas Maharashtra State 1973* (Bombay: Govt. of Maharahstra, 1973).

Sunderason, Sanjukta, "As 'agitator and organizer': Chittaprosad and art for the Communist Party of India, 1941–48," *Object* 13 (2011), 76–95.

———, "Shadow-lines: Zainul Abedin and the afterlives of the Bengal Famine of 1943," *Third Text* 31:2–3 (April 2017), 239–59.

Sunderland, Jabez Thomas, *The Causes of Famines in India* (Toronto, 1904).

Sweeney, Stuart, "Indian railways and famines, 1875–1914: Magic wheels and empty stomachs," *Essays in Economic & Business History* 26 (2008), 147–58.

Tauger, Mark B., "The Indian famine crises of World War II," *British Scholar* 1:2 (March 2009), 166–96.

Teaf, Jr., Howard M., "Origins of a Private Village Improvement Project," in Howard M. Teaf, Jr. and Peter G. Franck (eds), *Hands Across Frontiers* (Ithaca: Cornell University Press, 1955), 65–126.

Teed, Paul E., "Interfaith encounter and religious pluralism: J.T. Sunderland's mission to Brahmo Samajes of India, 1895–96," *American Studies* 50:1/2 (2009), 51–69.

Terracciano, Emilia, *Art and Emergency: Modernism in Twentieth-Century India* (I.B. Tauris & Co. Ltd, 2018).

Thakurdas, Purshotamdas, J.R.D. Tata, Ardeshir Dalal, Kasturbhai Lalbhai, A.D. Shroff, and John Matthai, *A Plan of Economic Development for India* (London: Penguin Books, 1944).

The Problem of Nutrition. Report on the Physiological Bases of Nutrition (London: The League of Nations. Technical Commission of the Health Committee, 1936).

Thompson, Michael G., "Sherwood Eddy, the missionary enterprise, and the rise of Christian internationalism in 1920s America," *Modern Intellectual History* 12:1 (April 2015), 65–93.

———, *For God and Globe: Christian Internationalism in the United States between the Great War and the Cold War* (Ithaca: Cornell University Press, 2016).

Tinker, Hugh, "A forgotten long march: the Indian exodus from Burma, 1942," *The Journal of Southeast Asian Studies* 6:1 (March 1975), 1–15.

———, "The India Conciliation Group, 1931–50: dilemmas of the mediator," *The Journal of Commonwealth & Comparative Politics* 14:3 (Nov 1976), 224–41.

Training Within Industry Report, 1940–1945. A Record of the Development of Management Techniques for Improvement of Supervision. Their Use and the Results (Washington: War Manpower Commission, 1945).

Trevithick, Alan, *The Revival of Buddhist Pilgrimage at Bodh Gaya (1811–1949): Anagarika Dharmapala and the Mahabodhi Temple* (Delhi: Motilal Banarsidass Publishers, 2006).

Trivedi, Lisa, *Clothing Gandhi's Nation: Homespun and Modern India* (Bloomington: Indiana University Press, 2007).

Trivedi, Lisa N., "Visually mapping the 'nation': Swadeshi politics in nationalist India, 1920–1930," *The Journal of Asian Studies* 62:1 (Feb 2003), 11–41.

Tschurenev, Jana, *Empire, Civil Society, and the Beginnings of Colonial Education in India* (Cambridge: Cambridge University Press, 2019).

Tsui, Brian, *China's Conservative Revolution: The Quest for a New Order, 1927–1949* (Cambridge: Cambridge University Press, 2018).

Tucker, Todd, *The Great Starvation Experiment: Ancel Keys and the Men Who Starved for Science* (Minneapolis: University of Minnesota Press, 2007).

Twentieth Annual Report of the Department of Land Records and Agricultural, Bombay Presidency for the Year 1902–1903 (Bombay: Government Central Press, 1904).

Twomey, Christina, "Framing atrocity: photography and humanitarianism," *History of Photography* 36:3 (Aug 2012), 255–64.

——— and Andrew J. May, "Australian responses to the Indian famine, 1876–78: Sympathy, photography and the British Empire," *Australian Historical Studies* 43:2 (June 2012), 233–52.

Tyrrell, Ian, *Reforming the World: The Creation of America's Moral Empire* (Princeton: Princeton University Press, 2010).

———, *Transnational Nation: United States History in Global Perspective since 1789* (New York: Palgrave Macmillan, 2015).

———, "Connections, Networks, and the Beginnings of a Global America in the Gilded Age and Progressive Era" in Christopher McKnight Nichols and Nancy C. Unger (eds) *A Companion to the Gilded Age and Progressive Era* (West Sussex: Wiley, 2017), 379–98.

———, "Vectors of Practicality: The Social Gospel, the North American YMCA in Asia and the Global Context" in Harald Fischer-Tiné, Stefan Huebner, and Ian Tyrrell (eds), *Spreading Protestant Modernity: Global Perspectives on the Social Work of the YMCA and YWCA* (Honolulu: University of Hawai'i Press, 2021), 39–60.

——— and Jay Sexton, *Empire's Twin: U.S. Anti-Imperialism from the Founding Era to the Age of Terrorism* (Ithaca: Cornell University Press, 2015).

Unger, Corinna R., *Entwicklungspfade in Indien: eine internationale Geschichte 1947–1980* (Göttingen: Wallstein Verlag, 2015).

Valdameri, Elena, *Indian Liberalism Between Nation and Empire: The Political Life of Gopal Krishna Gokhale* (Abingdon: Routledge, 2022).

Venkataramani, M.S., *Bengal Famine of 1943: The American Response* (Delhi: Vikas, 1974).

———, "The Roosevelt administration and the Great Indian Famine," *International Studies* 4:3 (July 1962), 241–64.

Verma, Vidhu, "Reinterpreting Buddhism: Ambedkar on the politics of social action," *Economic and Political Weekly* 45:49 (Dec 2010), 56–65.

Vernon, James, *Hunger: A Modern History* (London: The Belknap Press of Harvard University Press, 2007).

———, *Modern Britain: 1750 to the Present* (Cambridge: Cambridge University Press, 2017).

Vevaina, Leilah, "Good deeds: Parsi trusts from 'the womb to the tomb'," *Modern Asian Studies* 52:1 (2018), 238–65.

Visana, Vikram, "Vernacular liberalism, capitalism, and anti-imperialism in the political thought of Dadabhai Naoroji," *The Historical Journal* 59:3 (Sep 2016), 775–97.

Visram, Rozina, *Asians in Britain: 400 Years of History* (London: Pluto Press, 2002).

Visual Aids. A Guide for Peace Corps Volunteers (Washington: Peace Corps Information Collection and Exchange, 1976).

Ward, John M., "Stevens, Sir Bertram Sydney (1889–1973)" in J. Ritchie (ed.) *Australian Dictionary of Biography*, vol 12 (Melbourne: Melbourne University Press, 1990).

Ward, Robert, *Some Fruits of the Great India Famine* (Madras: The M.E. Publishing House, 1902).

Watenpaugh, Keith David, *Bread from Stones: The Middle East and the Making of Modern Humanitarianism* (Oakland: University of California Press, 2015).

Watson, Don, "The People's War and the People's Theatre: British Socialist Theatre 1939–45" in Colin Creighton and Martin Shaw (eds), *The Sociology of War and Peace* (Basingstoke: Macmillan, 1987), 159–77.

Watt, Carey A., "Education for national efficiency: constructive nationalism in North India, 1909–1916," *Modern Asian Studies* 31:2 (1997), 339–74.

———, *Serving the Nation: Cultures of Service, Association, and Citizenship* (New Delhi: Oxford University Press, 2005).

Watts, Tim J., "Watumull Foundation" in Huping Ling and Allan W Austin (eds), *Asian American History and Culture: An Encyclopaedia* (London: Routledge, 2015), 363–64.

Webster, David, "Development advisors in a time of Cold War and decolonization: the United Nations technical assistance administration, 1950–59," *The Journal of Global History* 6:2 (July 2011), 249–72.

Wedderburn, William, "The reorganisation of agricultural credit in India," *The Journal of the Society of Arts* 40 (Nov 1891), 646–54.

———, "The Skeleton and the (Jubilee) Feast," in *Speeches and Writings of William Wedderburn* (Madras: G.A.Natesan & Co., 1918), 284–322.

Weindling, Paul, "'Belsenitis': liberating Belsen, its hospitals, UNRRA, and selection for re-emigration, 1945–1948," *Science in Context* 19:3 (Sep 2006), 401–18.

Wieters, Heike, *The NGO CARE and Food Aid from America 1945–80: 'Showered with Kindness'?* (Manchester: Manchester University Press, 2017).

Williams, Cicely D., "Kwashiorkor: a nutritional disease of children associated with a maize diet," *The Lancet* (1935), 1151–52.

Wilson, Ann Marie, "In the name of God, civilization, and humanity: the United States and the Armenian massacres of the 1890s," *Le Mouvement Social* 227:2 (2009), 27–44.

Wiser, William Henricks, and Charlotte Viall Wiser, *Behind Mud Walls, 1930–1960* (Berkeley: University of California Press, 1971).

Wollons, Roberta, "Travelling for God and adventure: women missionaries in the late 19th century," *The Asian Journal of Social Science* 31:1 (2003), 55–71.

Wright, Robert A., *A World Mission: Canadian Protestantism and the Quest for a New International Order, 1918–1939* (Montreal: McGill-Queen's University Press, 1991).

Yang, Anand A., "Disciplining 'natives': prisons and prisoners in early nineteenth century India," *South Asia: Journal of South Asian Studies* 10:2 (Dec 1987), 29–45.

Zachariah, Benjamin, *Developing India: An Intellectual and Social History, c. 1930–50* (New Delhi: Oxford University Press, 2005).

Zanasi, Margherita, "Western utopias, missionary economics, and the Chinese village," *The Journal of World History* 24:2 (2013), 359–87.

Zimmerman, Andrew, "A German Alabama in Africa: the Tuskegee expedition to German Togo and the transnational origins of West African cotton growers," *The American Historical Review* 110:5 (Dec 2005), 1362–98.

Zook, Darren, "Famine in the Landscape: Imagining Hunger in South Asian History" in Arun Agrawal and K. Sivaramakrishnan (eds), *Agrarian Environments. Resources, Representations, and Rule in India* (Durham: Duke University Press, 2000), 107–31.

Zurbrigg, Sheila, *Malaria in Colonial South Asia: Uncoupling Disease and Destitution* (London: Routledge, 2019).

Selection of Newspapers and Periodicals

Allahabad Farmer, Allahabad
Amrita Bazar Patrika, Calcutta
Boston Daily Globe, Boston
Christian Herald, New York
Daily Worker, London
Dawn, Lahore
Free Hindustan, Vancouver and New York
Friends Journal, Philadelphia
Globe, Toronto
Hindustan Times, New Delhi
Hindustanee: The Official Organ of the United India League, Vancouver
Hindusthanee Student, Berkeley
Illustrated Missionary News, London
India Alliance, Akola
India To-Day, Allahabad
India Today, New York
Indian Express, Madras
Indian Messenger, Calcutta
Indian Social Reformer, Bombay
Issues and Events, New York
Lancet, London
Life, New York
Life and Light for Women, Boston
Missionary Herald, Boston
Modern Review, Calcutta
New York Times, New York
NewsIndia, London
People's Age, Calcutta
People's War, Calcutta
Servant of India, Poona
Social Scientist, Delhi
Socialist Appeal, London
Statesman, Calcutta
Times of India, Bombay

Town Crier, Birmingham
Tribune, Sydney
Vital Issue, New York
War Cry (Canadian), Toronto
War Cry (US), Missouri
West Australian, Perth
World Agriculture, Amherst
Young India, Ahmedabad
Young India, New York

Archives and Collections Consulted

Andover Theological Library, Harvard University, Boston
Bhowani Sen Library and Research Center, CPI West Bengal Headquarters, Kolkata
Bodleian Libraries, University of Oxford
British Library [BL], Asia, Pacific & Africa Collections, London [APAC]
Centre for Studies in Social Sciences, Kolkata
Dhananjayrao Gadgil Library, Pune
Haverford College Quaker & Special Collections, Haverford
Labour History Archive and Study Centre, People's History Museum, Manchester
Lamont Library, Harvard University, Boston
National Archives of the United States, Washington
Nehru Memorial Museum and Library, New Delhi [NMML]
New York Public Library [NYPL] Manuscripts and Archives Division, New York
P.C. Joshi Archives on Contemporary History, New Delhi
Sadharan Brahmo Samaj, Kolkata
UCLA Library Special Collections, Charles E. Young Research Library, Los Angeles
West Bengal State Archives, Kolkata
Working Class Movement Library, Manchester
Yale University Divinity Library, New Haven

Index

Kolibri 360 GmbH
Ferdinand-Jühlke-Straße 7
99095 Erfurt, Deutschland
produktsicherheit@kolibri360.de